Spanish Merino Sheep: Their Importation From Spain

Introduction into Vermont and Improvement Since Introduced

by Vermont Merino Sheep Breeders Association

with an introduction by Jackson Chambers

This work contains material that was originally published in 1879.

This publication is within the Public Domain.

This edition is reprinted for educational purposes
and in accordance with all applicable Federal Laws.

Introduction Copyright 2018 by Jackson Chambers

Self Reliance Books

Introduction

I am pleased to present yet another practical title on breeding and raising livestock.

The work is in the Public Domain and is re-printed here in accordance with Federal Laws.

As with all reprinted books of this age that are intended to perfectly reproduce the original edition, considerable pains and effort had to be undertaken to correct fading and sometimes outright damage to existing proofs of this title. At times, this task is quite monumental, requiring an almost total "rebuilding" of some pages from digital proofs of multiple copies. Despite this, imperfections still sometimes exist in the final proof and may detract from the visual appearance of the text.

I hope you enjoy reading this book as much as I enjoyed making it available to readers again.

Jackson Chambers

CONTENTS.

ILLUSTRATIONS.

INTRODUCTION.

THE Publishing Committee of the Vermont Merino Sheep Breeders' Association desire to offer no apology for presenting this volume to the public. They regret that the original importers and breeders of Spanish Merino Sheep could not have foreseen the importance of inaugurating a work which should have handed down, through accurate history and minute records, the pedigrees of the valuable animals which they introduced into this country and bequeathed to their posterity; and which have added so much to the prosperity and material wealth of the United States.

As the Committee have prosecuted the work assigned them in making their investigations, the thought has often occurred to them how much better for all breeders who are honestly endeavoring to pursue their calling in a legitimate manner, had Seth Adams, Chancellor Livingston, David Humphreys and William Jarvis, with such other breeders of pure Spanish Sheep, as had derived their blood from their importations, begun a work that could have been carried on and continued to us in comparative perfection.

Or, if we take a still later date, going back half the years that have elapsed since the great importations of 1810 and 1811, how much information, then in possession of the honored Jarvis and others who have since passed away, is now lost! And we must be content with surmising many links which they could, even then, have furnished to make the chain of evidence perfect and entire in the case of many flocks. And thus, with the departure of each year, have the difficulties of accurately tracing the pedigrees of the flocks of Merino Sheep increased, until at this day it would be presumption for us to claim absolute accuracy in every pedigree herein traced.

No farm stock except Merino Sheep has been affected by such extraordinary seasons of favor, amounting to mania, when prices

have been enhanced to such extraordinary fictitious values, when
large number without qualifications or due consideration have
embarked in the business of breeding them, until, by the action of
some unfriendly legislation it would become unprofitable, and the
most extraordinary reaction would immediately follow, depressing
prices as much below actual worth as the former ones had been
above, until, at times, many of the flocks of pure Spanish Merinos
would become crushed out, and all were threatened with annihila-
tion.

Immediately after Merino Sheep were introduced into the
United States, large numbers of them were brought into Vermont,
and, as it will appear in the account of their introduction into the
State, they were simultaneously introduced into many different
parts; and history shows that representatives from all the old Spanish Cabanas from which importations were made were among them.

The largest importer of Merinos located his home among the
hills of Vermont.   To that home he brought his choice of those
importations, and there established the largest flock of pure-bred
Merinos of that day.   These several causes served to awaken a
great interest in this famous race or breed of Sheep, and this interest,
early awakened, has been continued to this day, and its effect has
been to draw out and continue a line of breeders of acute judgment
and broad capabilities, whose taste, thought, and practical industry
seconded by a climate and location which have no superior for their
purpose, have enabled them to develope such an improvement upon
the breed of Merinos, as it came to their hands, as to attract the
attention of not only all our land, but of the most distant Sheep-
walks of the world.

The attention thus secured these great improvements, has
resulted in a general acknowledgment of their superiority over
those attained elsewhere; and thus has a demand been created for
the blood that represented these great improvements, from localities
less favored by natural advantages, or where the skill of breeders
had fallen short of developing as great improvements, to found
flocks of Merinos, or improve those already established, until nearly
all the Merino Sheep of the United States are tracing their pedigrees
in whole, or in part, to Vermont flocks; and these improvements
have been so well sustained, and their superiority so manifest, that
during a period not to exceed eighteen months past, contributions
have been furnished from them to Maine on the Northeast, Washington Territory on the Northwest, Texas on the South, with inter-
vening States and Teritories enough to make at least eighteen.

In short, every State where fine wool-growing is a practical industry, has received contributions of Merino blood from Vermont flocks to improve and enrich its own; and in the case of nearly, if not quite all these, contributions have been made by the car-load, and in those of some of the principal wool-growing regions like Ohio, by scores of car-loads.

Thus it is gratifying to find that the acme of improvement was not reached by a Rich or a Hammond, but that their mantle has fallen upon younger shoulders, apt scholars, who, profiting by the examples and teachings of those worthy masters, by careful thought, close study, patient application, intelligent observation, and sound judgment, have been able to develop improvements that may have been conceived by those masters, but were never practically accomplished by them.

To a State that has such a record, naturally belongs the work of investigating the histories and records of the past to learn the facts, as nearly as can be at this day, of the introduction of Merinos into our country, the histories of their breeding and improvements, as well as the pedigrees of all her flocks. Not only does this work belong to her, but the public everywhere these Sheep have been sent, have a right to ask that this Association publish the result of this labor to the world, and that we also publish as accurate histories and pedigrees of our flocks and their pedigrees as it is possible to gather, and that we show by actual facts, statistics and comparisons, what improvements her breeders have made in the Merino race of Sheep, to inspire confidence in wool-growers without her borders, that they will receive substantial benefit from the introduction of the blood of our flocks into their own.

The Committee have prosecuted their work in a spirit of loyalty to the fair fame of Vermont, well knowing that her good name and the permanent prosperity of her fine wool industry would be best promoted to stand upon a foundation of truth, rather than upon a fabric of fictions, made up of misrepresentations and claims that could not be substantiated by facts.

The labors of the Committee of Publication, as well as of the Committee on Pedigrees, have not been easy. Without any example or experience of others in like work as a guide, the Committee in both capacities have been obliged to feel their way slowly. It was found at an early stage of the work, that, to preserve accuracy, and decide justly, it was indispensable to make haste slowly. Long and patient investigation was absolutely necessary, and, as these investigations were being prosecuted to trace back to importations the blood of many of the flocks applying for registration, it was found that the published histories of some of the lines of blood and of their introduction into this country were not reliable, and that many of them were enshrouded in such a contradictory mass of neighborhood tradition and local theories as to their origin and blood, that the task of unraveling their mysteries seemed at times a hopeless one, and in a few cases has thus far proved impossible.

These investigations have been carried back in some instances to much greater extent that it was at first anticipated would be necessary ; but this labor has had its compensation in opening a field wherein our labors have gleaned many facts that had been lost sight of, and that seem to have been ignored or undiscovered by those who have published what they have supposed to be the whole facts with regard to the importation of Merino Sheep into the United States, and their history subsequently.

One of the greatest embarrassments to correct conclusions have been found to be the theories and damaging stories, started and circulated by breeders, not about their own, but of their neighbors' flocks. In some cases these imputations have been so long circulated, that the public have become prejudiced against some flocks to such an extent that their standing and value had become affected materially, while in reality their true standing was much better than that of those which belonged to the authors of the slanderous stories. It is confidently hoped that this practice may be made less popular after this Association has finished its work of investigating and publishing the histories of Vermont flocks.

In view of all the facts that have come to the knowledge of the committee, they have arrived at the conclusion that they will not use the names of any of the old Spanish Cabanas to distinguish or describe any of the lines of blood recorded in this work. While they insist that the origin shall be from some one or more of those celebrated Cabanas, unmixed with other blood, to entitle them to record, they would by no act of theirs indicate any preference for sheep that have descended from any one of them, over any other ; therefore, with their present light, they are not prepared to apply the name of any one of the old Spanish Cabanas to any flock of sheep to-day, as they have been unable to trace the blood of *any*, through a direct line, to one of those Cabanas unmixed with any other.

While making this a rule to guide their own action, they would not dispute the right of any breeder to call his sheep by any family name he chooses, and support its justice by any proofs he may be able to furnish the public in regard to it The only suggestion the committee have to offer is, that the excellence of the Merino sheep herein recorded consists as much in the very great improvements that have been made upon the best of the sheep imported from Spain, nearly or quite three-fourths of a century since, as in their having descended from any *one* of the old Spanish Cabanas, or upon their being named after some old Spanish convent, or Spanish grandee; indeed we would rather raise a question, whether those improvements in most flocks may not, in a great measure, be due to the sagacity of our best American breeders, as shown in their ability to unite the best excellencies of the different Spanish Cabanas, and weed out the imperfections of all, until the typical Merino sheep of our country is now an animal uniting all the good qualities of the different families, as they were imported, without the imperfections of any. Therefore, the committee have preferred to use the names of such American breeders as have proved, by their superior knowledge and success in improving the flocks they have bred and perpetuated, to be entitled to the honor of having sheep descended from their flocks named after *them*, in designating or giving names to families or lines of blood. Thus it will be found by the reader, that many names are used to designate the different lines of blood, but in reality they may all be traced back to a few importations, but the blood of these different importations having been united in most of the families or flocks of to-day, it seems necessary to give new names to the results proceeding from these crosses or unions of different importations; and what more appropriate than the the names that different localities soon learn to give sheep that have been stamped with the impress of an improvement conceived in the mind, and executed in the flock, of any one of the best of our American breeders.

As in the beginning we had Humphrey sheep, Jarvis sheep, etc., we soon, as a matter of necessity, had Atwood sheep, Blakeslee sheep, and, coming a little later, we have just as necessarily come to have Rich sheep, Hammond sheep, Cutting sheep, Robinson sheep; and to-day living breeders are just as surely giving their names to sheep that it will be as necessary to use when in the future breeders, historians or tracers of pedigrees, are mentioning the sheep that these breeders are stamping with the impress of their minds and practice, by selections within their own flocks, or by crossing the different families or sub-divisions of families or lines of blood; breeders also are living to-day who are entitled to and are enjoying the honor of this distinction.

After careful consideration of the subject, the committee have decided that the term IMPROVED SPANISH MERINO more fully comprehends a direct descent from the Spanish Merino, with the great improvements that have been made upon them by American breeders since importation, and as more appropriate than any other.

The term American Merino may be more flattering to the pride of American breeders, but it surely can be as truthfully applied to descendants of Merino blood from France and Saxony, as from Spain, and even to grades where their blood is made up by crosses of either of the three, upon common American sheep, or so-called natives

In carrying out the spirit and intent of Article 7 of the Constitution of this Association, and the Rules adopted by its Executive Board, the Committee have determined to admit Sheep for record in this Register, whose lineage can be traced through flocks of reliable breeders, from the importations from Spain made between the years 1800 and 1812, and by adopting the rule that the evidence presented with each application shall be sufficient to convince every member of the justness of the claims to be able to trace such line of ancestry through reliable flocks to such importations. They believe they have admitted no Sheep for record in this Register that are not entitled to the full confidence of the public as direct descendants from the Spanish Merino.

No register or record of flocks having been kept continually from 1812 to the present day, no perfect, authenticated line of ancestry or pedigree, which could be used as a calendar and guide, is known to exist; but history and tradition have become so far recognized in the case of many flocks as to leave no reasonable doubt in unprejudiced minds, that they had been continued down to us from importation, uncontaminated by any crosses, except such as have been going on among the divisions and subdivisions of Spanish Merinos.

The introduction of Saxon blood into nearly all the flocks of Merinos between 1822 (the time of the first importation of Saxon sheep), and 1840, has been a cause of embarrassment to the Committee, although there is no reason to doubt the statement of many of the leading breeders of those times, that they weeded out and sold all of that blood which they introduced.

The Committee have intended to rigorously exclude all sheep that had a trace of this blood, and they believe none have been admitted that have at any time in their ancestry been crossed with it.

Since 1840, breeders of Merino sheep in New England have been using rams from the heavier-wooled flocks of pure Merinos, and discarding the Saxon. Hence, if by any possibility the Committee have admitted sheep that had Saxon blood, previous to that time, there can be no appreciable amount of it left, as ten crosses by pure Merino rams will breed out all except less than a thousandth part, even when the original blood was all Saxon. It should be borne in mind that this Saxon blood has descended unmixed from the same Spanish stock from whence our best Merino blood was derived, and that the experience of large numbers of breeders is ample proof that by skilful selection and judicious crossing in the direction of the best Merino familes presenting the strongest constitutions and yielding the heaviest fleeces, it soon loses its peculiarities of weak constitution and light fleece, while the improvement has proved reliable and permanent. On the other hand numerous examples are not wanting to demonstrate that a taste for raising fine wool, with a lack of judgment that ignores and loses sight of constitution and weight of fleece on the part of the breeder, soon produces a sheep with the defective characteristics, and as little adpated to profitable wool-growing in the United States as those that were imported from Saxony.

In the absence of authenticated records and positive proof, which absence is universal, the Committee have been obliged to accept such circumstantial evidence and documentary proof following the histories of flocks, as tradition, the common judgment of men, and public opinion have come to regard as of nearly equal weight, and in all cases the Committee have insisted that these proofs should be sufficient to convince such member composing it, and in the case of any flock falling short of this, they have been rejected. Thus it has been necessary that the vote of the Committee should be unanimous to receive a flock, before it could be accepted for record.

The form of publication of the pedigrees and histories of flocks is thought to be one that will be the most useful to the public to understand the bloods and breeding of the flocks recorded. First, the pedigree gives all the lines of blood that have been introduced into the flock since the foundation, as well as the fountain at which the main stream commenced. Secondly, the purchase of ewes will show where the ewes for the original foundation were procured, and their blood, as well as the same facts in regard to subsequent additions of ewes to the flocks, if any. Thirdly, the purchase or use of rams will give the same class of facts in regard to the blood introduced into the flocks through them.

A careful reading of the last two will generally reveal the amount of each line of blood that has been introduced, which can not be determined by the pedigree.

It is hoped that this plan will be of more practical benefit to the mass of searchers after information in regard to the blood of Merino flocks than an elaborate pedigree of individual members of the flock, that would require an expert in pedigrees to trace through and understand, even if such a pedigree could be given ; while to pretend to give such an one of any Merino sheep now living, through the twenty or thirty generations of ancestry that must intervene between importation from Spain and the present time, is simply a farce, and, if pretended, may be used as a cover to defraud others than experts. In extreme cases a perfect authenticated line of of ancestry through individual sires and dams could not be given for ten generations, and the records of very few flocks have been so kept that such a one could be given for more than three generations. Pedigrees have been furnished this Committee containing a list of celebrated stock rams with excellent names, but omitting to give the names of dams at very important points in the pedigrees.

To one who wishes to understand the blood of a sheep it is much more important and satisfactory to know that a dam in any instance was bred in some flock of known, reliable blood, than that she was possessed of some very significant and pretty name.

It is hoped that the adoption of rules which require that all sheep that are registered shall be labeled with the owner's initials and with numbers—and that requires a flock register—shall be kept of all sheep recorded in this register, will make it possible from this time to trace the individual pedigrees of sheep and to facilitate this work, or even make it possible, the Committee cannot too strongly urge upon the members of this association the imperative necessity of complying literally with these rules.

Attention is called to the provision for recording the sheep by marks and individual numbers, and for notice to the Secretary of the sale or death of any sheep herein recorded. If these provisions are strictly observed, it will prove a great check upon sales of impure sheep for pure bloods or registered. While it will help to detect and punish fraud, it will prove a great protection to purchasers, and a help to honest dealers in Merino Sheep.

The committee venture to recommend a reform in giving names to stock Rams. The present practice of duplicating the names of celebrated stock Rams, excellent and apt though they may be, is a source of great inconvenience and confusion to all who are investigating and recording pedigrees. Occasionally some breeder selects a name for his Ram that no other breeder seems to copy ; but how many are the cases where the name of some stock Ram is duplicated, until they have an almost endless number of Old Blacks, Matchlesses, Greasies, Winkleys, Long-Wools, Sweepstakes, Consuls, Fortunes, Californias, Gold-Drops, Black-Hawks, Green-Mountains, Golden-Fleeces, Bull-Dogs, Hammonds, etc., etc., until the compiler of pedigrees often finds himself compelled to put the prefix of the breeder's name to the Ram, or confusion is sure to arise, and often is compelled to leave some link in the pedigree in doubt, because some one else in giving a previous one has failed to state what Golden-Fleece or Sweepstakes was the sire at some point.

The system of marking sheep by labels in their ears gives a ready and easy way to avoid this vexation and troublesome evil.

If breeders in the future will give their Rams no names but their own in the possessive case, with the flock number on the label in the ear of each Ram, the result, we think, will be satisfactory to all who are breeding, selling, or purchasing sheep, as well as to those who have to record or trace out pedigrees. If this plan is practiced hereafter, one breeder will not duplicate the name of another, and he will not duplicate the number in his own flock. Another great advantage of this system is, that the name of the Ram as given on his label will always be an advertisement for his breeder as long as he lives ; and a means by which he may be identified. We hope that breeders will see the importance and desirability of such a plan, and take common action by adopting it.

To the members of the Vermont Merino Sheep Breeders' Association and the public, we offer the results of our labors ; we claim only that we have by careful investigation and hard labor made as few mistakes as possible. We invite friendly criticism, that our successors may profit by them, and avoid our mistakes.

We gratefully acknowledge assistance and courtesies from many private individuals, historical societies, and public libraries, and to many authors from whose published works we have taken extracts, and for which we have intended to give in their proper place appropriate credit, when from time to time we have inserted those extracts in this work.

2

THE VERMONT

MERINO SHEEP BREEDERS' ASSOCIATION

Was organized at a meeting of the breeders of Merino sheep held at Middlebury, March 23, 1876. The meeting adopted the following Constitution :

ART. 1. This Association shall be called the Vermont Merino Sheep Breeders' Association.

ART. 2. The object of this Association is, in its own sphere, and in co-operation with other similar associations in other States, to preserve the purity of the Spanish Merino race of sheep, encourage their further improvement, and, as a means to these ends, to provide for the registration of the histories and pedigrees of the flocks of Spanish Merino sheep in Vermont and elsewhere.

ART. 3. Breeders of Merino sheep may become members of this Association by signing this Constitution themselves or instructing the Secretary to do the same, and paying into the treasury the sum of two dollars annually.

ART. 4. The officers of this Association shall be elected annually after this year, at the yearly meeting of the Association, to be held on the second Wednesday of January in each year, at such place as the president and executive board shall determine. Said officers shall be a president, two vice-presidents, a secretary and treasurer and four directors, and these nine officers shall constitute the executive board for the management of the affairs of the Association. There shall also be a committee of three on pedigrees. These officers shall hold their offices for one year, or until their successors are elected.

ART. 5. A majority of the executive board shall constitute a quorum for the transaction of business at any meeting of the board.

ART. 6. The president and vice-presidents shall have the power to call a meeting of the executive board by giving every member proper notice, and the executive board shall have the power to call a meeting of the Association at such time and place as they shall deem proper and just, by giving at least two weeks' notice, through not less than five newspapers published at prominent points in the State.

ART. 7. It shall be the duty of the committee on pedigrees to examine carefully the statements and evidence presented by persons who desire to have their sheep registered, and if, in their judgment, they shall deem the evidence sufficient to warrant, they shall admit their flocks to record ; but if they deem the statements erroneous, or the evidence insufficient, they shall reject the same. For the time actually employed in this work of examining evidence and pedigrees this committee shall be entitled to a compensation of two dollars and fifty cents per day and travelling expenses from those having the pedigrees of their sheep examined for record. This committee with the secretary shall constitute the publishing committee of the association, and they shall be entitled to like pay for time spent in preparing the manuscripts of pedigrees and histories of flocks for record and publication.

ART. 8. Any parties who may feel aggrieved at the decisions of the committee on pedigrees shall have the right to appeal to the executive board within sixty days, and their decision shall be final in the matter.

ART. 9. If at any time the funds in the hands of the treasurer shall be inadequate to meet the expense incurred in carrying on the work of the Association, the members and those having their sheep registered may be assessed to meet such deficiency in such manner and under such regulations as may be determined by the executive board.

ART. 10. The rules that shall govern the committee on pedigrees, and the prices by which sheep may be admitted to record in the Register, shall be established by the executive board.

ART. 11. By-Laws may be proposed and adopted at any regular meeting of the Association.

ART. 12. This constitution may be altered or amended at any regular meeting of the Association, provided notification of such amendment shall have been given at least three months previous to such meeting at a meeting of the Association, or by being published in at least three newspapers published at prominent places in the State.

The same meeting elected the following officers :

PRESIDENT :

N. T. SPRAGUE, *Brandon.*

VICE-PRESIDENTS :

S. S. ROCKWELL,   GEORGE CAMPBELL,
West Cornwall,   *Westminster West.*

SECRETARY :

ALBERT CHAPMAN, *Middlebury.*

TREASURER :

C. D. LANE, *Middlebury.*

DIRECTORS :

E. N. BISSELL,   J. H. MEAD,
East Shoreham,   *West Rutland,*
L. P. CLARK,   A. E. PERKINS,
Addison,   *Pomfret.*

COMMITTEE ON PEDIGREES :

W. R. REMELE,   J. T. STICKNEY,   J. J. CRANE,
Middlebury,   *Shoreham,*   *Bridport.*

COMMITTEE ON CORRESPONDENCE :

A. CHAPMAN,   E. N. BISSELL,
Middlebury,   *East Shoreham,*
CAPT. F. MOORE,   E. S. STOWELL,
Shoreham,   *Cornwall,*
E. G. FARNUM, *West Cornwall.*

The Executive Board, according to the provision in the Constitution, adopted the following Rules governing the Committee on Pedigrees in admitting sheep to the Register of the Vermont Merino Sheep Breeders' Association :

RULE 1.   Members of this Association only may enter their sheep in its Register.

RULE 2   Each breeder who wishes his flock registered must make application to the secretary of the society, or a member of the committee on pedigrees, furnishing a statement of the history of his flock, stating when and from whom the original breeding-stock was purchased, as well as of all individuals since introduced, giving as far as possible the families and breeders through which their lines of ancestry are traced back to importation from Spain.

RULE 3. If any rams have been used belonging to others, not comprised in the above, the applicant shall so state, giving the names of the rams, if they were named, the names of their owners, with their family or breeding, and the year or years such rams were used, so far as possible.

RULE 4. Such evidence and certificates as it is possible for the applicant to procure shall accompany the statements named in the two preceding rules. and he shall also solemnly affirm his belief in the truth of the evidence and certificates he shall present, as well as in the accuracy of his own statements.

RULE 5. In considering the application and evidence presented, the committee on pedigrees will be governed by Article 7 of the Constitution.

RULE 6. The price for admitting sheep to registry shall be, for every flock not exceeding fifty, $5.00, and for every sheep such flocks shall contain over fifty, ten cents each, besides such compensation to the committee on pedigrees as is named in Article 7 of the Constitution; and no applicant shall be entitled to a certificate of acceptance of his flock to registry nor to have them finally entered in the Register until all charges are paid.

RULE 7. As soon as the committee on pedigrees shall decide to receive or reject any flock to registry, they shall notify the applicant and the secretary of the Association of their decision, that any person interested may have an opportunity to appeal from the decision, according to Article 8 of the Constitution.

RULE 8. Applicants for registry of flocks must sign the accompanying blanks and affirmation, and attention of applicants is called to Article 9 of the Constitution

RULE 9. Any member of this Association who shall wilfully misrepresent the blood or breeding of his sheep for the purpose of procuring their registry or for the purpose of making sales, or sell any sheep as recorded that have not been registered, or shall change the labels from the ears of any sheep registered, or shall present any certificate or evidence to the committee, knowing them to be false, shall be expelled from the Association.

RULE 10. All sheep that have already been accepted for record in the Register of this Association shall be marked with metallic labels in their ears before the first day of June, 1877, and all those hereafter accepted shall be so marked within twenty days after they have been accepted. Said labels shall contain the owner's name or initials and his flock numbers; and such flock numbers shall be returned to the secretary for record, and to be published in the Register.

RULE 11. The owner of every flock registered shall keep a flock record after a form and in a book furnished by this Association.

RULE 12. If the owner of any sheep recorded in this Register shall sell the same, he shall give notice of the same to the secretary, giving the numbers of the sheep sold, and the name and post-office address of the purchaser, within ten days of the date of such sale. It shall also be the duty of every owner of flocks recorded to report all deaths by number to the secretary on the first day of January in each year, and it shall be the duty of the secretary to keep a record of the numbers and owners of such sheep as have been transferred or have died.

At the annual meeting, January 10, 1877, all the officers that were elected in 1876 were re-elected.

At the annual meeting held at Middlebury, January 9, 1878, the following were elected officers for 1878 :

PRESIDENT :

N. T. SPRAGUE, *Brandon.*

VICE-PRESIDENTS :

S. G. HOLYOKE,   ·   E. N. BISSELL,
    St. Albans,   *Shoreham.*

SECRETARY :

ALBERT CHAPMAN, *Middlebury.*

TREASURER :

CHARLES D. LANE, *Middlebury.*

DIRECTORS :

E. S. STOWELL,   V. RICH,
    Cornwall,   *East Shoreham,*
HENRY THORP,   H. BOTTUM,
    Charlotte,   *Shaftsbury.*

COMMITTEE ON PEDIGREES :

W. R. REMELE,   J. T. STICKNEY,   J. J. CRANE,
    Middlebury,   *Shoreham,*   *Bridport.*

COMMITTEE ON CORRESPONDENCE:

A. CHAPMAN,   E. N. BISSELL,   CAPT. F. MOORE,
   Middlebury,   *East Shoreham,*   *Shoreham,*

C. P. CRANE,   A. E. FULLER,   GEO. CAMPBELL,
   Bridport,   *Pomfret,*   *Westminster West,*

E. TOWNSEND,   B. W. COPE,   WM. BALL,
   Pavilion Centre, N.Y.,   *Smithfield, Ohio,*   *Hamburgh, Mich.,*

        C. M. CLARK,   E. PECK,
              Whitewater, Wis.,   *Geneva, Ill.*

DELEGATES TO MEETINGS OF THE NATIONAL WOOL GROWERS' ASSOCIATION:

ALBERT CHAPMAN,   E. N. BISSELL.

INTRODUCTION OF MERINO SHEEP INTO THE UNITED STATES
FROM SPAIN.

The committee of publication have not deemed it advisable to treat of the history of Merino Sheep previous to their being imported into this country from Spain. Numerous works are extant, in which compilations upon the subject have been published, and as nothing new upon the subject could be given here, it has not been thought necessary to make another compilation of these same facts which have been so many times published, but could occupy their time and space more profitably in treating more at length upon what the committee deemed more original matter.

It may not be of any great practical importance to know who imported the first Merino Sheep into the United States, but it is a question which has excited a good deal of interest. In 1798, Hon. William Porter, of Boston, imported from Spain two ewes and a ram, on Ship Bald Eagle, Captain Atkins. He gave these to Mr. Andrew Cragie, of Cambridge, who, not appreciating their value for breeding purposes, killed and ate them as mutton. This gentleman afterwards paid $1000 for a Merino ram. Mr. Dupont de Nemours, with a French banker, Delessert, imported four ram lambs to the United States from France, only one of which reached this country alive. This ram was used some years at Kingston, New York, but afterwards was taken to Delaware, and was the sire of some excellent flocks of grade sheep near Wilmington in that State. This importation was in 1801. Mr. Worrell in his "American Shepherd," says : "The sheep imported by Mr. Dupont de Nemours, two pairs, were selected from the Rambouillet flock in France." Dr. Randall, in his "Practical Shepherd," describes the shipment as before stated, four ram lambs, and he conjectures that the ram that lived was of original Spanish stock, not of the French.

The same year (1801), Mr. Seth Adams, of Zanesville, Ohio, imported in the brig Reward, Captain Hooper, which left Dieppe,

France, in August, 1801, a pair of Merinos, that arrived in Boston in the month of October following. Mr. Adams stated that these sheep were from France, from a flock imported from Spain by Bonaparte. Mr. Adams also claimed that he received a prize of $50 from the Massachusetts "Society for Promoting Agriculture," for importing the first pair of Merinos. Mr. Adams's account of this transaction is as follows :

"The Agricultural Society of Massachusetts having offered a premium of $50 for the importation of a pair of sheep of superior breed, General D. Humphreys imported a flock of Merinos, and sent some of them into Massachusetts, and he, or some one for him, applied to the Society for the premium. Knowing that his sheep did not arrive before the Spring season after mine, I applied at the same time for the premium, and after having examined the sheep and wool, and comparing with those of General H., the Society awarded to me the premium, and awarded to General H. a gold medal for having imported a larger number." We must think there are some inaccuracies in this statement, though perhaps not intentional.

The records of the Society for 1802, show that "a letter was received from Colonel Humphreys, late Minister to the Court of Spain, on the Merino breed of sheep, with a specimen of their wool, and remarks on the importance of propagating said sheep in the Northern and Eastern States." The letter mentioned that Colonel Humphreys had imported into Connecticut seventy-five ewes and twenty-five rams. "The subject was referred to a committee consisting of Mr. Lyman, Mr. Cabot, and Dr. Dexter, to consider the same, and report thereon."

"The introduction of Merino Sheep may be dated from this time, and so important it seemed, that at the next meeting of the Trustees, after Colonel Humphreys' letter was received, the question was raised, whether Colonel Humphreys should not receive the gold medal for his services, and at the following meeting it was awarded to him, not to exceed fifty dollars in value. A premium had already been offered to the person who should introduce Merino Sheep into the country, and the amount paid in this way was very considerable. The first claimant for this premium was Seth Adams, for the importation of two sheep of the Merino breed from France."

It would appear from this record, that there was no competition between Colonel Humphreys and Mr. Adams, but that the latter received the medal as a gratuitous reward, or recognition, for a very valuable public service, without having asked or applied for it.

Mr. Adams received his $50, if he received it at all, as one of the regular premiums, offered by the Society, "to the person who shall introduce into the State of Massachusetts, for the purpose of propagation, a ram or ewe, of a breed superior to any now in the State; if from a foreign country." We have not yet been able to find from the records of the Society that Mr. Adams received this premium. It may be possible that it was not finally awarded, in consequence of his not keeping the sheep in Massachusetts. Mr. Randall says he knows nothing of the subsequent history of these sheep, but we have found among the papers of the late Consul Jarvis, a letter written by Mr. Adams to Mr. Jarvis from Zanesville, in December, 1810, that says: "I have had the breed of sheep a number of years, and am continually applied to for the full bloods and know almost every person in this State or Kentucky, who is in want of them; and I have some conditional engagements for the next year. I imported in the year 1801 a pair of these sheep, the first pair imported into the United States, but I have but a small number of the full-blooded and I intend rearing of them, and I am known to have the stock; have a very great advantage over any person on this side of the mountain."

We do not know that there is anything on record relating to the character of these sheep. As will be seen by the following note* Mr. Adams afterwards took Merinos to the West.

This brings us to the importation of Mr. Livingston in 1802. This gentleman sent two pairs from France to the United States in that year; they were from the government flock at Chalons. Mr. Livingston subsequently imported a ram from the government flock at Rambouillet, and in one of the accounts of sales of the Jarvis sheep in 1810, rendered by one of the consignees, Mr. Livingston is mentioned as a purchaser; and he was a large purchaser about the same time of Merinos imported from the Duke of Infantado's flock, made by Mr Charles Henry Hale, of Harlem, New York.

In 1803, two pairs of Merinos were imported into Philadelphia, but as they were black and did not reflect any great glory upon their importers, history has taken no pains to hand down their names. Later there were some fifteen black Merino rams "selected from the best flocks in Spain" advertised in one lot for sale in Boston.

*Mr. Seth Adams has carried 176 Merinos to Kentucky and Tennessee.— [N. E. Palladium, August 3d, 1810.]

In 1802, Colonel David Humphreys, of Derby, Conn., made his celebrated importation from Spain. The following is his account of that importation, as published in his "miscellaneous papers."

"Convinced that this race of sheep, of which I believe not one had been brought to the United States until the importation by myself, might be introduced with great benefit to our country, I contracted with a person of the most respectable character, to deliver to me at Lisbon, one hundred, composed of twenty-five rams, and seventy-five ewes from one to two years old; they were conducted across the country of Portugal by three Spanish shepherds, with proper passports, and escorted by a small guard of Portuguese soldiers. On the 10th of April last, they were embarked in the Tagus, on board the ship Perseverance, of 250 tons, Caleb Coggeshall master. In about fifty days twenty-one rams and seventy ewes were landed at Derby, in Connecticut, they having been shifted at New York on board of a sloop destined to that river; the nine which died, were principally killed in consequence of bruises received by the violent rolling of the vessel on the banks of New Foundland."

For this importation, as before stated, Colonel Humphreys received a gold medal with the inscription, "Presented by the Massachusetts Society for Promoting Agriculture, to the Hon. David Humphreys, Esq., late minister to the Court of Madrid, as a testimony of respect for his patriotic exertions, in importing into New England one hundred of the Merino breed of sheep from Spain, to improve the breed of that useful animal in his own country, 1802."

Up to this time, no one has been able to learn from what Cabana in Spain Colonel Humphrey received these sheep, and from the account, and from the fact that he never in his writings says anything about, it we must conclude he himself did not know or considered it a matter of no consequence. Jarvis in his account of importations says: "They were probably Transumantes, which is the most material fact worth knowing."

The following is a copy of MSS. in the collection of the New Haven Historical Society, copied by Prof. W. H. Brewer. It proves Mr. Jarvis' conjectures were correct:

"To all concerned to whom these presents shall come, I hereby certify and make known that the flock of Merinos belonging to me, and entrusted this day to the care of Mr. Elihu Ives, to be disposed of for my account at his best discretion, consisting of full and high degree of mixed bloods, are the genuine descendants and offsprings of those pure bloods extracted by me from Spanish Estramadura, in the beginning of the year 1802, and ascertained by their pedigree to be of the purest and best race in Spain.

I moreover declare that it has been proved by the best of experience, in almost every part of the United States, that they have not in any respect degenerated, but on the contrary, that the breed of whole-bloods has, in some points, much improved in this country.

To prevent any apprehension on the part of purchasers of such frauds and impositions as have frequently been attempted to be practiced with sheep pretended to be descended from the Humphreyville flock by speculations (sic), the said Merinos described in this certificate are branded with the letter H in the following manner, to designate their different grades of blood, to wit:

One-half Bloods,	H on the left side.
Three-fourths Bloods,	H on the right side.
Seven-eighths Bloods,	H on rump.
Fifteen-sixteenths Bloods,	H on right hip.
Full Bloods,	H on left hip or both hips.

In testimony whereof I have given this instrument, at Humphreyville, this 8th day of July, 1812.

D. HUMPHREYS."

In the same package with the above was another document, with all the legal signs, signatures, seals, witness, etc., in which Elihu Ives, of New Haven, is delegated agent and attorney to dispose of all such mixed blooded Merino rams as shall be forwarded to him at Pittsburgh, or elsewhere, from Col. Humphreys' flocks, together with such articles of domestic and other manufacture as may be forwarded, etc. This document, in short, makes him his agent to buy wool, sell and buy for him. This is dated Humphreysville, Derby, Ct., March 5th, 1810. So far as the other signatures are concerned, D. Humphreys' signature is dated Philadelphia, April 2, 1810.

With the above is also a letter from Mr. Ives, dated Pittsburgh, August 13, 1813, in which he writes that he has closed a bargain for 38,000 lbs. wool from the Province of Texas, and also writes of a brother who has gone west to Kentucky with 42 rams, alludes to rams sent last season, gives some details of business, etc.

The success that had attended the introduction of Merinos by Messrs. Dupont, Livingston and Humphreys, together with the high price of wool, had prepared the way for a very hearty welcome to the first of the Jarvis importations. We read that "at the fair of the

Columbus Co. Agricultural Society at Georgetown, D. C., on the 10th of May, 1809, nearly all the gentlemen present wore clothing of domestic manufacture.   President Madison sported his inauguration suit, the coat made from Merino wool of Col. Humphreys' flock, and the waist-coat and small-clothes made from the wool of the Livingston flock at Clermont."   At the same fair, "two Merino rams were exhibited, sired by 'Don Pedro,' owned by Mr. Dupont, of Wilmington."

Thus we find that by the cloth made from the wool, or by live animals, three importations made seven or eight years before, were represented at that fair in 1809; and the letter of Mr. Adams, alluded to before, was evidence that the four importations, made in 1801 and 1802, had been preserved, made useful in improving the sheep and wool of the country, and had become such favorites as to command very large prices, as was proved at Clermont in 1810 by the sale of four lambs at $1,000 each.   (In a letter written to Consul Jarvis by Messrs. Cornelius Coolidge & Co., of Boston, they speak of Col. Humphreys repurchasing for $1,000 one which he had sold, and of his selling two pairs for $6,000.)

There is ample evidence that the sheep imported by Col. Humphreys were rapidly disseminated and made great improvements in the flocks in the States where they were taken ; and we have evidence that the improvements that have reached such a high standard at the present day, commenced immediately after they were first introduced into this country.

Col. Humphreys wrote to the Massachusetts Society for Promoting Agriculture in 1807 as follows :

BOSTON, Nov. 28th, 1807.

DEAR SIR:—More than five years having now elapsed since the introduction into New England of the flock of Merino sheep, in consequence of which the Society for Promoting Agriculture in the State of Massachusetts were pleased to present to me a gold medal, it will doubtless be acceptable to that respectable and patriotic body, to learn that their hopes and expectations concerning the utility of this interesting species of animals have not been disappointed.   The attempt to propagate the pure Merinos in this country has been attended with complete success.   The extent of the experiment insures the duration of the unadulterated breed.   Instead of degenerating in the quantity or quality of their fleeces, the identical sheep which I brought to this country yield, on an average, half a pound more of wool apiece than they did at the first shearing after their arrival.   Nor, on the nicest and most candid examination, is it found that there is any finer wool produced in Spain than that which is annually shorn from these same imported Merinos and their full-blooded offspring. The rams born in America are, however, generally preferred to those born

In Spain, by persons who now make application to my agent for Merino rams, to cross the blood of their flocks, in breeding from them by American ewes. It is the opinion of all the farmers in Connecticut who have been acquainted with the original flock and its descendants, both of the pure and mingled blood, that they are hardier, better adapted to our climate, and more easily nourished, both in Summer and Winter, than the common breed of American sheep. They are likewise remarkable for being more gregarious and less disposed to stray or get over fences than the others. Finally, it may truly be asserted that they preserve the entire character, shape, features and qualities of the best Merinos in Spain.

The mixture of the Spanish with the American blood has succeeded in ameliorating the pile of the fleece beyond my most sanguine expectations. As a proof of the superior value of the wool of the half-blooded Merinos, it is a well-known truth that it has been sold for a dollar a pound in Connecticut, and still dearer in New York, the present season, while the best common wool has been sold for about half that price. The half-blooded Merinos produce more wool than the common sheep, and they ordinarily attain a larger size than the Spanish or American breed, from which they are descended. The facts here stated agree in substance with those established by experience in every country of Europe in which I have traveled, where this breed of sheep has been introduced. In England and France, the greatest care and expense are now bestowed under royal and imperial protection, for its extensive propagation.

A difficulty was experienced at first in carding the wool by the common carding machines. This has been overcome.

Some farmers, who early introduced a mixture of this blood into their flocks, have made in domestic manufacture, for sale, five or six pieces of cloth from this wool, during the present year. I shall have several hundred yards, fabricated entirely by machinery, from pure Merino fleeces. Several thousands made by the same process, from the common sheeps' wool of the country, have already been sent to market. Samples of both kinds, with the prices, are enclosed.

How long a period must pass before the prejudice against the fabrics of our country can be extinguished is not for me to decide. If any suitable means for their extinction could be devised and adopted perhaps an essential service would be thereby rendered to the real prosperity and independence of the United States.

With sentiments of great respect and esteem, I have the honor to be, dear sir, your most obedient and most humble servant,

D. HUMPHREYS.

Dr. Aaron Dexter, one of the Vice-Presidents of the Society for Promoting Agriculture in the State of Massachusetts, etc., etc., etc.

We also add another letter written a few days afterwards :

FACTORY (Renomon Falls), DERBY, Dec. 10th, 1807.

DEAR SIR—The importance of rightly understanding the best means of multiplying and improving the fine-wooled breed of sheep, derived from a cross of the pure Merino blood with that of the common flocks of the country, must be my apology for offering a few observations in addition to those which I had the honor of communicating to your Agricultural Society, on the 28th of last month.

To facilitate the extension of this improved breed, and to confirm its superior excellence in point of wool, it is conceived, are objects which have a peculiar claim to the public attention.

A mixed breed being first produced from our finest-wooled Ewes by full-blooded Merino Rams, it is still desirable that the Spanish blood should be renewed for three or four generations, through the medium of sires of that race. Then the system of *breeding in and in*, as it is technically called, and as it has been ably explained by Dr. Parry, of Bath, in his late "Essay on the nature, produce, origin and extension of the Merino breed of Sheep," proves decisive for the accomplishment of the objects proposed, in the shortest time, at the smallest expense, and with the greatest certainty, of any other plan hitherto suggested.

It is judged by the farmers in this neighborhood, who are best acquainted with this confirmed mixed breed, that, aside of their superior excellence with respect to wool, they have a greater tendency to fatten, on the same keeping, than any other sheep within the compass of their knowledge. Although this disposition to fatten is of little consequence so long as they are bred for the fleece only; yet it may be well, that those farmers who may hereafter propagate them for the sake of the carcass should not be ignorant of the fact.

From my further inquiries with regard to the weight of the fleeces of my Merinos, I learn that they have increased somewhat more than I stated in my letter of the 28th of last month. One of the rams born here has produced, this season, seven pounds and five ounces of washed wool.

This wool, would, it is presumed, be worth one dollar and a half per pound in England. I have the united testimony of all the people engaged in, or acquainted with, its fabrication into cloth, to prove that it has not deteriorated, by reason of its augmented quantity, in any respect whatsoever.

I take the liberty of inclosing four more specimens of cloth. Nos. 1, 2 and 3 were made from the wool of the pure Merinos; and No. 4, from that of the half-blooded race.

I beg you will receive the assurances of the real and great esteem, with which I have the honor to be, dear sir, your most obedient and very humble servant,

D. HUMPHREYS.

To the Hon. Dudley A. Tyng, Corresponding Secretary to the Society for Promoting Agriculture in the State of Massachusetts.

Dr. Randall has noticed, in his works, an importation of a small number of Merinos from Hesse Cassel into Philadelphia in 1807, and in a note says: "These, crossed with Col. Humphreys' sheep in the flock of Mr. William Caldwell, of Philadelphia, were the origin of the formerly highly celebrated flocks of Wells & Dickinson, of Ohio."

It is a matter worthy of note just here, that Mr. James Caldwell, of Philadelphia, in September, 1810, purchased of Mr. Levi Hollingsworth & Son, one hundred and nine Merino sheep, consigned to Messrs. H. & Son by Hon. William Jarvis, paying therefor the sum of one hundred and fifty dollars each, or total amount $28,500.00.

Taking the locality, the name of Caldwell, and the numbers purchased, it is possible, that, the Wells & Dickinson flock sprung more largely from the Jarvis importations, than the other two mentioned. (Besides this purchase, Mr. Caldwell purchased more of Mr. Jarvis and others of the importations of 1810 and 1811.)

The following is from the Records of the Massachusetts Society for Promoting Agriculture for the year 1809 :

"Capt. William Bartlett received a premium of fifty dollars for the importation of a Merino ram."

In 1809 the Hon. William Jarvis, at that time Consul at Lisbon, through the intervention of George W. Ewing, Esq., then U. S. Minister at Madrid, purchased of the Mayoral of the Escurial Cabana in Spain, 200 of the Escurial breed, and it was claimed by Mr. Jarvis that these were all that were ever imported into the United States from the Escurial flock.

This brings us to that extraordinary exodus of most of the celebrated Cabanas of Merino sheep from Spain, that followed the second invasion of the French under Joseph Bonaparte in 1809, and caused by the confiscation of four of the principal Cabanas of Spanish sheep by the Spanish Junta in consequence of their proprietors joining the French. These were the Paulars ; which, although it had numbered thirty or forty thousand, was by the vicissitudes of confiscation and war, reduced to about seven thousand and five hundred ;* the Negrette, which had numbered nearly or quite as many as the Paulars, reduced by the same causes to six thousand. The Montarcos formerly numbering thirty thousand, reduced to four thousand, and the Aguirres, numbering before the wars thirty thousand, were reduced to three thousand. If the estimate of Mr. Jarvis was correct there were one hundred thousand of the finest sheep in Spain sacrificed to the devastation of war, leaving only a little over twenty thousand, out of a total of about one hundred and twenty thousand.

Besides these four Cabanas there were probably as many more that had about the same number of fine Transhumanta sheep, including the Cabana of the Duke of Infantado, which numbered about forty thousand. The Duke not joining the French, his flocks were not confiscated, but in consequence of the permission given to export the confiscated flocks that were sold, there is no doubt that large numbers of those Cabanas that were not confiscated were also sold, and

---

* We here use an estimate of the numbers left in these flocks, made by the Hon. W. Jarvis in 1837.

many of them came to this country. We have evidence that over two thousand of the Infantado, and large numbers of the Guadaloupe Cabanas were also purchased and brought to the United States in 1810 and 1811. As there were a number more Cabanas than those named, it is easy to see where the large numbers of pure Merino sheep that were imported in the years 1810 and 1811 could have been obtained, and we do not share the opinion of many who have written upon the subject, that a large portion of the importations of those years were not of the noted and best flocks of Spain.

As we shall show by actual statistics, previous writers upon these importations have not nearly comprehended the large numbers of sheep that were imported in 1810 and 1811. Some of the first writers for want of the proper information seemed to have made erroneous estimates and statements, and later ones without taking pains to make proper investigations have copied their errors. Those who were engaged in the enterprise at that time had reason to suppress the facts in regard to the numbers that were constantly arriving, and thus it is not so much a cause of wonder, that the real numbers should not have been generally known to historians who relied upon published accounts, rather than upon records and facts.

Soon after we commenced these investigations, we noticed in a file of old papers, that a sea captain estimated the number that had been shipped from Spain to the United States at 15,000, which we supposed an exaggeration ; but we now believe that the old salt hardly comprehended the numbers enough to do full justice to the subject. He understated, instead of exaggerating the numbers. It is evident that Mr. Jarvis writing in later years must have relied upon his memory and upon his own personal observation in giving the numbers that were purchased and sent by others to this country in 1809-10 and 11 ; and that he was somewhat deceived or did not comprehend the full numbers ; for among his own papers we found a letter from Lisbon written to him December 24, 1810, (a few months after he left there with his family,) in which the writer advises Mr. Jarvis that Cochran Johnson had " sent about three thousand Merinos, principally Aguierres, to New York," and that others had shipped large numbers, so that altogether, the writer supposed that near five thousand head had gone, or were about going, since Mr. Jarvis had left. The same letter says : " Mr. O'Neil sends on his own account 120 Merinos by the brig Ann to Georgetown." It will be remarked by those who have read Mr. Jarvis's account of the importations, whence the sheep came, and where they were sent, that he makes no mention of any Aguierres having been sent to the United States except those that he sent himself, yet here is evidence that we cannot doubt, that nearly three thousand were purchased and sent to New York by one person. Again he makes no mention of those sent by Mr. O'Neil to Georgetown; yet we have found other evidence in Mr. Jarvis's letters that these sheep did probably arrive at Georgetown.

Again in 1847 Consul Jarvis gave it as his opinion that, as the Cabana of the Duke d'Infantado was not confiscated, none of his sheep were ever sent to this country, but the letter of Mr. Charles Henry Hall, which is published elsewhere in this work, will show that over two thousand were purchased of the Duke and sent to this country. The noble character of Consul Jarvis, his well-known reputation for honest integrity, forbid at once our ascribing to him any other than good intentions, when he gave these accounts, which prove to be inaccurate, and irreconcilable with the facts; the only theory that we can entertain is, that his memory was at fault, and his knowledge did not embrace all the facts; although there is no doubt that he thought he was well informed concerning them all, and no doubt was the best of any one of that day.

We have spent much time in looking up and gathering statistics at the different ports to which Merino sheep were shipped in 1810 and 1811. The list which we publish is very far from including all that were shipped to the ports south of New York as only those that appear to have been reported through Northern ports, or through the correspondence of Consul Jarvis, are included in the lists. There is plenty of evidence that there were large arrivals at Philadelphia, and ports south of that, which are not included in the list we give. We think the list of arrivals at New York, and the New England ports are nearly accurate That for Boston was kindly furnished by the Hon. J. H. Simmonds, collector of that port, and gives the number entered at the Custom House in that city, and of course is exclusive of those which had died on the passage. At the other ports the numbers that were shipped were generally given and include nearly three thousand which died on the passage. In collecting these statistics we acknowledge with pleasure, valuable and efficient assistance we have received from Prof. W. H. Brewer, New Haven, Conn., E. I. Mulchahey, Providence, R. I , and E. H. Libby, formerly editor of the Scientific Farmer, Boston, Mass.

These arrivals commenced early in 1810 (we have not found

any record of any arriving in 1809); a few scattering animals
were reported from time to time during the Spring and Summer of
1810, but by far the larger part of the arrivals were between September
1st, 1810, and July 1st, 1811.

At the port of New York between September 8th, 1810, and
April 26, 1811, fifty-two vessels arrived from Spain on which were
shipped over nine thousand Merino sheep, and during the same time
eight more vessels on which were shipped Merino sheep, but we
have not been able to learn the numbers. If the numbers shipped
on these eight averaged as great as on the other fifty, then over ten
thousand were shipped to New York alone, during a period of eight
months and twenty days.

In the accounts which Consul Jarvis gave of his shipments, he
says, that he sent to Portland and Wiscasset; but we find that those
he sent to the latter place, and a part if not all of those sent to Portland,
were entered at the Boston Custom House, and appear in the
numbers for that Port.

NUMBER OF VESSELS ARRIVING AT DIFFERENT PORTS IN THE UNITED STATES
FROM SPAIN, DURING 1810 AND 1811, WITH MERINO SHEEP AND
THE NUMBER OF SHEEP.

Port	Vessels		Sheep	
Portland, Me.,	1	Vessel,	43	Sheep.
Newburyport, Mass.,	3	"	280	"
Gloucester, "	1	"	50	"
Boston, "	29	"	2049	"
Providence, R. I.,			1235	"
Newport, "	1	"	60	"
New London, Conn.,	1	"	111	"
New Haven, "	4	"	275	"
Sag Harbor,	1	"	30	"
New York, N. Y.,	52	"	9349	"
Philadelphia, Pa.,	4	"	389	"
Baltimore, Md.,	2	"	300	"
Alexandria, Va.,	3	"	330	"
Georgetown, D. C.,	1	"	120	"
Norfolk, Va.,	3	"	401	"
Richmond, Va.,		"	525	"
	106		15767	

One vessel sailed from Lisbon with 220, but we have not been
able to find at which port she arrived. She is not in the above list.

In addition to the above, during the same time there arrived from Spain with Merino sheep, one vessel at Gloucester, Mass., eight at New York and two at Alexandria; but we have not been able to ascertain the numbers sent on them. We could not learn the number of vessels that arrived at Providence, R. I., or at Richmond, Va. A fair estimate for the eleven that arrived without numbers being reported, would be the average of numbers by the one hundred and seven that arrived at the different ports where the numbers are given, which was one hundred and thirty; this would add 1,430 to the 15,767 reported, which would make a total of 17,197; and we would again call attention to the fact, that this number, large as it is, does not include nearly all that came to ports south of New York.

IMPORTATIONS OF CONSUL JARVIS.

We will now consider more in detail the importations of Hon. William Jarvis. Mr. Jarvis wrote in 1837 that he sent about 3850, and he also wrote Dr. Randall, a few years later, that he sent that number. In 1844 he wrote Deacon Gregory that he sent 1400 Paulars, 1700 Aguierres, 200 Escurials, 100 Negrettis, and about 100 Montarcos. Of this number about one hundred were sent to Wiscasset and Portland, one thousand one hundred to Boston and Newburyport, one thousand five hundred to New York, three hundred and fifty to Philadelphia, two hundred and fifty to Baltimore, one hundred to Alexandria, and two hundred to Norfolk and Richmond. This account differs from the preceding two, in the numbers being less by 200.

In none of Mr. Jarvis's published statements that we have found does he state to whom he consigned the sheep he sent to this country, but by the kindness of his daughter, Mrs. Hampden Cutts, now living at Brattleboro, we have been permitted to examine a large portion of his correspondence of those days, and by her we have been materially assisted in collecting the facts, whereby we are enabled to throw much light upon the arrival of these sheep in this country, their dissemination and the fate of a large portion of them.

The consignees of the Jarvis sheep at the different ports, were at Wiscasset, Mr. J. F. Wood, at Portland, Messrs. Newhall & Watson, Newburyport, Jacob Little, Boston, Cornelius Coolidge & Co., and Dr. George Bates, New York, Hicks, Jenkins & Co., Baltimore, Smith & Buchanan, Alexandria, J H. Hooe, Norfolk, Moses Myers & Son, Richmond, M. & B. Myers.

There is evidence in these letters that Mr. Jarvis sold some of the sheep that he purchased before shipping them. Messrs. Cornelius Coolidge & Co. in some of their letters complain that their market was damaged by the sales he had made to others, mentioning the captains of several vessels and Mr. Crowningshield. This as well as other testimony, leads us to believe that Richard Crowningshield procured at least a part of his Merinos through Mr. Jarvis.

The sheep sent to Mr. J. F. Wood, at Wiscasset, were sent on board the ship America, the number being twenty-two ewes and five rams, landed alive. In a long letter Mr. Wood relates the result of the venture. They were sold late in the Fall, and brought low prices, Mr. Wood himself being one of the principal purchasers. The sheep sent to Messrs Newall & Watson at Portland, were sent on the same vessel ; thirty-one were landed alive, but some died soon after ; none were sold and before the next Spring they had dwindled to seventeen old sheep and six lambs ; and they were moved elsewhere, probably added to the flock that was finally taken to Weathersfield, Vt.

It is evident from the numbers accounted for by Mr. Jacob Little at Newburyport, Mass., that the numbers that arrived there must have been greater than the numbers, two hundred and eighty, reported in the list we have prepared, for he sold thirteen for $2,312.-50 ; he delivered two hundred and seventy-four to the agents of Mr. Jarvis, otherwise accounts for four and still says : "The remainder of the sheep shipped to my care must of course be accounted for by the skins of those that have died." The sheep that were sent to Mr. Little were more fortunate, or had better care than those sent to Portland. On one of the vessels a shepherd with his dog came to Newburyport. This shepherd afterwards went up to Claremont, from which place we hear complaints of his want of reliability.

Some of the first shipments of Mr. Jarvis were to Messrs. Cornelius Coolidge & Co., of Boston, and were a part of a lot of sixty-five that were purchased in Spain on joint account of Mr. Jarvis, Nicholas Gilman, Esq., of Portsmouth, N. H., Messrs. C. Coolidge & Co., and Mr. Charles O'Neil, of Lisbon. Mr. O'Neil subsequently disposed of his interest to Mr. Jarvis.

The following, from the records of the Massachusetts Society for Promoting Agriculture will show that Messrs. C. C. & Co. received a premium for importing the first Merino ewes from Spain :

"1810.
The premium of two hundred and fifty dollars was awarded to Cornelius Coolidge, of Boston, for the first ten ewes imported from Spain."

A few of the sheep sent to Boston were consigned to Dr. George Bates. Seventy-nine were entered in his name at the Custom House, about thirty-seven were entered to Mr. Jarvis, and three hundred and twenty-seven to Cornelius Coolidge & Co. We have found reports of a number of sales by the last-named gentlemen. At one sale of rams, probably a part of the sixty-five before-named, eleven sold for $10,902.66—nearly $1,000 each. Eleven more were rented at this time for $4,440, or a little more than $400 each. In another account this firm report the sale of a few sheep received by ship "Hamlet" and a few by ship "Three Brothers," neither of which were reported in the Custom House list received, and consequently they are not given in the number of arrivals at Boston.

This, with the case at Newburyport, is evidence that the arrrivals at the New England ports are probably somewhat larger, even, than those we have given. Other sales reported by Messrs. C. Coolidge & Co. were one hundred and three for $12,771, averaging nearly $124 each, and one hundred and seven for $13,671, an average of $128 each.

This firm, in several of their letters, make complaints that Mr. Jarvis did not consign all his shipments to them, and enable them to keep up the price by controlling a monopoly of the sales. It is very evident, from the tenor of many of their letters, that Messrs.C. Coolidge & Co. had a very imperfect knowledge of the vast numbers of Merino sheep that had been confiscated and were being sold in Spain at that time.

This firm also complain in several of their letters, that they were obliged to meet and counteract certain slanderous stories that different parties interested in other importations had circulated in regard to their Merinos, that they were not of the pure breed, etc.

Thus it will be seen that this detraction commenced at an early day, and gentlemen who are engaged in it at the present time will please take notice that they are not entitled to a patent for discovering a process for bringing their neighbors' swans down to a level with their own geese.

In the letters alluded to, this firm charge very emphatically that Col. Humphreys had taken every occasion to indirectly injure their sheep.

We hope that those gentlemen now owning flocks descended directly from the Colonel's will not feel it their duty to commence defaming their neighbors' flocks, because their illustrious predecessor was charged with doing the same.

As a matter of curiosity we publish the following advertisement from the Boston Gazette, of August 16th, 1810:

MERINO SHEEP.

CORNELIUS COOLIDGE & Co., having been repeatedly informed that certain characters have industriously propagated reports tending to impress upon the minds of the community that the Merino sheep recently imported by them, are not of the genuine blood and breed, and considering it due not only to themselves, but to the agricultural and manufacturing interests of their country, to take every possible measure to defeat the object of such misrepresentations and falsehoods, give this public notice, that all the sheep imported by them were obtained through the agency of William Jarvis, Esq., U. S. Consul-General for the kingdom of Portugal, and are warranted of the genuine, full-blooded Merino breed of the first class, as appears by a Spanish certificate accompanying them, signed by the Secretary of the Supreme Junta of the Province of Estramadura, and at least equal to any ever brought into this country.

It is a fact that, upon comparing the wool of C. C. & Co.'s sheep with that of Chancellor Livingston and Col. Humphreys' flocks, the preference was decidedly given the former.

To those who are acquainted with this valuable race, the appearance of the sheep imported by C. C. & Co. is the best evidence of their blood, and as there may be many persons desirous to purchase a Merino who are influenced by the reports alluded to, and to whom it might be too inconvenient and expensive to visit Boston, upon an uncertainty, they are referred, for full satisfaction upon the subject, to the following gentlemen, living in those parts of the country where the false reports are the most prevalent, and who are possessed of sheep imported by C. C. & Co., viz.:

Geo. Fitch,	279 Pearl St., N. Y.
Ezra Reed,	Hudson, "
Roger and Solomon Tainter,	Hampton, Conn.
David Watkinson,	Hartford, "
Frederick Butler,	Weathersfield,"
William Kinne,	Plainfield, "
Joseph Bellows, Jr.,	Walpole, N. H.
Aaron Tufts,	Dudley, Mass.
James M. Lellan,	" "
Hon. Samuel Lathrop,	Springfield, "

With others that may be known upon application to Cornelius Coolidge & Co., No. 53 Long Wharf, Boston, who have still for sale or hire a number of the *pure blood*.

The printers in the United States who are friends to the manufacturing interests of their country, and wish to correct misrepresentations calculated to check the present spirit of enterprise, are requested to give the above a place in their respective papers. The printers in New Haven and its vicinity, where the envious reports originated, are particularly desired to publish the above.

Of the fifteen hundred sheep that Mr. Jarvis shipped to New York, we find mention in his papers or letters of only two or thre hundred shipped to Messrs. Hicks, Jenkins & Co. From the notices of arrivals, and the consignees named, together with the facts related above, we have more evidence that a large portion of the fifteen hundred Mr. Jarvis sent to New York were sold to other parties before they landed at that port. We have heretofore intimated to whom a part of these may have been sold, and shall again mention other circumstances that strengthen our belief in that opinion.

Messrs. Hicks, Jenkins & Co. were a company of gentlemanly Friends, and it is a pleasure to read their letters, and we think Consul Jarvis must have been pleased when he read the one of which the following extract is a part:

NEW YORK, 6 M., 7, 1810.

WILLIAM JARVIS & Co:

ESTEEMED FRIENDS:—We have now the pleasure to acknowledge receipt of your favor by the ship Maria Theresa, Capt. Dickson, with a consignment of seven Merino sheep, six of which we have to advise of having sold for seven thousand five hundred dollars, at two and four months' credit. The other not looking quite so well, from his being affected with the scab, have not yet sold it; and have tried it at auction, but found no bidders. We therefore had it struck off, after running it up ourselves to $750, but are in hopes of closing sales soon, when you shall be furnished with account sales, etc.

It will be seen that these six sheep sold at an average of $1,250 each. Among all the accounts of sales we have found by the agents of Consul Jarvis, these six sheep sold for the highest average price. Indeed in the best sale reported by Messrs. Coolidge & Company the highest price obtained was $2166 66 for two sheep, or $1,083.33 each; thus it will be seen that the average obtained for the six sold by Messrs. Hicks, Jenkins & Co. was $166.67 more than that obtained for any one sheep by the other agents of Consul Jarvis of which we have any account.

In the same letter the firm warn Mr. Jarvis that he must not expect so large prices for any further shipments he may make to them. We are unable to learn much about the large portion of the sheep sent to New York. One letter mentions their having been sent to Brooklyn, L. I., for keeping.

We are able to account for many more of the sheep that were sent to Philadelphia to Messrs. Levi Hollingsworth & Son, than of those sent to New York. Mr. Jarvis in his letter to Dea. Gregory says he sent "about three hundred and fifty to Philadelphia." From letters we found from Messrs. Levi Hollingsworth & Son, we should judge that nearly or quite that number were consigned to them, but

5

that many died on the passage. They render accounts of sales of one hundred and ninety to Mr. James Caldwell at one time, for $28,500, $150 each, and of $5,960 worth with no numbers given to Mr. Caldwell, Mr. Warner and others, and of fourteen at another sale, for $75 each. At the sale for $5,960, Mr. Caldwell made a $3000 purchase. The last cargo arrived at Philadelphia September 20, 1810, and in consequence of large arrivals found a low market. There were one hundred and forty-two shipped; but a number died on the passage and more soon after. December 24th, following, only one hundred and eighteen were left. They were in bad condition and diseased when they arrived.

In January, 1811, Michael Kippley, Esq., offered $50 each for the whole flock. May 13th, they were offered Judge Griffith, of New Jersey, for $60 each. May 14th, they were offered at $45 each for the entire lot. May 20th, Messrs. Young, Dupont and Warner offered to take the lot on shares.

We have been unable to ascertain the final disposition of these sheep. The last letters discuss the question of sending them to New York by sea or over land, as was ordered or proposed by Mr. Jarvis, but there is no light upon the final disposition of them.

The two hundred and fifty Consul Jarvis sent to Baltimore were consigned to Messrs. S. Smith & Buchanan. In one letter they speak of previously sending advices of sales of sheep that brought $22,159.69 nett. The number sold is not given but the letter states: "They were sold at a most fortunate period for your interest." It was written November 27th, 1810. These sheep arrived in October previous. In a letter December 18th following, they wrote to Messrs. Myers at Richmond at the request of Mr. Jarvis: "We are persuaded that a further sale at this time would be impracticable on almost any terms." Mr. Jarvis gave the numbers sent to Alexandria at about a hundred. The numbers must have exceeded that. They were consigned to Mr. J. H. Hooe, but one cargo was received and, in the absence of Mr. Hooe they were sold by one Mr. Muncaster. Mr. Hooe sold one cargo of a little over one hundred to Mr. George Fitch for $10,864.11, and later a sale of seventeen was made for $1,250, and still later one ram and two ewes for $300. March 22d, 1811, five rams and forty-three ewes, with fourteen lambs, were brought from Richmond.

From the sheep sent to Alexandria, Mr. Jarvis ordered a pair selected and presented to President Madison, and another pair to ex-President Jefferson, which orders were executed. The last arrivals at Alexandria were unfortunate. They were diseased, and the dogs killed a number. April 14th, 1811, Mr. Hooe proposed to take all that were left upon shares.

April 28th, 1811, there were still on hand thirty-seven old sheep and fifteen lambs, which is the last information we have of them.

Mr. Jarvis's account gives about two hundred as the number of those shipped to Norfolk and Richmond. Here again the numbers must have exceeded that, for his correspondents and consignees account for more. They were consigned to Messrs. Moses Myers & Son, who sent a part, one hundred and twenty-five, to Messrs. M. & B. Myers, at Richmond.

Very few were sold at Norfolk. We have an account of a sale of one ram and five ewes for $750, to a gentleman of South Carolina. Evidently the Merinos were not appreciated at any of the Southern ports, and the business was very dull, the sheep badly diseased, and the consignees ignorant of the care and best management of them. The large numbers arriving, their poor condition, falling prices, no sales, and the need of great care, with poor results from their best efforts, must have been very depressing, as Messrs. M. Myers & Son wrote January 20, 1811, to Mr. Jarvis, that they "see no prospect of success, more having arrived, and we look for a ship which has more on board to our address. Heaven knows what will become of them."

October 20th, 1810, they were offered $100.00 each for forty. November 14th, more dying, no more sold. November 20th, same story. Offered to let a gentleman from South Carolina pick ten at $130 each. In February the weather was such as had not been experienced for years. Sixteen died, and the remainder were put into a house where a fire protected them from the cold. In March there were seventy-eight survivors, but during that month the dogs destroyed eighteen in one night. March 12th, twenty were shipped to Messrs. S. Smith & Buchanan at Baltimore. April 19th, 1811, only thirty-five sheep and four lambs survived, and we are without information of the final fate of these. Altogether the shipment to Norfolk was most unfortunate. An incident was the detection and trial with a public whipping of a slave for stealing a number of the skins from those that died. In one of the latest letters Messrs. M. M. & Son say : "Never have we had a more unpleasant consignment. We were never more anxious than for your order to ship the residue ; they perplex us much."

The sheep that went to Messrs. M. & B. Myers at Richmond, as before stated, were one hundred and twenty-five; fifteen died on the passage, and eight soon after, and further losses followed.   They were taken to a farm near Richmond belonging to a Mr. Temple, who seemed to exercise a watchful and benificent supervision of them, for they improved at his place.   Six of these, a Paular ram, three Paular and two Aguierres ewes, were sold to A. B. Venable, President of the Virginia Bank, for $1200.   Two rams, eight ewes and two lambs were sold for $862.   Two coops containing one Paular and one Negrette ram, two Paular and fourteen Negrette rams were shipped to Messrs. Hicks, Jenkins & Co., New York   March 22d, 1811, the remainder, five rams, forty-three ewes and fourteen lambs, were sent to Mr. Hooe at Alexandria.

Of the sheep imported by others in 1810, that were probably procured through Mr. Jarvis, we have found in an old file of Providence papers, an advertisement of a cargo for sale at the farm of David Buffum, in Newport, R. I., consisting of " seventy-four Merino rams and ewes, warranted of the pure Merino breed, shipped by William Jarvis, Esq , American Consul at Lisbon, etc."

The advertisement was signed by Capt. Paul Cuffe and Isaac Cory and dated 9 M. 7, 1810.

Some of the older citizens of Newport remember Capt. Cuffe as a sea captain of tall, commanding appearance, who came to the meeting of the Friends when in port ; and they also remember that he was generally accredited with importing into Newport from Spain the Merino sheep from which the flocks of David Buffum, William Bailey and others principally sprung, though there is a probability that blood from the Humphreys importation may have been also introduced into some of the flocks at Newport.

As intimated before, there is reason to believe that Mr. Richard Crowningshield procured from Consul Jarvis at least a part of the sheep which he imported into New York.

Some sheep that were driven into Addison County, to be kept for Mr. Crowningshield, were said to have been Jarvis sheep.   In the account given by Judge Lawrence, of the origin of the flocks of Andrew Cocks, of Long Island, he states that Mr. Cocks' first purchase was of two Escurial ewes of Richard Crowningshield ; which, according to Mr. Jarvis's statement, he could not have procured in any other way than through himself.   In the list of arrivals at New York of vessels with sheep from Spain in 1810 and 1811, we find that three full cargoes were consigned to R. Crowningshield, the numbers being given on two of them at two hundred and ninety. In the other case the numbers are not given. Another vessel also arrived with Merino sheep consigned to R. Crowningshield, Bailey & Swan, and Buckley & Somerdyke.

As a matter of interest we give the consignees of sheep at New York, and as far as can be ascertained the numbers consigned to them in 1810 and 1811:

S. Hathaway,	6
Wood & Skinner,	30
Post Grinnell and Minturn,	410
Hicks, Jenkins & Co.,	220
Wm. P. Viger,	150
J. Creighton,	206
Wm. and S. Craig,	356
N. L. and G. Griswold,	138
R. Crowningshield,	290
E. Dexter (Providence, R. I).,	44
John Jubel,	320
Green & Lovet,	21
Ingraham, Phœnix & Nixen,	16
Wm. Codman,	1118
J. Murray & Sons,	1729
Lawrence & Whitney,	216
Captain Hayes,	5
W. Osborne (owner),	148
To order,	150
A. Cranston & Co.,	151
Taylor & Hamilton,	80
Hall, Hull & Co. (owners),	184
M. Smith,	70
A. Barker,	100

Besides these, consignments were made to the following persons, where the numbers to each were not stated, but the cargoes were consigned to more than one person or firm:

Leroy, Burnell & McEvers, Rossier & Bowley,	300
Isaac Clawson (owner), and Howard,	200
Hall, Hull & Co., and Charles Dickinson,	269
Captain (owner), and Gilbert Robinson,	
N. T. Allcott and R. Dickey,	248
J. Clawson (owner), and A. Barker,	210

A. Gracie & Son (owners), Wm. Codman, and
    A. G. Thompson,   300
S. Whetmore and others,   50
William Watkinson, Lawrence & Whitney,   136
John Murray & Sons, and the Captain Morgan,   270
John D. Miller, and G. S. Mumford,   75
D. Hadden, Col. Thorndike, and Wm. Codman,   85
N. L. Griswold, E. Leavenworth, Strong & Haven,   330

There were also consigned to the following eight cargoes of sheep where the numbers were not given :

One to E. Leavenworth, one to A. Ruden, W. Codman, Smith & Hubbell, P. & G. Havens, Capt E. Townsend and Charles Dickinson, one to R. Crowningshield (owner), one to J. Murray & Sons, one to J. Barker, one to J. Robertson, one to R. Crowningshield, Bailey & Swan, and Buckley and Somerdyke, and one to no name.

The consignees at Boston, were :

Cornelius Coolidge and Co.,   377
Eben Parsons,   156
Goodwin and Whitney,   100
Samuel Ruggles,   7
Gideon Snow,   149
Henry Bass and Co.,   23
William Hewes,   30
Ignatius Sargent,   45
Dr. Bates,   79
Newell and Watson,   30
J. S. Wood,   33
Jona Allen,   80
John W. Rich,   127
Nathaniel Curtis,   25
William Story,   67
William Jarvis,   37
Charles Heard,   100
Thomas Thaxter,   67
Brigham and Bigelow,   75
J. B. Kettle,   15
E. Snow,   3
Jona. Buffington,   198
Oliver Keating,   8
Elisha Ayer,   200
John Doak,   3

The following extract of a letter was published in 1844. Mr. Hall's residence at that time was at Harlem, New York. It shows a large importation of Infantado sheep from Spain, during the years we speak of: "The Duke del Infantado it is true joined the patriot cause, and went ambassador to England from the cortez at the time Ferdinand was detained in France, and returned to Cadiz, when that city was in a state of siege. There I was introduced to the Duke by the United States ambassador, Mr. Ewing. His flocks he informed me were in positions of safety from the contending armies, in various parts of Spain, some of them in Andalusia. The result of my interview was a purchase from the Duke of a flock of four hundred sheep by myself and associates, which were shipped to Virginia, consigned to Messrs. Brown and Rives at Richmond. Subsequently there were obtained from the Duke two thousand more sheep, having this mark (a brand of Y, upon the side of the face of the sheep), which were shipped to New York and Philadelphia, for account of Commodore Charles Stewart, Consul Richard Hackley, myself and others. Of one of the cargoes, Chancellor Livingston had a large lot of my Infantado sheep which he purchased of my agent, Mr. Henry Ward, and I think in some of his writings, he speaks of the high estimation in which he held the flocks of the above-named Duke. The invoices of these sheep and the result of the shipments, I have among my papers and will select them out hereafter for the inspection of Mr. Allen.

CHARLES HENRY HALL."

Some things in this letter are corroborated by other circumstances. H. Ward is given as one of the consignees of Merino sheep that arrived at New York on board the Maria Theresa from Cadiz, and 38 days from Villa Real, Spain. She arrived Oct 19, 1810. The brand on the face of the sheep is the same as was upon the sheep imported into New Haven in 1810, as described by Jacob N. Blakeslee, a more full account of which we shall publish hereafter.

Another coincidence is that both vessels sailed from Villa Real, not from Lisbon, where most of the confiscated flocks—perhaps all— were shipped. There are several reasons for believing that this importation into New Haven, which we will now notice, was from the Infantado flock, and probably a part of the purchase described by Mr. Hall.

In the Marine News of the Connecticut Journal, of Dec. 27, 1810, will be found the following:

"Arrived,—Brig Ceres Capt. William Fairchild, from Villa Real, Spain, 45 days, with 150 Merino sheep, to the Captain C. Peck and F. Woodward."

Upon the day in which they arrived, seventy-eight of these sheep were advertised to be sold at auction, Jan. 9, 1811, by Joel Atwater, auctioneer.

The advertisement reads :

"These sheep were selected from the Duke Infantado's flock of 1500 by a person who went from this country for the purpose, and are said by judges to be superior to any that have been imported."

The issues of the same paper for Jan 10th, 17th, 24th and Feb. 7th, 1811, also have an advertisement by Capt. William Fairchild, of "sixty full-blooded Merino sheep, selected from the Duke Infantado's flock, imported in the brig Ceres, Capt. Wm. Fairchild, direct from Spain, to be sold at private sale. Inquire of the subscriber in Wooster St."

It is evident from these advertisements and others, that Joel Atwater acted as auctioneer for several flocks and importations, but probably was not an importer himself. Acting, however, as auctioneer or agent for those importing, it was very natural that he should have been supposed to be an importer himself at that time, and as such various authors and historians of importations of Merino sheep have recorded him.

Another importation into New Haven was the schooner Elizabeth Little, via Turk's Island and New York. They were auctioned Nov. 30th, 1810. The same date their arrival was noted, as "60 genuine Merino sheep, selected from the best flocks in Spain," etc., for sale by Prescott & Sherman and Norton & Bush, to whom they were consigned.

The last importation of Merino sheep from Spain into New Haven of which we find any account, was what was called the Heaton importation.

They arrived at New Haven, Jan. 3d, 1811, on the brig Bellona, Capt. White, 43 days from Lisbon, and were forty-two in number. They were consigned to Abraham Heaton, and *others* (not Abraham Heaton & Co., as some have given it). The following advertisement will explain from whence these sheep came. It is from the Connecticut Journal of Jan. 10, 1811 :

To be sold at auction, in this city, on the 17th instant, January, 42 MERINO SHEEP, imported in the brig Bellona from Lisbon. Thirty of these sheep are the improved breed, the most unquestionable documents accompanying them, proving them to be of unmixed race of Leonese Merinos, of the flock termed Guadaloupe, they were purchased of the Prior of the Royal monastery of Guadaloupe, in Spain, and are warranted genuine.......

The remaining twelve are of the Negretti breed, with certificate attesting their genuineness. Sale to commence at 10, A. M.

JOEL ATWATER, Auctioneer.

From the importations by the Ceres and the last-mentioned by the Bellona, sprung the flock of Jacob N. Blakeslee, and perhaps some others, in Connecticut. The statement of Mr. Blakeslee, made years afterwards, with regard to these importations, although varying in a few unimportant particulars, is in the main corroborated by the facts of the case. His memory, as in the case of Consul Jarvis, was in fault in some respects, though in some cases where it varies from the facts it seems more a confusion of them in regard of the two importations, than any addition to or suppression of the facts in the cases.

We may give here as well as anywhere, Mr. Blakeslee's statement, as he related it to Prof. W. H. Brewer, of New Haven, in 1868 or 1870. We do not give the whole statement as given by Mr. Blakeslee, but such parts as are of material interest to understand the subject:

"There were two importations into New Haven.

In 1810, an importation by Peck and Atwater, of New Haven. In 1811, another by Abraham Heaton & Co. John DeForest was supercargo. Merino sheep fell in price about that time. These were sold at auction—at least some of them were. I saw all the sheep of both these importations. Both had the same brand on the nose, a V or Y, fork upwards ; no hair where the mark was put on. I was then told that this was the Infantado brand. Capt. Peck told me that they were the best flock in Spain, and were called the Infantado.

When the Heaton importation was sold there was one particular ram that I wanted. He was unlike the others ; he had a peculiar fleece. Several wanted him. He was sold when some of us were away, getting something to eat. Daniel Bacon, of Woodbury, got him. I afterwards got two rams of his get, but I never owned that identical ram."

He then stated he took some of both these importations upon shares. Some he took of Peck & Atwater, and some of Mr. Woodward. In speaking of the ram which Mr. Bacon bought, he said,

6

"That ram of Bacon's was the best of all the original importations I ever see." Elsewhere Mr. Blakeslee has stated that this ram was an Escurial, and that he was sold to Mr. Bacon for $275, who, after using him for five or six years, sold him again to Mr. William K. Sampson for $1,130. An item taken from a Connecticut paper of that time notices the sale, but gives the price at $1,300. (Further notice of this ram will be given in treating of the introduction of Merinos into Vermont.)

The critical reader will observe some differences in the above statements from some other acconnts Mr Blakeslee has published of the origin of his flock; but there are a number of points, and those the most vital ones, which are adhered to through all the accounts he has given. In all that we have seen, he gives the first foundation as from Peck and Atwater. In 1844 he stated that a part of these were Negretti and a part Montarco. Here there is some confusion in his memory, but no more than we find in the statement of Consul Jarvis at the same time; and as the latter was a gentleman of fine education, and as we are obliged to ascribe his mistakes to bad memory, why may we not extend the same measure of charity to Mr. Blakeslee, a man of infinitely inferior mind, much less culture, and comparatively limited experience to Mr. Jarvis's?

Only a few months before he died, Mr. Blakeslee wrote a friend of ours that his "sheep were imported by Peck & Atwater, of New Haven; they were selected by Capt. Peck from the Infantado flock, the best in Spain. I took them of Capt. Peck in 1815 with the letter V branded on the nose of every sheep."

We have thus given such extracts from Mr. Blakeslee's statements as will show their variation, and still we shall find that mainly they agree in some important particulars, and where they vary the differences are such as most men would make in describing events that happened in their early youth.

Taken altogether, we think them strong affirmatory evidence that the statement of Mr. Hall as to Infantado sheep having been imported, and their having the brand of the Infantado flock upon their faces, was true.

Somewhere in our researches we have seen it stated that the sheep that were imported into New London were from the Infantado flock, but as our memorandum of the matter is lost we can only give it as a rumor, without authority.

We give the following extracts from a letter written by Mr. H. H. Green, from Lisbon, to Mr. Jarvis, December 14, 1810. It throws light upon some matters connected with the importation of Merinos at that time, some of which have been already alluded to :

"We have bought from Mr. O'Neill 100 sheep of the Cabanas of the Paular and Montarco at $30 per head, 1 ram to 7 ewes, for account of Messrs. Smith & Buchanan, to go by the Scioto. Cochran Johnson, as already advised, has sent about three thousand Merinos, principally Aguierres, to New York. Several others say Mr. Grant, Mr. Edwards, etc., have shipped. In fact almost every vessel bound for the United States since you left, has carried sheep, so I suppose near five thousand head have gone and are going. Mr. O'Neill sends on his own account, 120 Merinos of the Montarcos, by the brig Ann, to Georgetown."

Mr. O'Neill, here spoken of, seems to have been a very extensive purchaser and dealer. He was one of the four who were engaged in the first purchase and shipment of Merinos, but afterwards sold out his interest to Mr. Jarvis. In the life of her father, by Miss Cutts, we see that through the intervention of Mr. Jarvis, he made a sale of one thousand Montarcos to General E. H. Derby, of Salem, Mass. These sheep were shipped on one vessel, contrary to the advice of Mr. Jarvis, and their history shows that they were very unfortunate. The voyage was long and tedious, with adverse winds, and nearly half of them died on the passage. They were kept in very muddy yards on their arrival in New York, and many more died there. Only about two hundred and fifty were left to sell in the Spring, besides less than two hundred which Gen. Derby reserved to put on his farm at Salem.

In Consul Jarvis's account of the importations he says : "The Guadaloupes, Paulars and Montarcos, which were shipped to Boston by others, were for the account of Gorham Parsons, Esq., Gen. Sumner, D. Tichenor, and E. H. Derby, Esq." We think Mr. Jarvis's memory must have been at fault in this account, for we can find no account of any entry of sheep of E. H. Derby at the Custom House, at Boston, and Mrs. Cutts's account gives New York as the place of the arrival of Gen. Derby's sheep. We think the only vessel that could have brought Gen. Derby's sheep was the ship "Mount Hope," that arrived at New York Nov. 28, 1810, with seven hundred Merinos, having lost three hundred and fifty on the passage ; the captain's name was reported as Derby, and the sheep were consigned to William Codman. No other vessel at any of the ports was reported as having half as many sheep on board as were reported on this.

The following letter written by Gen. Derby will show that misfortune and disease followed these sheep to their new home:

SALEM, January 7th, 1811.

WILLIAM JARVIS, ESQ.,

DEAR SIR:—Permit me to felicitate you upon your arrival in Boston after so long an absence, and thank you for your politeness to me while in Lisbon. I have been very unfortunate with my sheep, owing to a disorder very similar to the small pox, and which I am since told the shepherds called the pox. I fear that our friend O'Neil knew of it when he sold me, for all those affected at the nose invariably died; as did almost all that were much affected about the head, and he scolded much at my rejecting such ones. I am losing a number, and amongst them some of the most healthy, owing (I fear) to their being littered on with some of the ship litterings; the smell of it I fear is injurious and infectious to the lambs as soon as born. I fear much I shall lose almost all my lambs. If you think of anything to relieve my difficulty, I will thank you for a hint. Pray inform me what you have done with your wool, or what prospect you have. Is it not adapted to hatting as well as for the clothier? what is the price? Hoping for the pleasure of seeing you, I am with respect, your

Very humble servant,

ELIAS HASKELL DERBY.

P. S. If you are not better provided, I will sell you my Ten Hill farm in Charleston. If you have a shepherd to spare, or know of one, may it not be best for me to send one to New York.

The disease here spoken of, clauean or sheep pox, affected many of the importations that were arriving in 1810 and 1811, and no doubt was the cause of some of the fearful losses on ship-board at that time.

There were at least six vessels that arrived at New York in those two years which lost over half of the numbers that were shipped upon them, namely the Brooks, which lost 104 out of 194, the Lydia, 139 out of 269, the Gen. Coleburn, 191 out of 300, the Ann, 244 out of 304, the Orion, 40 out of 75, the Fox, 93 out of 128. Besides these, there were a number that lost one-third or more of the number shipped on them. The same disease raged fearfully among a part of the flock Mr. Jarvis sent to Claremont.

This disease is very accurately described in a small volume published by Samuel Bard, M.D., at New York, in 1811, upon Merino sheep, their management, diseases, etc. It is a very interesting little work and gives excellent advice in regard to breeding and keeping Merino sheep, with a very good description of the diseases which sheep-flesh is heir to. From it we learn that the foot-rot came to this country with the importations from Spain previous to its publication in 1811, instead of first with the importations of Saxony sheep many years later, as was supposed by Mr. Jarvis and others. It is noticeable that the sheep of Gen. Humphrey are only alluded to once in this work, while those of Chancellor Livingston are spoken of several times ; his opinions and experience are also often quoted.

The same may be said of the files of papers that were published about the same time. We find much more information of the sheep of Chancellor Livingston than of Col. Humphreys', the records of the Massachusetts Society for Promoting Agriculture having more information (chiefly through the communications of Col. Humphreys) than we get from any of the papers that were published in Connecticut even at that time.

Before we leave this matter of importations, we wish to again express our opinion that few of the vast numbers of the sheep imported in 1810 and 1811 were impure. It is evident that, at that time, it was about as easy to procure sheep of the pure-blooded, Leonese Transhumate flocks as any ; and, as thousands were being slaughtered and eaten by the armies of Spain, it is but reasonable to suppose that they could be procured at low rates. It is true that shrewd men and good judges would then, as always, take advantage of the occasion to buy superior selections, but it is more than probable that the most of the sheep that were imported at that time, were pure-blooded and of the best flocks of Spain, although there is no doubt that there was much difference in the individual excellence of the different importations. In the midst of these importations the New York Gazette published the following :

"The number of Merino sheep imported from Lisbon during the last month may be justly considered as matter of astonishment by those who recollect the difficulties which were stated to exist in procuring those animals, etc.   *   *   *   The recent importations, it is believed, are all accompanied with a variety of well-authenticated documents, so as to leave no doubt of the breed being as represented, etc."

In nearly all the advertisements of imported sheep for sale at that time, it was stated that these certificates attesting the purity of the sheep accompanied them.

While there is no doubt that these certificates were numerous, it is probable that most of them were English copies of the original ones given in Spanish, and most likely attested to be genuine by our Consuls at Spanish and Portuguese ports. The great efforts made in later years to revive recollections of what were in the *original Spanish certificates*,—now supposed to be lost—as they were remembered by many persons, and even surmising that the original Spanish certificate given Consul Jarvis at the time of his large purchase from the Paular Cabanas was probably one of these documents that has been lost, are simply ludicrous, in view of the fact that the persons certifying to what the certificate contained probably could not understand a word of Spanish, while the original certificate is still in the possession of Mrs. Cutts, daughter of Consul Jarvis, and that the English translation of it, by Gen. J. W. Phelps, was first published in 1869, in her memoirs of her father.   In an interview with Mrs. Cutts we called her attention to this statement, when she wrote the following note :

" I am not aware that any of my father's certificates are missing.   He took great care of them during his life, and I have done the same since his death, and had all that came into my possession translated by Gen. J. W. Phelps, and the translations inserted in the life and times of William Jarvis.   I think I was mistaken in saying some were lost in that memoir."

INTRODUCTION OF MERINO SHEEP INTO VERMONT.

We have found no certain record of the first introduction of Merino sheep into our State. Some were introduced from the flock of Col. Humphreys previous to the Jarvis importations in 1810. As early as July 25, 1810, Elias Gallup, of Woodstock, advertised in The Washingtonian, rams, from Col. Humphreys' flock, probably, though they were grades from 2-8 to 6-8 wooled. In the same advertisement he states : "Said sheep can be had by applying to the subscriber on Woodstock Green, Samuel Dumar, Esq., of this town, Judge Keys, of Stockbridge, Elisha Hotchkiss, of Chelsea, Oliver Lathrope, of Sharon ; Freeman Leavitt, of Hartford, Doct. Phineas Parkhurst, of Lebanon, N. H., and Samuel Montague, of Bridgewater, where samples of wool and cloth made from the said wool can be seen, etc. * * * The public are cautioned against imposture, etc. * * * All who possess higher blooded than the above can shew it by certificates from the subscriber, or Col. Humphreys, who has an accurate account of the whole of said sheep in this part of the country."

Oct. 6, 1810, the same paper contained the following advertisement :

MERINO RAM TO LET.

"The famous full-blooded Niles ram will be kept for use this season, on such terms as shall be agreed upon.

Enquire of

CAPT. PETTIS."

Just what the term "Niles ram" means here, is not certain, but we have some evidence in the following statement, furnished us by Jerome Holden, Esq., Westminster, Windham County, that he was probably bred by Col. Humphreys, and the statement will also account for the introduction of more Merinos into our State. Says Mr. Holden :

"I consulted Hon. W. C. Bradley at Brattleboro' in September, 1865. Mr. Bradley was a former resident of this town, and one

whose word was always considered reliable. At the interview with
Mr. B., I wrote down the following :

I first met Col. Humphreys at a tavern in Medway, Mass., I
think in 1808 or 1809. I could tell exactly if I were with my papers
in Westminster. He said he had some full-blood Merino rams that
he wished to get some one to take and use, as he wanted to get the
blood introduced The rams were at Hartland, Vt., and the man
that had them was not doing anything with them. I think they
were called the Niles rams. He valued the largest one worth
$1,000 and the smaller one $950. Said if any one would take them
and keep them well, and use them, they might have the use of them
for nothing. I told Col. Humphreys there was a man in Westminster by the name of Mark Richards, a wealthy farmer, that I thought
would take them, that he had a nephew living with him by the name
of Luther Richards, that took a great liking to sheep. I informed the
Richardses, and they sent for the rams, and they bred their ewes to
them, and it made such an improvement in the quality of fleece they
concluded to purchase some full-blood ewes, and attended the auction sales of Capt. Nathan Dorrs, of Roxbury, Mass., and purchased
several ewes. They had rams of Col. Humphreys (and, perhaps,
some ewes) for several years. The auction sales were advertised in
the Boston papers of that day....

The Richardses bred Merino sheep for several years, purchased some of Consul Jarvis. I incline to think the Jarvis sheep
were the best I think the Niles rams sheared three or four pounds
of washed wool each.''

We give the statement somewhat at length, as throwing light
on two or three points. One, that so-called Niles rams were bred
and owned by Col. Humphreys ; second, that Col. Humphreys did
send Merino sheep into Vermont before Consul Jarvis imported
Merinos, and that these were bred (and by the testimony we have)
were disseminated among the flocks of Vermont. Mr. Eldad Harlow,
of Westminster, "remembers going into a pasture when a boy, with
his father, to see Merino sheep. They looked black as muddy hogs.''
The Harlows purchased some of these Merinos of Mr. Richards,
"and bred them for a long term of years,'' but the Richards flock of
a thousand or more were ruined by the introduction of Saxony
blood, like so many more of the fine flock of Merino sheep of that
day.

Early in 1811, T. W. Perkins, Esq., of Boston, purchased of
Consul Jarvis a Nigretti ram, and fifteen Aguierre ewes, that he sent

to his agent, Amos W. Barnum, of Vergennes. As Mr. Jarvis would not allow the sheep to be selected, the following plan was agreed upon for the purpose of getting an average : " It is understood that the flock is to be put into a yard, the flock being driven together, a gate then to be opened and the first fifteen sheep which escape are to be mine."

Some of the older citizens of Vergennes remember these sheep, and that they were kept and bred there for a number of years, and were scattered through the adjacent towns and counties. Rowland T. Robinson, Esq., of Ferrisburgh, remembers that ex-Gov. Chittenden and his brother, Truman Chittenden, of Williston, with the assistance of his father, Thomas R. Robinson, selected some Merinos from this Barnum flock, and returned to Ferrisburgh with them in a sleigh. Mr. Robinson says : " After spending the night with us I think the three went together to Barnum's, and returned the next evening with their purchase. I do not remember the number, but it could not be *many*, as the whole were stowed into a common sleigh with the three men."

Mr. R. T. Robinson also gives the following account of the purchase and result of another introduction of Merinos into Addison County:

" In the autumn of 1811, my father, in company with his nephew, Jonas Minturn, of the City of New York, purchased from off shipboard, direct from Spain, three Merino ewes and one ram ; the latter and one of the ewes were of the Paular family of Merinos, the other ewes where one Escurial and one Aguierres. As I remember them, the Paular ewe was the largest of the three, the wool finer, and somewhat shorter than that of the other two, the weight of fleece five to seven pounds washed on the back, I mean the average of the three. We kept them and their progeny, together with the various crosses with our old flock of natives long enough to greatly improve our own flock and those of the surrounding region, until the introduction of the Saxony sheep, when, like most other keepers of large flocks, we took the fatal disease which put an end to our hopes of success in the line of wool-raising. Many years afterwards, I made some amends for my folly, by introducing a flock of ewes from Rhode Island, from which I obtained a good flock."

We have not been able to find a definite date of another early introduction of Merino sheep into Addison County, but it must have been about the same year of the above. The facts, as nearly as we can ascertain, are, that a number of Merinos that were from one of the importations by Richard Crowningshield were brought into

7

Weybridge and kept there in a pasture for a few months. They were badly diseased and finally sold to Hon Horatio Seymour (from their hopelessly diseased condition, probably for a very low price).

The sheep were bred for a number of years by Mr. Seymour, and were disseminated somewhat among the farmers of the vicinity. The manuscript records of the *old* Addison County Agricultural Society, for a number of years from its formation in 1819, are now in our possession. Esquire Seymour is often mentioned in the premium list as having been awarded premiums on Merino sheep. As several others, John N. Hunt, Samuel Crafts, William B. Sumner, A. Catlin, and Daniel S. Potter, are also mentioned as receiving premiums on Merino sheep, the supposition is that Mr. Seymour must have sold some of the blood, or that they procured it from some other source.

The impression these sheep of Mr. Seymour's produced upon the mind of the late Edwin Hammond, when a boy, was what caused him to look for the Atwood sheep, which he said to the writer were the first he had ever found that looked like the Seymour sheep—the peculiarities spoken of that gave them the resemblance, being mainly in the appearance of the ends of the wool, or surface of the fleece. Although there is no record what persons procured these sheep, there is no doubt but that in the case of all three of these flocks of Merinos, the Barnum, the Robinson and Seymour, did become disseminated and produced great improvements among the flocks of Addison and adjacent Counties.

William Samson, of New Haven, purchased a Merino ram of Danel Bacon, of Woodbury, Conn., at an early day, but the exact date of this purchase we are unable to give. In an account given of the sale in a Connecticut paper, the sum paid for the ram was $1,300. Mr. Blakeslee gives the sum as $1,100. The ram was imported at the time of the Heaton importation, and was the cause of much competition at the time that importation was sold. In Mr. Bacon's hands he became celebrated as a very superior stock ram, and obtained a wide notoriety as an improver of the flocks of the neighborhood.

A gentleman, then a resident of New Haven, but now living at the West, write us, that he remembers looking at this ram when a boy; that he was very black, and his appearance was a great contrast to that of other sheep. As New Haven has always been celebrated for her flocks of fine-wooled sheep, this ram probably made his impression upon them.

In April, 1811, the sheep Consul Jarvis had reserved from his different importations, and gathered at Claremont, New Hampshire, during the previous Fall and Winter, were moved across the Connecticut river, and settled upon the farm he had purchased a short time previous as his future home, and from whence so many of the flocks of Eastern Vermont, and many of those in the Western part of the State, derived the blood for their foundations, and procured other additions in after years, to recuperate and improve the quality of the first, or to improve that derived from other sources.

As several accounts of this flock have been published, it will be unnecessary to give any very extended account of it here. The whole flock numbered about four hundred, probably short of that number. Three hundred were selected by the shepherds who came with the Paular flock to Lisbon. About half of them were Paulars, one-fourth were Aguierres, an eighth Escurials, and the other eighth Montarcos and Negrettis. The other hundred were some that remained over after the sales of those sent to Boston.

The following advertisement, from the Washingtonian, of Aug 10, 1811, will show that five rams and four ewes were brought into Wilmington, in this State, in May, 1811 :

"MERINOS FOR SALE OR TO BE LET

by the subscriber, on reasonable terms. Nine full-blooded Merino sheep, viz : five bucks and four ewes, which were imported from Spain to New York in the month of April last, and were brought to this town in the month of May last. A certificate of their importation will be shown by the subscriber, and of the flock from which they were taken. A liberal credit will be given, etc.

LINUS AUSTIN.

Wilmington, August 10, 1811."

In 1814, Chief Justice Skinner brought from Watertown, Connecticut, a number of Merino sheep that were said to have been of the Humphrey importation. Their descendants afterwards passed into the hands of the late Hon. J. S. Pettibone, of Manchester.

In the Spring of 1816, Zebulon Frost and Hollet Thorn purchased from Judge Effingham Lawrence and Andrew Cocks, Flushing, Long Island, a flock of Merino sheep, and brought them to Shoreham, shearing them on the way at Poughkeepsie, New York. We know of no record of the numbers or of the particular Cabana blood of these sheep. Some of them were kept pure, and their blood has come down to us without other than Merino blood being crossed with it, to this day. About the same time some parties brought some full-blooded Merinos into Bridport. One of our veteran sheep-breeders, Prosper Eithorp, remembers something of them, that they were good sheep, and made their mark upon the flocks of the neighborhood, but we have been unable to learn more of them.

In 1822, the late Judge Pettibone, of Manchester, bought of John Nettleton, of Watertown, Conn., twenty full-blooded Merino ewes. They were bred from the flock of Jacob N. Blakeslee. The produce of these, united with the purchase of the Skinner flock, laid the foundation of the large flock that Judge Pettibone bred so many years, until his death, with credit and profit.

In 1823, Jehial Beedle, Elijah Wright and the Hon. Charles Rich, of Shoreham, through Leonard Beedle purchased of Andrew Cocks, of Long Island, his flock of Merino sheep, consisting of about one hundred. Of the bloods of this flock, the most reliable account we have is a letter received by Dr. Randall from Judge Effingham Lawrence, and published in the Albany Cultivator in 1844. It was as follows :

" Yours is duly received, in which you refer to a conversation we had on the subject of Merino sheep, and particularly of the quality and purity of the flock of Andrew Cocks, who was my near neighbor. We were intimate, and commenced laying the foundation of our Merino flocks about the same time I was present when he purchased most of his sheep, which was in 1811. He first purchased two ewes at $1100 per head. They were very fine, and of the Escurial flock imported by Richard Crowningshield. His next purchase was thirty of the Paular breed, at from fifty to one hundred dollars per head. He continued to purchase of the different importations, until he run them up to about eighty, always selecting them with great care. This was the foundation of A. Cock's flocks, nor did he ever purchase any but pure-blooded to my knowledge or belief. Andrew Cocks was an attentive breeder, saw well to his business, and was of unimpeachable character. His certificate of the kind and purity of blood I should implicitly rely on. I recollect of his selling sheep to Leonard Beedle, of Vermont.

    (Signed,)   EFFINGHAM LAWRENCE.
Flushing, Long Island, Oct. 19, 1844."

After the arrival of this flock in Vermont, it was divided by the owners, the Hon. Charles Rich receiving one-fourth, Mr. Beedle one-half, and Mr. Wright one-fourth.

A few of the Beedle flock have descended to us unmixed with other than Merino blood. We do not know that any of the Wright flock have been kept pure, but from that portion of the flock kept by the Hon. Charles Rich, much of the pure Spanish blood of to-day has descended. In 1824, upon the death of Hon. Charles Rich, this flock descended to his two sons, J. Thurman and Charles Rich. The rich pecuniary recompense, and the meed of fame these men and their heirs have since received, is but a portion of what they deserve as a reward for the judgment and firmness exhibited by resisting the popular mania for Saxony fineness and blood. When Jarvis, Atwood, Blakeslee, and almost all gave way, John Thurman and Charles Rich stood firm. It is true that the first three named, with a few others, discovered their error in time to retrace their steps, and save to us much of the good old blood; but their judgments were fascinated and bewildered by the mania for fine wool that swept over the land between 1824 and 1836, vitiating the blood and constitutions of nearly all the flocks of fine-wooled sheep, depleting the pockets and destroying the hopes of their owners. To-day let us all especially revere the memory of Thurman and Charles Rich, whose firmness and judgment were not bewildered, and who have left unto their heirs and the land, the goodly heritage of the Rich flock without even the smell or rumor of Saxony upon its outermost skirts.

The Charles Rich branch of this flock was bred pure and unmixed with other blood until 1836, when a portion was sold to Erastus R. Robinson, and the remainder to Tyler Stickney, thus laying the foundation of two of our most justly celebrated flocks, and giving two more names that this with succeeding generations will remember and honor.

In 1824, N. H. Bottum, Esq., of Shaftsbury, purchased and brought to that town twenty-three full-blooded Merino sheep from Connecticut. We have not been able to learn the name of the breeder in Connecticut from whom these sheep were purchased, but they were bred with care and made improvement upon the flocks in the neighborhood of Shaftsbury.

In 1827, Dea. Frederick Button, of Clarendon, bought of Mr. Stephen Atwood, of Connecticut, two lots of Merino sheep which he brought to Clarendon, and from which he bred up a flock, afterwards breeding in blood from the flock of Consul Jarvis. At the time Mr. Button made one of these purchases, he was accompanied by Mr. David P. Holden, of Wallingford, who also purchased some sheep of Mr. Atwood, and brought them home with him. These are the first Atwood sheep that we have been able to trace into Vermont.

About the year 1835, Eber R. Murray, Augustus Munger, and a man by the name of Bundy, purchased a flock of about 150 Merino sheep of breeders living at Newport, Rhode Island, mainly, if not all, from the flock of Joseph J. Bailey, and brought them to Whiting, where a portion of them went into the hands of S. T. Baker, from his into those of James M Ormesbee, and from his to D. & G. Cutting, where and when the foundation of the Cutting flock was laid.

The evidence is conclusive that these sheep were pure descendants from one of the Jarvis importations brought over by one Capt. Paul Cuffe, a Quaker, and probably sold him at Lisbon, by Mr. Jarvis.

About the year 1838. Merrill Bingham and Zenas Skinner of Cornwall, purchased of a breeder by the name of Buck, and his neighbors in Lanesboro, Mass., one hundred and ten Merino ewes, and brought them to Cornwall, which was the foundation of the original flock of Mr. Bingham. They were said to have been bred from the Atwood flock.

In 1838, A. L. Bingham, of Cornwall, bought of Messrs. Buck & Atwater, of Connecticut, twenty ewes, which he added to a small number that he had already purchased of Messrs. Cuttings. In 1841 Mr. Bingham, purchased of Jacob N. Blakeslee, of Connecticut, sixteen ewes, and added them to his flock, with eight that he had purchased of Mr. C. Atwood, of Connecticut, and twenty-seven that he had purchased of Mr. Joseph I. Bailey, of Newport, Rhode · Island. In 1842 Mr. Bingham purchased of Mr. Bailey, forty ewes, and the following year fifty-one, being the entire flock Mr. Bailey had left.

Thus it will be seen that Mr. Bingham purchased and brought into the State at different times, one hundred and sixty two sheep descendants from the importations of Spanish Merinos.

Mr. Bingham was also one of the first to introduce French Merinos into Vermont. Between 1847 and 1853 he purchased of John A. Taintor, of Hartford, Connecticut, one hundred and sixty-one, for which he paid $37,500.

But these purchases of Merino sheep were surpassed by those of Mr. S. W. Jewett, who imported a large number from France, at a cost of of over $50,000.

David Cook and his son, Charles B. Cook, of Charlotte, bought in October, 1841, of Stephen Atwood, of Woodbury, Conn., twenty-three ewes and a ram, and brought them to Charlotte. In Jan'y, 1845, they made a further purchase of six ewes, and Mr. Atwood's best three ram lambs, and of Chauncey Atwood, five ewe lambs.

In 1847, in company with Prosper Elithorp, Esq., of Brid. port, Mr. C. B. Cook purchased a few ewes of Mr. Atwood, and eleven ewe lambs of Chauncey Atwood.

In January, 1844, Messrs. W. S. and E. Hammond, of Middlebury, and R. P. Hall, of Cornwall, made their first purchase of Atwood sheep, three rams and a few ewes of Mr. Atwood himself, and ewes enough to make twenty-one in all, of one of his, Mr. Atwood's, neighbors, Mr. Northrup, of the same, or Atwood blood. Subsequently, or within a period of three years, Messrs. Hammond made of Mr. Atwood four or five purchases more. Soon after this Ward M. Lincoln, of Brandon, bought of Mr. Atwood two rams and a few ewes, and also a few ewes of Mr. Blakeslee. About the same time Messrs. Cuttings bought a ram of Mr. Atwood.

In the fall of 1844, Mr. W. C. Wright and S. L. Bissell, of Shoreham, and S. W. Jewett, of Weybridge, purchased a ram of Mr. Atwood, while at the New York State Fair at Poughkeepsie. After bringing him to Shoreham they re-sold him the same season to Messrs. Prosper Elithorp, of Bridport, and L. C. Remele, of Shoreham, in whose hands he became famous. Soon after these purchases Hon. Joseph Marsh, of Hinesburgh, C. W. Brownell, of Williston, and Mr. William Gage, of Ferrisburgh, purchased a few sheep of Mr. Atwood, and each bred a flock of pure Atwood sheep for many years. That of Judge Marsh was bred pure to about the time of his death in 1877. About the same time Judge Penniman, of Colchester, purchased a number of pure bred Merino sheep, it is believed of Jacob N. Blakeslee, of Watertown, Conn., brought them to Colchester and bred them there.

In 1844 or 1845, Hon. W. R. Sanford, of Orwell, purchased the Ram, "Old Black," of Mr. Atwood, and in 1849 he purchased of the same twelve ewes, and later, eight or ten more.

Previous to this Mr. Sanford had purchased of Mr. John Nettleton, of Watertown, Conn., some ewes of Blakeslee blood, and bred them for a number of years, but after his purchase of Atwood sheep he sold out his Blakeslee blood. In 1846, Rev. L. G. Bingham, of Williston, purchased of Jacob N. Blakeslee a number of Spanish Merino ewes. If we mistake not, this was in the same year that Mr. Bingham purchased the Collins flock of Rambouillet sheep, and the first French ram of the Taintor importation ever introduced into Vermont, though Mr. Jesse Hinds, of Brandon, had previous to this, introduced a very few of the Rambouillet sheep from Mr. Collins' importation.   In 1849, Alfred Hull, of Wallingford, bought of Mr. Atwood thirteen ewes and a ram, and they were bred for many years in company with Hon. N. T. Sprague, the President of this Association.   Mr. Hull had previous to this bred an excellent flock of Jarvis Merinos.

In 1846, Philo Jewett, of Weybridge, bought of some of the Messrs. Atwoods ten or twelve Atwood ewes.   Soon after this A. A. Farnsworth, of New Haven, bought all the yearling ewes Mr. Atwood raised in one year.   In 1863, E. N. Bissell, of Shoreham, purchased five ewes and a ram of Stephen Atwood, three ewes of Chauncey Atwood, twenty-nine ewes and one ram of George Atwood, and six Atwood ewes of Jerry Smith.   These, with those before mentioned, are all the Atwood sheep I am at present able to trace into Vermont.

We have no doubt there have been many introductions into the State of Merinos from different flocks, especially from those of Jacob N. Blakeslee, of Connecticut, J. I. Bailey and David Buffum, of Rhode Island, the Shakers of Lebanon, New Hampshire, and others. Indeed we have evidence that there have been many more Merino sheep brought into different parts of the State than we have given in the foregoing narrative, but not being able to give any reliable facts in regard to the breeders, the purchasers, or the time of their introduction, we cannot give data that would be either valuable or interesting.

It is believed that an account of the origin, blood and introduction of all the ancestors of Merinos herein recorded has been given, as it has been the purpose of the Committee to admit none of which at least as much of their ancestry is not known.

IMPROVEMENT OF MERINOS IN VERMONT.

In a public discussion of the question of adaptation and profit of the different breeds of sheep for different localities, Prof. Agassiz once remarked that "the Merino was the sheep of the hills and the mountains," and if we trace the history of the race, or breed, we shall find that, in hilly and mountainous countries, they have proved susceptible of the highest development and have reached their highest grade of improvement.

In lower altitudes, warmer climates, and more level countries, breeders of Merinos have been able to increase the size of these sheep, but we think the examples and facts of history will bear us out in saying, that it has not been so easy to retain the hardiness, and the large proportion of wool to the size of the carcass.

The herbage of the hills, although very nutritious and strengthening, is scattering, and the exercise necessary to obtain a sufficiency to satisfy the appetite, stimulated by the pure, health-giving air and water usually found in near proximity to the mountains, develops a strong, well-knit frame and vigorous constitution.

As with the frame so with the covering. In high altitudes and cold climates nature supplies animals with the thick, fine, warm coat necessary to enable them to withstand the excessive cold and the violent and extreme changes in temperature inseparable from such localities Skins of the fur-bearing animals bring much higher prices when taken from the natives of the mountains and hills of the Northern States, than they do when taken from the same species of animals in the Southern States.

If it is possible to retain the dense covering on animals taken from scant feed and cold climates to luxuriant pastures and warmer localities, the inevitable law of nature comes in to affect the progeny from them, and not more than one or two generations will have passed before it will be found that a deterioration from extreme density, finer qualities and strength of fibre will have taken place. By increase of the size of the animal the aggregate amount of covering may be maintained or even increased, but the relative amount of quantity will be found to be decreasing, and we think it will be erroneous to suppose that the increase of size denotes a corresponding increase of vigor or constitution. It will be safer to presume that with this increase has taken place a deterioration from the greatest vigor and constitution, corresponding to the depreciation in the quantity and highest qualities of the covering.

All these natural laws and their effects we hold to be immutable, and should be properly considered to appreciate the great advantages Vermont possesses in her high latitude and mountainous conformation, to second the skill, enterprise and good judgment of her breeders, and enable them, not only to reach the high standard of improvement of the Merino race of sheep already accomplished since their importation from Spain, but to encourage them to go on, with the assurance that by the exercise of the same enterprise, skill and good judgment still greater improvements may be accomplished in the future, with all these advantages within the border of our State, and that the typical Vermont Merino will represent the acme of improvement, in the future, as it has in the past years.

It is hard to find any very concise facts as standpoints from which to estimate the improvements that have been made in Merino sheep since their introduction into this country from Spain, and so many contingent variations in feed, management and other causes are to be considered that it seems almost impossible to arrive at a very close estimate of the true amount of improvement. Still, by careful study and close consideration of such facts as we have, we think we can arrive at the truth approximately. It is fair to suppose that sheep that were held in such high esteem, and brought such prices as the first Merinos that were imported, would be as well cared for, and have as high feed as their owners were capable of giving them, with the experience and light they then had.

It is the opinion of close observers in the laws of breeding that the animal that has been developed to a high standard of perfection in a useful direction, by artificial means judiciously used, has the power to transmit in a measure the excellencies thus developed. If this opinion has its foundation in fact, then it is certain that the development that is obtained by such artificial means as are judiciously employed by the enterprising and intelligent breeder is a part of the improvements themselves.

Probably no flocks were so badly kept and as sadly neglected at the time of importation as some are to-day, yet there is little room to doubt that the poorest-kept and worst-neglected flocks of pure-bred Merino sheep in Vermont now will shear a larger quantity of wool on the average than the best-kept and best-managed flocks at that time and we will show by well-contrasted facts, that the average Merino to-day shears more than twice the per centage of fleece to live weight as did the very best specimens of which we have any account at the time they were imported from Spain.

In a work on Merino sheep, published in England, 1869, by an experienced breeder, Charles H. Hunt,* we find an account of the yield of wool in 1804, from a single flock of Spanish Merinos, belonging to a Mr. Follet. This flock consisted of sixteen rams, thirty-two ewes and eight shearling ewes, or fifty-six in all. The yield from the rams was eight pounds two ounces each, the ewes five pounds fifteen ounces each, and the shearling ewes four pounds five

---

*This work was kindly sent us by Prof. Brewer, of New Haven, Conn.

and one-half ounces each. The total average of the flock, six pounds six ounces each, an aggregate of three hundred and fifty-six and one-half pounds of unwashed wool, which, when washed according to the Spanish method, gave one hundred and eighty-four pounds washed wool, or a trifle over three and one-fourths pounds per head. It was *estimated* that this wool would shrink further by scouring, to one hundred and fifty-two pounds, or four-ninths of its original weight, an average per fleece of a trifle more than two pounds eleven ounces of scoured wool.

This gives us the best basis from which to estimate the improvement we have made in the amount of wool produced before the importations of Spanish sheep to this country we can find, and, as the flock was small and probably well-kept, it would seem to be a fair case to use in making comparisons.

The accounts of the weights of fleeces yielded by the Merinos, previous to their importation from Spain, is given by different writers as from four and one-half to five pounds for the ewes, and from eight to eight and one-half for the rams, the wool being unwashed.

The records of the weights of fleeces shorn from the first Merinos imported, give that of the ram Don Pedro, imported in 1801 by M. Dupont de Nemours, as eight and one-half pounds of brook-washed wool. This was a French Merino and of large size, weighing one hundred and thirty-eight pounds. Although this ram was from France, Mr. Randall conjectures he was of the original Spanish stock, and says that the fleece was the heaviest of any borne by the earliest imported Merinos. In order to make a proper comparison between this fleece and those of the present day, we must add the third supposed to have been lost by washing, which would make the amount of unwashed fleece twelve and three-fourths pounds, or a per centage of wool to live weight of nine and two-tenths. If we deduct sixty per cent. from the unwashed fleece, allowance for shrinkage by cleansing, we have five and one-eighth pounds of cleansed wool; a per centage to live weight of three and six-tenths.

At Chancellor Livingston's sheep-shearing in 1809, a ram, fourteen months old—Clermont—sheared nine pounds six ounces unwashed wool, his live weight being 126 pounds, a per cent of 7.4.

Another ram — Rambouillet—five years old, sheared nine pounds, his live weight being 131 pounds, or 6.8 per cent of unwashed wool to live weight. Two other rams—Columbus and Hornless—weighing 123 and 122 pounds, sheared six pounds seven ounces unwashed wool each. The four gave 6 2 per cent. to live weight.

At the same shearing, seven ewes gave thirty-six pounds unwashed wool, an average of five pounds two ounces, the ewes weighing fifty-three or fifty-four pounds each, or 9.5 per cent. of wool to live weight. The rams and ewes together gave an average of 7 7 per cent. of unwashed wool to live weight.

When these sheep were shorn at Clermont in 1810, the first two named rams—Clermont and Rambouillet—each weighed 146 pounds and each sheared nine pounds unwashed wool, a per cent. of 6.1 to live weight. At the same shearing the ewes gave an average of five pounds thirteen ounces unwashed wool, the heaviest fleece weighing eight pounds twelve ounces. The live weight of these ewes was not given, but we are told they exceeded those of the year before, as did those of the fleeces, the sheep having been better kept.

At this shearing a fleece was shorn from a yearling ram that weighed eleven pounds eleven ounces, but as his live weight after shearing was 145 pounds, it is hardly possible that his age was much less than a year and a half, and we think we are not justified in using his fleece or weight to make a comparison with our Merinos. Rams of just one year do not equal in weight of carcass and fleece those of mature age of the same breed.

It is reasonable to suppose if the weights of fleeces of Chancellor Livingston's sheep had been equalled or exceeded by those shorn from the sheep of Colonel Humphreys' and others breeding Spanish sheep at that time, we should find published accounts of them, but a careful search through the files of papers published in Connecticut, Massachusetts, Vermont and New York at that time, fails to give us any information of the fleeces shorn from any but Mr Livingston's. We regret we cannot find reliable data of the early shearing and history of Col. Humphreys' flock. Once he mentions that a ram raised on his farm yielded seven pounds five ounces of washed wool. This would give, by applying the same rules as applied above, nearly eleven pounds of unwashed wool, or four pounds six ounces cleansed.

Of Consul Jarvis's flock, we learn by his letter to Mr. Morrell, author of the " American Shepherd," that his flock from 1811 to 1826 averaged four pounds washed wool. By applying the same rules, we have six pounds unwashed, or within a fraction of two and one-half pounds cleased wool. At the same time Mr. Jarvis wrote Mr. Morrell that his best stock rams sheared six and one-half pounds washed wool, which by the same rule would give nine and three-fourths pounds unwashed, or a fraction less than four pounds of cleansed wool each fleece.

It is noticeable that Mr. Jarvis, writing the above letter nearly or quite fifty years after he imported his Merinos from Spain, does not claim any improvement in weight of fleece during that time, but says of the Merino part of the flock, " it will not vary materially from the original weight."

In mentioning examples from the flock of Consul Jarvis, we have given only published accounts of the weight of his fleeces, but Prosper Elitharp, Esq., of Bridport, has a letter written to him by Mr. Jarvis, April 9, 1844, in which he says he had had bucks in high case that had sheared as high as seven and one-half pounds each. This would make by the rules applied in the other cases eleven and one-fourth pounds unwashed, or four and one-half pounds cleansed wool.

We have thus somewhat at length analyzed all the fleeces from the early Merinos, the weights of which we have found recorded. We have been obliged to make these reductions ascending and descending in most cases for want of actual recorded facts as to cleansed wool ; the deduction of sixty per cent. is that given by as good authority as we can find, for the average shrinkage of Merino fleeces by scouring in those days; we believe it to be very liberal towards the sheep of that period. The heaviest rams' fleeces would probably shrink more.

The first point of advance or improvement from these weights in cleansed wool that we are able to find at this time is in 1846, and in a descendant of the Humphreys flock. In that year Prosper Elitharp, Esq., of Bridport, sheared a fleece from a ram purchased of Mr. Atwood in 1844 that weighed fifteen pounds unwashed, which, when thoroughly cleansed, gave six pounds of scoured wool   Later, Mr. Atwood reports a somewhat heavier fleece, comparatively, than this one, or twelve pounds four ounces washed, equal to eighteen pounds six ounces unwashed wool.   Whether the amount of cleansed wool was as great, we are in doubt.

At the same time Mr. Atwood gave the average weight of ewes' fleeces in his flock as five pounds for his ewes, lambs the same, weathers six pounds, and rams seven to nine pounds.   The heaviest ewe's fleece was six pounds six ounces, and the heaviest ram's fleece (the one named above) twelve pounds four ounces—all washed as clean as he could in the river, and sheared in six or eight days after.

The next fleece we shall notice is one shorn from a ram raised by Mr. Hammond from the Atwood stock, sold Messrs. Baker and Harrigan, of Comstock's Landing, N. Y., from which was shorn in 1865 twenty-three and three-fourths pounds unwashed wool, that weighed seven pounds when scoured.   We have not the live weight of this sheep.

The same year a ram was shorn at Canandaigua, N. Y., that weighed ninety-five pounds, sheared unwashed twenty pounds one ounce, that cleansed six and one-fifth pounds scoured wool—a per cent. of 21 of unwashed wool and of 6.5 of scoured to live weight.

At the same shearing there were ram lambs and ewes shorn, of small size and thin condition, that yielded larger proportions of wool, but as we have been citing rams of larger size and more nearly matured condition, we do not use them in comparison.   The largest per cent. of wool to live weight at that shearing was 23.9, and the average per cent. from all shorn on that occasion was 19 3.   At the shearing held by the same association at Hemlock Lake, Spring of 1878, there were six sheep shorn that exceeded the per cent. of wool to live weight over the best one shorn at Canandaigua in 1865 ; the average per cent. of the six being 25.3 ; the highest per cent. given by one was 29.5.   Throwing out a few yearlings, the growth of wool on which exceed thirteen months, the average on the thirty left at that was 20.4, or 1.1 per cent. more than at Canandaigua in 1865. At this shearing the days' growth of fleeces given show seven cases out of thirty where the growth of wool was over one year, the average days' growth of the seven being 377 days, the highest being 395, the average of the thirty being 362½ days.   The average length of staple of the thirty was two and eight-fifteenths inches ; the longest staple reported was three and one-half inches.*

The next step in improvement we are enabled to notice was the case of Col. E. S. Stowell's ram Red Leg, that at two years old (in 1868) sheared a fleece of twenty-eight pounds from a live weight of 110 pounds, a percentage of fleece to live weight of 25.4 per cent. This fleece when thoroughly scoured weighed eight pounds one and one-half ounces, of which three pounds fifteen ounces was No 1 wool. This is a percentage of 7 3.

The large percentage of wool to live weight that has been produced for the seasons of 1877 and 1878 in Vermont is very justly a cause for astonishment, and were we able to give a few selected cases only, might with reason be supposed to be accidental. Fortunately an unexpected opportunity has come to our hand for estimating the per cent. of wool to live weight shorn from a larger number of sheep than in any case we have cited. In the fleeces and samples of wool collected and forwarded to the Paris Exposition from Vermont, we have sixty-seven cases where the weight of wool, weight of sheep after shearing, days' growth, and length of staple are all given, and as these samples and fleeces were selected with a view to showing fine quality and style of wool as well as weight of fleece, the comparison must be fairer than a few cases selected to show great weight or large percentage of wool, without regard to quality, or even such sheep as it would be fair to suppose would be selected to be shorn at a public shearing. The precautions taken by the Association, requiring that the sheep be shorn and the fleeces and sheep weighed before witnesses, make these cases and their results reliable as examples. The committee from personal knowledge are aware of a number of instances where the samples selected with a view to showing extra quality and style of wool instead of weight of fleece, and that the per cent. of wool to live weight was decidedly less than the average of the fleeces shorn from the flock. There were also eight of the sixty-seven that were seven years old and over, the average age of the eight being over ten years. One of the eight was sixteen years old, another was fifteen years, and all the eight had been used for breeding purposes Forty-one of the whole number had also been used for the same purpose. These several causes must be acknowledged to have reduced the percentage of wool to live weight materially. The longest time any fleece of the sixty-seven was growing was 385 days. There were eight that were growing over a year, the average of fleece of the eight being nearly 371 days. The average age of the sixty-seven was nearly 361 days, or over four days less than a year. The per cent. of wool to live weight of the whole sixty-seven was 22; of the best thirty, 25 2 ; of the best six, 30.1. The best single one was 36.6. The fleece of the last was 362 days' growth. Those of the best six averaged 365 1-3 days growth. Twenty-one of the whole number were rams, yielding

---

*This wool, we believe, was measured in the natural condition, as it grew on the sheep, without drawing out the crimp.

an average per cent. of fleece to live weight of 22.8, or a little above the average of the whole.   Of these twenty-one rams, three were one year old, seven were two years old, seven were three years old, one four years old, one five years and two seven years old.   The average live weight of the twenty-one was 108 1-3 pounds.   The eleven rams that had attained mature age averaged 118 pounds live weight.

Discarding from this last class the two seven years' old rams that had past their prime, and had somewhat declined from their greatest weight of fleece, the remaining nine sheared an average of a fraction over twenty-nine pounds; their average live weight was 120½ pounds, thus making the average per cent. of fleece to live weight 24.1.

In the Practical Shepherd published in 1864, at page 76, Mr. Randall says that twenty-seven pounds had probably never been excelled for gross weight of fleece, and that 21 per cent. was probably the greatest proportion of unwashed wool to meat that had been yielded to that time.   It is but fourteen years since that book was published, but in that time our breeders are able to show the great improvement of two pounds of unwashed wool in nine fleeces over the best in 1864, and among the nine we have one fleece that exceeds the best then by ten pounds.   In the matter of per cent. of wool to live weight, the best of the nine fleeces shows 28 9 per cent., or very nearly eight per cent. better than the best of 1864, while the whole nine average three per cent. better than the best at that time.   Nor is this all.   We are willing to affirm from personal knowledge that at least nine more rams can be found in Vermont that will yield equally as good gross weight of wool and per cent of fleece as the nine named.

If we go back to our starting point, we find that the average per cent. of unwashed wool to live weight of the five fleeces shorn from the Dupont and best two Livingston rams in 1809 and 1810 was 7.1. or a little less than one-third that shorn from the whole twenty-one rams we have brought forward as a class from which to make comparisons, and this percentage does not exceed the percentage of wool shorn from the stock rams of the best flocks of Merinos of Vermont, but probably will fall something short of it.   It would be very interesting to know the amount of cleansed wool the fleeces of these twenty-one rams would yield after scouring. but the fleeces of only three of them have been scoured separately to test this question. We will compare the three with the ram of Mr. Dupont and the best two of Chancellor Livingston.   The three rams we refer to are "Banker," bred and owned by Mr. V. Rich, Shoreham; "Patrick Henry," bred and owned by Mr. L. P Clark, Addison, and "Stub," bred and owned by Mr. H. C. Burwell, Bridport.   The fleeces from these three rams at three years old weighed respectively thirty-one, thirty-seven, and thirty-five pounds, their live weights being one hundred and eight, one hundred and forty-seven, and one hundred and twenty-one pounds each.   Their fleeces after scouring weighed eight pounds six and one-half ounces, nine pounds, ten ounces, and eight pounds thirteen ounces, the aggregate live weight of the three being three hundred and seventy-six pounds, the aggregate of un-washed wool one hundred and twenty-three pounds, or 27.3 per cent. The aggregate of cleansed wool was twenty-six pounds thirteen ounces, or 7.1 per cent. to live weight, or the same per cent. that we find was given by "Don Pedro," "Clermont," and "Rambouil-let," of unwashed wool, while the latter was but little more than one-fourth the per cent. that "Banker," "Patrick Henry," and "Stub," gave of wool in the same condition.

We will mention one more example of large yield of cleansed wool as remarkable in showing what has been accomplished by one breeder. H. C. Burwell, of Bridport, whose flock is herein recorded, has clipped in one year from three stock rams, all bred by himself, three fleeces that weighed on an average over thirty-one pounds each, which, after being thoroughly scoured, weighed over eight pounds each. One of these rams was two years old, the others four years.

Of the forty-six samples, or fleeces, from ewes, sent to the Paris Exposition, all weighed over nine pounds, only four weighed less than ten pounds, the percentage of wool to live weight of the four being 156. One of the four was a yearling, one fifteen years old, and one sixteen years. The whole forty-six ewes averaged over fourteen pounds unwashed wool, and gave a per cent. of wool to live weight of 21 1

This comparison shows that the lightest fleeces out of all these ewes (where the selections were largely made for fineness and style of wool, rather than for a large per cent of fleece to live weight) exceeded the heaviest fleece shorn from any ewe at the Livingston shearings by half a pound ; that the average of the forty-six fleeces was nearly double that from the best ewe at the same shearing, and that the per cent. of wool to live weight of the forty-six is more than double that of the fleeces of the Livingston ewes in 1809. We cannot make the same comparison of the ewes shorn in 1810 as we have not the live weight of the Livingston ewes that year, but the average per fleece of five pounds and thirteen ounces is more than twice exceeded by the forty-six in 1878 approximating three times as much wool, rather than for a large per cent. of fleece to live weight) ex-ceeded the heaviest fleece shorn from any ewe at the Livingston shearings by half a pound ; that the average of the forty-six fleeces was nearly double that from the best ewe at the same shearing, and that the per cent. of wool to live weight of the forty-six is more than double that of the fleeces of the Livingston ewes in 1809. We can-not make the same comparison of the ewes shorn in 1810, as we have not the live weight of the Livingston ewes that year, but the average per fleece of five pounds and thirteen ounces is more than twice ex-ceeded by the forty-six in 1878, approximating three times as much.

Had the selection of these last been made to show large weight of fleece, the average would have been greater, and the per cent. of wool higher : flocks of two and three years' old ewes, containing as many or larger numbers, have shorn heavier fleeces on the average than these forty-six.

The average length of staple from these fleeces and samples was three and one-fourth inches, the shortest two and one-half inches. There were twelve that measured less than three inches, nine that measured four inches or over, and two that measured full four and one-fourth inches. This length of staple is remarkable, and probably longer than the average of the Merino wools of Vermont, and is easily accounted for by the fact that in making selections of wool for exhibiting, length of staple would be one of the qualities sought for.* It may be interesting to know that the three stock rams, Banker, Patrick Henry, and Stub, gave samples respectively, four, three, and three inches. It is certain that the increase of length of staple is one of the improvements that have been made in the fleeces of our Merinos, and will account for a part of the increase of weight of fleece, but the density that has been added should probably be credited with more of it. As a rule, the longest stapled fleeces do not show the largest percentages of wool to live weight.

The large increase of the average weight of fleeces in Vermont and the United States during the twenty years from 1850 to 1870 is a subject worth noticing, and a strong argument in favor of the substitution of the blood of the heavy shearing Merino sheep of Vermont flocks for the lighter wooled Saxony sheep so popular forty to fifty years ago, and it will be shown to have added very largely to the material wealth of our country.

In 1850 the average weight of fleece of the sheep of Vermont was three pounds five and one-half ounces, of the United States, two pounds six and three-fourths ounces. In 1860 the fleeces of the sheep of Vermont had increased to four pounds two and one-half ounces, an increase of over thirteen ounces per fleece. During the same time the average of the fleeces of the United States had increased to two pounds ten and two-eighths ounces, an increase of only about four ounces per fleece. In 1870 the fleeces of Vermont averaged five pounds five and two-thirds ounces, an increase in ten years on the average per fleece of one pound and three ounces. In the same year the fleeces of the United States averaged four pounds seven and one-fifth ounces, an increase in the ten years of one pound and thirteen ounces on the average.

The decade from 1860 to 1870 was the one during which Vermont sent so many sheep to the West. Carload after carload of the prime blooded sheep of Vermont were taken from her best flocks, and it is but reasonable to suppose that the increase of nearly two pounds in the average of the fleeces of the United States was largely due to this importation into other States of the superior sheep of Vermont and the infusion of that blood into the flocks of those States.

The number of sheep in the United States in 1870 was 28,477,-951. This increase of a pound and thirteen ounces in ten years would make an increase of the wool product for 1870 alone of 51,616,286 pounds, which, at forty-six cents, the average price of

---

*These measurements were for the full length of the fibre with the crimp drawn out.

9

wool in 1870, would add to the total value of the wool produced that year $23,643,591.56 It is but fair to claim that a large proportion of this increase is due to the preservation of the heavy wooled Merinos in Vermont and Connecticut, and the great improvements made in them by their breeders, between 1840 and 1870. That this improvement is still appreciated, the large numbers that are still drawn from the flocks of the Merino sheep-breeders of Vermont and sent to the wool-growing States and Territories West is ample proof. During 1877 no less than twenty-nine and one-half carloads of blooded Merino sheep were sent from one station alone. In the first six months of 1878 the shipments from the same station were sixteen carloads, against eleven for the same time in 1877. Of these forty-five and one-half carloads, twenty-four and one-half went to Ohio. In these cars were shipped an average of over one hundred each.

SIZE.

In treating of the question of size it is necessary to take into consideration the opinions of breeders at different periods as to the most desirable or most profitable size. For wool-growing alone, there is no doubt but the small size has the greatest proportionate surface, and, other things being equal, will yield the greatest per centage of wool to live weight; and this fact has at times led breeders into the practice of breeding too much of this large per centage of wool to the sacrifice of the size that is most profitable, and often small rams have been used that have yielded a much larger percentage of wool than their size and constitutions would warrant their sustaining ; consequently at some periods when wool was very high the size of our Merino sheep as a whole has declined slightly. Another cause that has sometimes seemed to give an impression that the size of our Merinos was decreasing, has been the fact that most excellent stock rams have been used that from some accidental cause when young have failed to attain their full natural size. In such cases, while the progeny may have been affected to a limited extent by the accidental causes, the effect has not been serious, provided the aim of the breeder has not been to encourage the depreciation in size.

Insufficiency of keep and excessive service when young and growing has been the cause of many stock rams not reaching their full size and proper development, and this practice continued in a stock inevitably prevents improvements in size, and often causes a marked depreciation of it.

Notwithstanding these several causes operating in the contrary direction, our Merino sheep have materially increased in size since the period of their importation from Spain, though not nearly in the same ratio as in the weight of fleece.

We find the same difficulty in estimating the live weight of the Spanish sheep at the time they were imported from Spain that we do in finding a starting-point from which to estimate the improvements made in weight of fleeces : there are very few records of weights of those sheep at the time of their importation, if there are any. The live weight of the Livingston and Dupont rams was much larger than the average of the Merino rams imported direct from Spain ; this may be attributable to the taste of those who selected the animals from the Spanish Cabanas with which to start the French flocks ; certain it is that all writers describing the French flocks of Merinos, even at a comparatively early period after they were imported, speak of them as being much larger in carcass than those or any of the Spanish Cabanas. To attain large size, by selection, breeding and stimulating feed, has always been a leading object of the breeders of Merinos in France, and, it is probable, influenced those who selected the original stock to found the first flocks of French Merinos. For these reasons, we must conclude the Livingston and Dupont rams would not fairly represent the average weight or size of Merino rams as we received them from the Spanish.

The great difference between the weights of the ewes and rams of the Livingston flock is very noticeable, and not easily accounted for. It was much greater than in other flocks at that time, according to the best authorities we have.

Petri, a French writer upon Merino sheep, of acknowledged authority, who visited the Cabanas of Spain in the fore part of the present century, previous to the great importations of 1810, made a careful estimate of the size of rams and ewes of several of the Spanish Cabanas, and gave weights and measurements, as he took them from personal inspection of some of the largest of the Merinos of Spain. The table he gives shows but little variation in the sizes of the mature individuals of the different Cabanas, the difference in three of them being only three pounds in the ewes, and three and a half in the rams. According to that writer, the best Merino rams of that day weighed about one hundred pounds, and the ewes a little less than seventy pounds at maturity. These weights were of sheep in the best condition in their fleeces, and this should be borne in mind when we compare them with those of the present day, the weights of which are taken after shearing.

The mature ewes of the present day, which we take for comparison, having been used as breeders, and suckling their lambs when their weights were taken, their condition and weights must be much lower than if taken from the same sheep in November or December ; and here it is proper to state that the proportion of lambs to mature ewes raised now in our own country much exceeds that raised in the old Spanish Cabanas. Seventy-five years since it was, as reliable writers inform us, the practice of Spanish shepherds at that time to kill all the smaller, weaker lambs, and give the remainder of the lambs two ewes each to suckle. This practice would very materially increase the weight of the sheep ; the draft being much less upon the ewes, they would be in much better order and weigh more, and the lambs become much better developed and reach a larger size.

With all these advantages in favor of the original Spanish sheep at the time Petri made his measurements and took their weights, we find that the ewes that had reached three years or over, from which we sent samples to Paris, weighed as much without their fleeces as the old Spanish ewes with theirs on. We find a larger increase in the size of the rams than in that of the ewes in the same time. an increase of eighteen to twenty pounds in grown rams, taking from the same class as before, and giving the old Spanish rams the advantage of having been weighed with their fleeces on and ours without, which would make from six to eight pounds more. It would be safe to claim that our rams now weigh at least one-fourth more than the old Spanish rams did, seventy-five years ago—a very creditable increase to make : while we have increased the fleece-bearing proportions to nearly or quite three times its former capacity during the same time.

There is no doubt that this addition to the weight of our sheep from the original Spanish has been effected so as to much improve their appearance and beauty ; they are shorter in the leg, broader in the chest and across the hips, and altogether much more stocky, heavier-boned, and giving evidence by these changes of a better constitution, which is needed to sustain the largely increased percentage of wool our present Merinos are called on to grow. We think this improvement in constitution is also sustained by increased longevity. We believe the Merino sheep of this day live longer and maintain their usefulness as wool-producers and breeders more years than when they were first imported. Youatt says they had been known to breed in extreme cases at fifteen years of age, but we have at least two cases in Vermont of Merino ewes continuing to breed until after they were twenty years old. John H. Mead, of West Rutland, had a pure Merino ewe that brought a lamb when she was twenty-one years old, and L. P. Clark, of Addison, had an ewe that raised a lamb at twenty-four years, and died the fall after.

One of the members of this committee has an ewe bred by P. Elithorp, Esq., of Bridport, that is at least thirteen years old This ewe has brought twins nearly every year for the last ten years, always raising one of them and sometimes both. better than the average of Merino ewes. Notwithstanding this great draft upon the vital energies and constitution of this ewe, her fleeces for the last three years have averaged about ten pounds each. Within the last year she has been driven twice with a small flock of ewes, over twelve miles each drive, performing the journey easily in about six hours, and leading the flock nearly the entire distance. At this writing (July, 1878), she is raising one of a pair of twin ram lambs she brought last spring (both of which were at least full average size for Merino lambs) ; the lamb is growing finely, and the old ewe is in better condition than she has been for a number of summers past. Since she came into the possession of her present owner she has had during the winter a very small allowance of grain and roots, but during the time she has been at pasture and through the Fall grass only. She has been housed during the Winter, but not at all during the time she has been at pasture. There is nothing in her present appearance that indicates a decline of her powers for usefulness for some years longer, though the ability to withstand such drafts for even this number of years is ample evidence of constitutional powers of extraordinary magnitude, and it would seem they could not be continued much longer without exhaustion. Scores of examples could be cited of ewes breeding at an older age than this one, but probably not many that would show the ability to answer such large drafts for so many years without exhaustion.

Mr. William H. Cook, of Shoreham, has an ewe fifteen years old that has brought a lamb and raised it for fourteen years, and is still healthy. Until she was ten years old her fleeces never fell below twelve pounds, and one fleece has reached fifteen pounds. Her fifteenth fleece, taken off last spring, weighed nine pounds. She shed her teeth when she was twelve years old.

These cases, with the large numbers more that could be cited, are abundant to prove that the increased percentage of wool-bearing capacity has not been achieved at the expense of constitutional strength of vigor, but that the latter must have been increased in full proportion to the other to enable the Merino sheep of to-day to show so many examples of constitutional vigor and longevity while yielding so much heavier fleeces than were shorn from their ancestors, seventy years ago.

Returning to the matter of size, we think breeders in Vermont at this time are generally breeding to increase the size of their Merinos. Small sheep yielding very large per centages of wool have served their purpose—a most useful one—to increase the wool-bearing capacity of our flocks, and they seemed for the time almost indispensable to develop to the present extent this quality ; but the power having become fixed without sacrificing any other good qualities—even in size an improvement having been made—breeders are now able to direct their efforts to this particular point of size, and will be able to make great improvements in that direction, without sacrificing the other good qualities of form, constitution, and the ability to produce large fleeces of fine wool. Stock rams having already been bred weighing about 150 pounds, of the most vigorous constitution, and yielding twenty-five per centage of unwashed fleece, or over six per centage of cleansed wool to live weight, we are encouraged to believe that increase of size is within easy reach of breeders of Merino sheep in Vermont, and that while it is being attained all the improvements that have been made in other directions can be carried along with it : though just how far it is desirable to carry the increase of size, or how far it may be profitable to carry it, is probably at this time an open question, and must be for some time to come, and it is without the proper sphere of the committee to give any opinion upon the subject; the good judgment and experience of the breeders of Vermont will enable them to arrive at a practical solution of the question.

QUALITY OR FINENESS OF FLEECE.

It is important to consider the quality of the fleeces borne by the Merino sheep of to-day, with the quality of the fleeces borne by their ancestors when the improvements of which we have been treating were commenced. The committee have made investigations upon this point that must be valuable, and must, we think, be a subject of satisfaction to the breeders and improvers of Merino sheep.

These investigations were commenced with the hope that we might be able to establish the fact that the great improvements we had made in the weights of fleeces, the beauty of forms, and the constitutions of Merinos, had not been at the sacrifice of the quality of wool, with some little fears that in this instance the facts might not favor us to realize our hopes; we certainly were not prepared to expect anything like the improvements that have been made in this direction.

Wishing to give this investigation as much practical value as possible, and make the measurements and comparisons as comprehensive as we could, we procured samples of wool from the leading wool-producing countries of the world and from the leading breeds in this country, and had samples from them measured under a powerful microscope by Dr. H. A. Cutting, of Lunenburgh, Vermont, an accomplished expert, who will be acknowledged as authority in scientific investigations. The very exhaustive and accurate tests he employed will give the results of his work great value and cause them to be received as authority.

The samples of Australian, Mauchamp, Silesian, Hungarian, Cape of Good Hope, and Saxon wools were the best selections we could make from a quantity kindly sent us from the extensive collection of George William Bond, Esq., of Boston, collected by him at the Centennial wool exhibit at Philadelphia and elsewhere, as well as from the large collection he himself contributed to the Centennial show.

The South Down wools are furnished by C. K. Gray, Esq., of East Montpelier, from some of his flock that received prizes at the last show of Vermont State Agricultural Society. Our request to furnish samples of Cotswold wool from the flocks of Vermont breeders not having been complied with, we sent to Dr. Cutting some samples that were given us at St. Louis last fall from a Leicester ram, a Cotswold ram and a Cotswold ewe, that received prizes at that show. For samples of Vermont Merino wool, we selected mainly from our heaviest fleeces and best-built rams and ewes, such at this time as are regarded in the estimation of our best breeders as among our best representatives of the heaviest-fleeced and strongest-constitutioned Merinos; and in no case were any samples sent where they were from sheep without the other good qualities, nor from any that do not stand well in the estimation of the breeders for their other good points besides the fineness of their wool. It will readily be seen that had they been selected to show quality alone, or had a number of such samples been sent without any regard to the other good qualities of the sheep, the average fineness would be much greater; but it was the object of the committee to show how fine the fleeces of the *best* Merino sheep of to-day are, rather than how fine samples could be shown from the finest fleeces, without regard to their other and more profitable qualities.

Good breeders will be gratified to learn that in securing so great improvements in the wool-bearing capacity and constitution of our

Merino sheep, no sacrifice of the fine qualities of the fleece has been necessary, but rather that the improvement in fineness has been going on, and keeping pace with the other good qualities; indeed, we may raise the question whether this improvement in the fineness of fibre may not be inevitable, while the wool-growing power has been so greatly increased, as from the poor demand for, and inadequate compensation for, wools of very fine grade, we believe very few breeders have been encouraged to improve the fineness of their fleeces, or cared to do so; while there are some few that do pay attention to and endeavor to breed sheep that will produce wool of good quality, most of them prefer and select rams bearing wool of strong masculine staple, believing that they will be more sure to get stock with strong constitutions, and that will yield heavy fleeces; as they express it, they do not like a ram with a ewe's fleece, and as rams of this strong masculine character of fleece are generally used and patronized, the samples used for the measuring test were mostly selected from such ones.

We would state here that this masculine appearance or want of crimp when viewed by the naked eye is not a safe guide to judge of the fineness of the wool. The unerring measurement of a powerful microscope is necessary to determine that. The results of some of the measurements of the samples sent Dr. Cutting were unexpected. The absence of crimp in some of the samples indicated, to the naked eye, greater size of fibre than in some samples that had more crimp, but the measurements proved a larger size of fibre in the latter samples than in the former.

In 1856, under the direction of Mr. William Youatt, with a microscope of 300 linear power, made by Mr. Powell, of London, the latter gentleman made very accurate and interesting measurements of the size of fibre of wools from several different breeds of sheep. As the results of these measurements Mr. Youatt gives the size of fibres from a few breeds as follows:

Merino wool................$\frac{1}{750}$ of an inch | South Down..............$\frac{1}{355}$ of an inch
Saxon wool................$\frac{1}{740}$ of an inch | Leicester............ ......$\frac{1}{565}$ of an inch

Mr. Randall and Mr. Morrill both acknowledge these measurements as accurate, and refer to them as good authority. From the latter gentleman's *American Shepherd* we give the following as the French and German methods of grading and sorting the different qualities of wool, and the size that each quality measures:

SORT.	NAME.	MEASURE.
............1....................	Super-electa..............	$\frac{1}{818}$ of an inch
............2....................	Electa.....	$\frac{1}{755}$ of an inch
............3....................	Prima....................	$\frac{1}{785}$ of an inch
............4....................	Secunda Prima............	$\frac{1}{717}$ of an inch
............5....................	Secunda............	$\frac{1}{517}$ of an inch
............6....................	Tertia....................	$\frac{1}{515}$ of an inch

It will be noticed by comparison with Mr. Youatt's measurements that the first quality, or super-electa, is the size Mr. Y. gives as the size of Saxon wool fibre, and the second as very nearly that of Merino wool fibre.

Mr. Morrill in the same work says : "The fibre may be considered coarse when it is more than the five-hundredth part of an inch in diameter, and very fine when it does not exceed the nine-hundredth part of an inch, as exhibited occasionally in choice samples of Saxon-Merino wool." It is said there are animals which have a wool underneath a covering of hair, the fibre of which is less than the twelve-hundredth part of an inch."

With these examples to give a proper appreciation and understanding of the subject, we give Dr. Cutting's report and tables of measurements of the samples of wool sent him. These samples were sent Dr. C. by numbers and a careful record of the numbers kept, so that he did not know what the samples came from until he had reported his measurements. As the samples were sent as they were gathered, without any regard to classification, we have changed the numbers to different positions to give the rams and ewes in separate classes, but the measurements are in every case as he gave them.

LUNENBURGH, Essex Co., Vt., March 25, 1878.

Messieurs :

I send you herewith statement of measurements of wool. The measurements were made with a micrometer of one millimeter in length, divided into one hundred parts, and those divisions acted upon by a micrometer eye is as a vernier, dividing each division of the stage micrometer into ten parts. The objective used was one-sixteenth of an inch, consequently you will see that great accuracy was attainable. The fibres of wool are none of them of equal size through their entire length. I consequently made ten measurements on each fibre, and thus measured ten fibres out of each lot, and give the average for the result. One millimeter is .03937 of an inch. The measurements are decimal fractions of this millimeter. As a standard of comparison I use human hair, the largest one being from the coarsest head of hair I could conveniently find, the medium from an ordinary head of hair, and the other the finest I could get in this immediate vicinity. The wool as a whole is good fibre, yet of course much filled with the gum so common to wool—especially fine wool. Of the peculiarities visible to the eye I say nothing, as you know those much better than I do. A peculiar structure, visible only under magnifying power, I mention.

H. A. CUTTING.

MEASUREMENTS.—HUMAN HAIR.

Coarse.. .098
Medium... .051
Fine.. .042

WOOL.

Number	RAMS.	Age of Sheep	Weight Fleece. lbs.	oz.	Live Weight.	Size of Staple in Millimeters	Size of Staple in fractions of an inch
1	Banker	3	31	......	108	.0215	
2	Patrick Henry	5	37	......	147	.0235	
3	Stub	4	35	......	121	.0255	
4	Stock Ram	3	33	......	128	.027	
5	" "	2				.027	
6	" "	5				.0275	
7	" "	7	21	8	91	.024	
8	" "	5	32	8	132	.0285	
9	" "	2				.024	
10	Ram Teg	1				.018	

Average Rams........ ......
Highest, or Finest.......
Lowest, or Coarsest......

Number	EWES.	Age of Sheep	Weight Fleece. lbs.	oz.	Live Weight.	Size of Staple in Millimeters	Size of Staple in fractions of an inch
11	Breeding Ewe	6	19	3	......	.015	
12	" "	12	9	11		.0135	
13	" "	10	12	......	80	.024	
14	" "	4				.015	
15	" "	6				.021	
16	" "	10	13	12		.0205	
17	" "	13	10	......		.0265	
18	" "	4	12	12		.0245	
19	" "	12	12	8		.021	
20	" "	2	20	......	86	.024	
21	" "	3	19	......	64	.021	
22	Ewe Teg	1				.02	
23	" "	1				.0145	
24	" "	1				.0225	

Average Merino Ewes...
Finest............
Coarsest............
Average of both Rams and Ewes

25 Sample of very fine Australian wool........ .0175
26 Sample of very fine Australian Lustre wool........ .015
27 Sample of Mauchamp Merino wool........ .024
28 Sample of Silesian cleansed wool........ .0245
29 Sample of Saxon Pan Handle, J. Brown's........ .015
30 Sample of Hungarian cleansed wool........ .014
31 Sample of Hungarian unwashed wool........ .0145
32 Sample of Cape of Good Hope wool........ .012
33 Sample of Cotswold Ram's wool........ .059
34 Sample of Cotswold Ewe's wool........ .055
35 Sample of Leicester Ram's wool........ .052
36 Sample of South Down Ram's wool........ .036
37 Sample of South Down Ewe's wool........ .033

We append some remarks made by Dr. Cutting upon some of the samples. Of No. 1 he remarks: "The evenness of the fibre is just superb, you can stretch it out and see its superiority with the naked eye, and with a hand magnifying-glass can see it very well." Of No. 2 he says: "This is a very even structure and remarkable strength, not showing the usual changes in growth that occur in most specimens, I should presume the sheep to have an iron constitution, not easily affected by any change in climate or food."

Of the samples from the ewes, he says of No. 12: "This is one of the best in its structure in the lot, if not the very best."

Of No. 24 he says: "This is an excellent structure, free from bad places, and of very good strength."

He has disparaging remarks to make of two samples only, and they were taken from sheep that had experienced a period of sickness during the growth of the samples sent from them.

To show still further the pains taken to secure accuracy, and the amount of work the Doctor was willing to perform to secure it, we copy a portion of the letter to us apologizing for the delay in furnishing the report of his investigations:

"It may seem a long time that I have been about this work, but I have worked all I could at it, as it is very close work and two or three hours a day is all my eyes would stand, and I believe all any person would do. Then it has required over three thousand adjustments, and of course those adjustments, made to measure with a micrometer, must be perfect and in line, or no measurement could be made. I have consequently got it off as fast as I could. It has at least been twenty times the work I anticipated, as there is such a difference in the size from root to tip, that no single measurement of a fibre could be depended upon."

It would seem that we could add very little to this report to show its force and value, but we venture to call attention to a few points in it, and to make a few comparisons to show the fineness of fibre borne by our Merinos to-day, when compared by other standards or by the most critical scientific tests. First, it will be noticed that not a lock measured from our heavy-fleeced Merinos proves quite as as coarse as the Saxon Standard, Super-Electa or $\frac{1}{10}$ of an inch. We find the average size of the wool from the ten rams to be less than four-fifths the size of Super-Electa. We find the average of the wool from the fourteen ewes to be less than two-thirds the size of Super-Electa, while out of only fourteen samples we have four bearing wool only half as coarse as Super-Electa, and these four are all as fine or finer than the Saxon or Australian samples, except the Lustre wool. Our best sample is equalled in fineness only by the exquisite sample of very light wool from the Cape of Good Hope. The Hungarian Wools are the only ones except the last named, that prove as fine as a few of our best, though neither sample of that quite equals our very best in fineness. We give the names of the rams from which samples 1, 2 and 3 were taken, that our readers may understand how the three rams we have used in making comparisons heretofore stand the test as to this last particular when

brought into comparison with the finest sheep, by the most accurate
tests.   In one case we have a ram that shears 37 pounds unwashed
wool (that cleansed weighs 9 pounds 11 oz.), a pick lock from
which measures fully one-fourth finer than Super-Electa, and one
yielding a fleece of 31 pounds (or 8 3 per cent. of cleansed wool to
live weight), a pick lock from which measures more than a third
finer; the average size of the samples from the three is fully one-
fourth finer.   Among the ten rams we find five that shear an average
of 33 pounds 11 ounces of unwashed wool, or 26.4 per cent. to live
weight, pick locks from which measure on the average about one-
fifth finer than Super-Electa, while the coarsest of the five is finer
than that high standard   Among the ewes we have one that yields
20 pounds unwashed wool ; a lock taken from the fleece after it was
done up measures about one-fourth finer than the Saxon Standard,
and one which yields 19 pounds (or 29.3 per cent. to live weight),
a pick lock from which measures over one-third finer.   The very
coarsest ewe of the lot yields wool over one-seventh finer than
Super-Electa.   But the most remarkable case of high excellence in
large weight of fleece and fine quality is in ewe No. 11 that sheared
19 lbs. 3 oz. unwashed wool, a pick lock from which measures less
than half the size of Super-Electa ; in the case of this ewe the lock
was preserved from one of her heaviest fleeces, when she was in
her prime, it is believed from her heaviest, 19 lbs 3 oz., but it is not
certain that it may not have been from one a year or two later ; the
samples from 11 and 12 are from the same ewe at different ages. This
test would seem to prove the general opinion that the same sheep
yields wool of finer quality as it increases in age is correct. This
ewe died at the age of thirteen years ; she was the dam of a family
remarkable for weight of fleece and fine, even quality.   Judged by
the standard of Mr. Morrell, quoted at page 76, these measurements
give us the right to claim that our heavy-fleeced Merinos, as improved by our Vermont breeders, bear wool of very fine quality.

We would call attention to the measurements of Leicester and
South Down wool fibres ; the former was grown in Kentucky, the
latter in Vermont ; while the Leicester wool measures a trifle
coarser than the size of that wool given by Mr. Youatt, the South
Down wool is finer, showing either quite an inprovement in the
quality of that wool or the effect of climate and altitude ; if the latter is the case, then it is another fact that goes to show the natural
effect of Vermont's latitude and location upon the sheep grown
within her borders.

There are other peculiar points to be made and lessons to be
learned that could be adduced from these measurements, weights
and comparisons, but we have already dwelt much longer upon this
subject than we intended.   It is to be hoped that the Association
may induce Dr. Cutting at some future time to continue these investigations, and take measurements from the different parts of the same
fleece in several instances, and thus determine whether there has
been advance or deterioration in the evenness of the fleece, and throw
light upon that and many other points for the benefit of our breeders.

As to the evenness of fleece, we would call attention to the fact hereinafter stated, that in all the nine first-class, and in two of the special awards at the Philadelphia Centennial Exhibition upon Merino sheep from Vermont, they were commended for evenness of fleece, and in the other two cases of special premiums both lots were commended for the same excellence in other awards.

The facts we have presented upon the high standard attained by our breeders of Merino sheep in great weight, fineness and evenness of fleece, combined with great constitutional development and hardiness, would seem sufficient to prove a surety to breeders in those countries where the finest, lightest wools are grown, that they may very materially increase the wool-bearing capacity and profits of their flocks by introducing liberal amounts of the blood of our heavy-wooled Merinos, without fear there would follow any decline in the quality of their fleeces, or depreciation of the value of their wools.

There are but two more points which we wish to notice under this head of improvement of Vermont Merinos, both of which go, we think, to sustain positions we have taken in this paper, and both having a direct bearing upon the subject.

The first is the list of rams and ewes recorded in the United States Sheep Register, published under the auspices of the Ohio Wool Growers' Association In this list we find eight rams and ninety-nine ewes out of a total of 340 that were bred in Vermont, submitted to an ordeal of scaling by points by a committee appointed by the association to test their comparative excellence. In the same list were twenty-seven rams and one hundred and thirty-seven ewes that were bred in Ohio, five rams and thirty-seven ewes bred in Pennsylvania, and twenty-six ewes bred in Virginia. Of those that were bred in Ohio, we find the sires and dams of twelve were bred in Vermont, and in forty-six cases either the sire or dam was bred in this State. In the case of the remainder of the sheep bred in Ohio, where any authenticated line of ancestry is traced by sire and dam, the line ends in both sire and dam bred in Vermont, before many generations are passed.

The following table will show the results of the scaling by points by the committee, no one of which was a resident of or a breeder in Vermont :

Rams and Ewes.	Whole No.	Average of points.	No. that scaled 81, the lowest admitted.	Number that reached 85.	Number that reached 90.
Rams.					
Vermont	8	$85\frac{1}{4}$	0	5 or 62 per ct.	0
Ohio	27	$83\frac{5}{27}$	9 or 33 per ct.	6 or 22 per ct.	1
Pennsylvania	5	$87\frac{1}{5}$	0	5 or 100 per ct.	0
Ewes.					
Vermont	99	$84\frac{22}{99}$	16 or 16 per ct.	39 or 39 per ct.	4 or 4 per ct.
Ohio	137	$82\frac{113}{137}$	59 or 43 per ct.	26 or 9 per ct.	1
Pennsylvania	37	$83\frac{26}{37}$	5 or 13 per ct.	0	0
Virginia	26	$82\frac{4}{13}$	10 or 38 per ct.	2 or 7 per ct.	0

It would be erroneous to suppose that the sheep named as bred in Vermont were the very best in the State at the time they were sold from it, or even as good as the very best we had at the time; nor do we believe they were as good as could be selected from those sold to go to Ohio within the last year, though the tests show they averaged better than those to which the same tests were applied that were bred in the other States.

This scale of points was not made by Vermont breeders, nor were their principles applied by them nor by a committee of their selection. The result of their application by the committee of the United States Register, selected by the Ohio Wool Growers' Association, is an impartial tribute to the excellence of our Vermont Merinos, for which our breeders have reason to be grateful.

The other point of evidence tending to prove the advanced stage of improvement reached by our breeders upon their Merino sheep, is the awards made upon this class at the Centennial at Philadelphia in 1876.

This class of sheep received at that show thirteen first-class awards and fifteen second-class ditto; of these, nine of the first-class and four of the second were received by Vermont breeders. These were received in common with the breeders of Merino sheep from other States. Besides these, there were four other special competitive prizes or awards where the prizes were given for the best only.

There was an award for the best two rams and fourteen ewes selected from those exhibited from each State. This prize was taken by the Vermont State Wool Growers' Association.

There was a prize of $100 offered by the Pennsylvania State Agricultural Society, for the best flock of Merino sheep, to consist of one ram and four ewes, bred and exhibited by one breeder. This prize was taken by Mr. Joseph T. Stickney, of Shoreham, Vermont.

There was a sweepstakes prize or award for "The Best American Merino Ram of any age." This award was made to the Ram Bismarck, bred and owned by Mr. H. C. Burwell, Bridport, Vermont.

There was also a sweepstakes prize for the three best American Merino ewes of any age. This prize was won by Mr. A. E. Perkins, of Pomfret, Vermont. So far as we have been able to learn, all the special competitive prizes on Merino sheep at that exhibition were awarded to Vermont breeders that have their flocks recorded in this Register.

Of the nine lots of sheep that received first-class awards, all were commended for "high excellence in quality," "evenness of fleece," "large constitutional development," and for being "superior specimens of the breed to which they belong." Eight of them were commended for their "density of fleece," seven of them were commended for their "symmetry," four of them for "uniformity," and two of them for "length of staple." The fourteen ewes and two rams selected from the lots exhibited from Vermont were commended for all these excellences: "High excellence in quality, uniformity of symmetry, density, evenness of fleece, and length of sta-

ple, large constitutional development, and for being very superior
specimens of the breed to which they belong."

A short account of the way this exhibit of sheep from Vermont
came to be made will show that the awards they received there were
won on their natural, sterling merits, and were not due to early
shearing or long fitting and preparation. Up to September 1st there
was no sheep breeder in Vermont that intended to make an exhibition of sheep there, or had made preparation for it, the expense
being greater than the prospect of any adequate return from honors
or premiums, as they were announced by the Centennial commis-
sioners.

During the time of the State fair, held at St. Albans, September, 1876, a number of consultations were held among the sheep-
breeders, at which great disapointment was felt and expressed that
no Vermont breeders were intending to exhibit sheep at the Centennial Exhibition, or had any sheep that had been shorn early enough,
or properly fitted to compete with the Merino sheep from other
States. This feeling was so wide-spread that a number of breeders
expressed a willingness to place their sheep at the disposal of the
State Agricultural Society and Wool Growers' Association—late-
sheared and imperfectly-fitted as they were—provided that Association would appropriate money to defray the expense, and appoint
some one to take charge of and make an exhibition in the name of
the State.

This offer was magnanimously accepted on the part of the State
Association, and the exhibition was made, with the results as stated
above. There is no doubt that had the sheep from Vermont been
shorn earlier, with a view to making that exhibition, more of them
would have been commended for length of staple; but as all but one
lot were commended for density, a more important excellence, Vermont breeders would be unreasonable not to be satisfied with the
laurels they won upon the Merino sheep they sent to the Centennial.

The Jury were governed by the following points in judging for
Merinos, which are identical with those that governed the committee of the U. S. Sheep Register in scaling the 340 sheep noticed
heretofore.

1st, pedigree. A perfect, authenticated line of ancestry, extending to one or more of the importations of fine-wool Merino sheep
from Spain, made prior to 1812, without admixture of any other
blood—20 points.

2d, constitution : This is indicated by a healthful countenance ;
expanded nostrils ; short, strong neck ; deep chest ; round barrel ;
short, strong back ; strong loin ; heavy bone, of fine texture ; mus-
cle fine and firm ; skin thick, soft and of a pink color—18 points.

3d, fleece : This includes the quantity, quality and condition
of the wool, as shown in the weight of the fleece, the length and
strength of staple, crimp, fineness and trueness of fibre, evenness
throughout, freeness from gare, and the fluidity and amount of yolk
—13 points.

4th, covering : The extent and evenness of the fleece over the whole body, legs, belly, neck and head; the quality, lustre, crimp, density, and length of wool, and the quantity and kind of oil or yolk —13 points.

5th, form: The shoulders should be broad and well placed; back broad; quarters long and well filled up; head short; folds on the neck, shoulders, flanks, belly, thighs and the tail; all parts in just proportions—9 points.

6th, size: Rams at full growth, in breeding condition, should weigh 130 pounds or upwards; and ewes about 100 pounds—8 points.

7th, head: Should be medium size; muzzle clear; nose covered with short, furry hair; eyes bright and placid; forehead broad; ears soft, thick, and set wide apart; horns, on the ewe none, on the ram well turned (set not too closely to the head and neck, nor yet standing out too widely from them), and free from black or dark-colored streaks—6 points.

8th, neck: Should be short on top and long below; strongly set to the head and shoulders, becoming deeper towards the shoulders; folds heavier underneath and extending up the sides of the neck, including heavy dewlap or apron, and cross-folds higher up, but diminishing in size over the top of the neck—5 points.

9th, legs and feet: Legs short, straight, well spread apart and bone heavy; hoofs clear in color and well shaped—4 points.

10th, general appearance: Good carriage; bold, vigorous style; symmetrical form, and proper complexion of the covering— 4 points.

Making a total of one hundred points. The whole number of Merino sheep exhibited was 218; of this number Vermont sent 61, selected from the flocks of ten different breeders, and nine of these secured first-class awards, besides the special prizes named. The tenth was a single ram exhibited by his purchaser, not by his breeder.

We give a part of the report of the committee of jurors upon Merinos:

The United States was the only exhibitor in the Merino class. The display, though small in numbers, was very fine in character: numerous animals in each class being perfect specimens in their respective breeds.   *   *

Fully three-fourths of the territory of the United States is well adapted to wool growing, and wool is destined in the near future to become one of the leading staples of export.

The question, then, as to which is the best sheep for general purposes in our country, assumes an importance of large proportions. In awarding this position to the "American Merino," we think we are fully borne out by the following facts and conclusions:

First. The climate of our country is a variable one, subject to great extremes of temperature, requiring animals of great hardiness of constitution to enable them to endure its frequent and sudden changes.

Second. Our agriculture is still in a rude- and transition state; few of our citizens being in posession of the necessary facilities required for the successful rearing and keeping of the large mutton breeds. Again, that portion of our country best adapted to sheep and wool-growing is still almost a wilderness, a grand expanse of nature, and must remain such for many years to come unless utilized by the culture of some race of animals so hardy as to be in a measure self-sustaining and at the same time profitable to their owner.

Such an animal we claim is the "American Merino" as now bred among us—a sheep of fair size (weighing 100 to 160 pounds), compact in carcass, symmetrical in form, possessing a strong constitution, and carrying a fleece of great density, weight and value, covering the sheep in every part where wool ought to grow, thus protecting every part of the body from climatic influences, a sure breeder, and a fair nurse. As a working sheep, it is capable of living on short pasturage and scanty fare, easily herded, and the only good sheep which can be profitably kept in large flocks—a sheep which fills more nearly all the requirements which the condition of our country and people demand than any other known breed.

As a further proof of the great value of this breed, and the correctness of our position, we may mention that three prizes—two first and one second—were awarded to an entry of twelve animals of this breed, at the International Exhibition held at Hamburg, Germany, in 1863. They competed with two hundred and ninety-two rams and fifty-five pens of ewes. Only eight prizes were awarded.*

The great improvement which has been wrought upon Spain's type of Merinos, as imported into this country by Col. Humphreys, Consul Jarvis and others, between 1802 and 1812, through the enterprise and skill of our breeders, is but an earnest of the further advances which the same skill and enterprise are destined to effect during the next half century.

Your committee, in conclusion, would further state that there were a number of very good sheep on exhibition, which would have received more honorable mention had they not, in our judgment, been unfairly shorn, and artificially dealt with.†

Signed,   GEORGE CAMPBELL,
                                  O. H. P. BUCHANAN,
                                  M. STOKING.
                                  J. S. MAYNARD.

International Exhibition, October 19, 1876.

---

*The twelve sheep were all bred in and taken from Vermont to the Hamburg Exhibition, by Hon. George Campbell, of Westminster, Vermont.

†This did not apply to any of the Merino sheep exhibited from Vermont.

AIMS OF BREEDERS OF MERINO SHEEP IN VERMONT.

It is deemed advisable to devote a short chapter treating of the subject of the further improvement of Merino sheep in Vermont, and in explaining some of the characteristics of the typical Merinos of our State, as exhibited in some of the best specimens of the race as we have them to-day, and incidentally in considering some of the objections to them, as urged by breeders and wool-growers in other sections of the country.

In discussing this subject we shall aim as far as possible to give the views, as we understand them, of the majority of our best breeders and *improvers* of Merino sheep, those that have taken a leading position, and, by their success in breeding, proved that their opinions and views are entitled to the greatest weight.

There is no doubt that the opinions of these breeders have undergone something of a change in the last ten to fifteen years, and though perhaps they are still far from a unit in their opinions on all points it is desirable to improve most, they still do not differ as much in the main as they probably did in 1865, and we believe the different flocks, families, or lines of blood, are gradually becoming more uniform as to general characteristics and points of excellence.

There are some breeders in Vermont who do not agree with the majority in regard to the most desirable type of sheep for us to breed, and perhaps it is best it should be so. If, in a term of years, experience proves that our breeders as a class are not breeding the the best, most desirable and profitable type of Merino sheep, those who differ from the majority will have proved public benefactors by breeding something more desirable and more profitable, and will thus enable our breeders to retrieve their lost ground.

It is certain that our breeders, as a class, are breeding Merino sheep of strong and marked characteristics. They believe that wool-growing simply cannot be profitably carried on without we can produce heavy fleeces of wool of a fair quality. As we have heretofore observed, they do not, as a class, breed for *fine* wool. They certainly are breeding for heavy fleeces, and believe they have accomplished in this line more than has been accomplished elsewhere by the breeders of any state or country.

They believe it is easier to increase the size of their sheep, and grow a less proportionate amount of wool, than it is to retain this

11

strong, wool-producing capacity. Even in Vermont those flocks that are neglected soon decline in their wool product, and they some times (not always), become finer in the fleece.

It is believed that the great improvements that have been made have been largely due to a close attention to the subject of selecting sires and mating them with dams, with the view of improving and remedying defects where they exist. Since this practice of selection and close study to couple every sire and dam to improve and remedy defects has been practiced, the improvement has been very rapid, compared with what had been accomplished in the first fifty years after importation.   The last twenty has produced most wonderful results, and we think we run no risk in asserting that even what has been accomplished in the last twenty years will be much exceeded by what will be reached in the next twenty.

In accomplishing what they have, Vermont breeders have developed a large amount of oil in the fleece, and encouraged an increase of the looseness of the skin and consequent development of wrinkles over the carcass.   In the opinion of some breeders, within and without the State, we have an excess of these two qualities that is not desirable.   In regard to the amount of oil, we do not believe there has been a real increase of the natural amount in excess of the increase of the amount of wool.

Our best sheep, as they are cared for and sheltered from the Fall rains, have apparently more oil in proportion to amount of wool than their ancestors thirty-five years ago, but we believe it is only the preservation of what is naturally produced that makes the proportion seem greater.

The Atwood sheep, brought from Connecticut by Messrs. Cooks and Hammond were more oily than are their pure bred descendants in Vermont to-day, when the latter are not housed from the Fall rains.   We know of no stock ram among the class we have selected as examples to make comparisons of heavy fleeces, that has as much oil in proportion to amount of wool as did Old Black, Wooster, or Old Greasy, Nos. 9, 16 and 18, of our list of stock rams.

Be this as it may, our best breeders do not believe we can dispense with any of the average amount of oil without running a risk of serious loss from the deterioration of the amount of wool, as well as the strength, fineness, and evenness of fibre ; they believe it is absolutely necessary to retain these most desirable qualities of the Merino fleece. The majority of them believe it is not so easy to retain these qualities as it is to breed dry, light fleeces with twisted dead ends with an uneven fibre, that may not waste so much in the scouring tub as in the cards, and devoid of that elastic strength and felting quality that gives Merino wool its great value.   Certain it is, that the very heaviest fleeces of scoured wool we have produced thus far, have in no instance been from our bulky dry ones ; our experience proves that such fleeces disappoint our expectations in yielding cleansed wool.   That this development of oil may be carried to such excess in an occasional individual as to impair its constitution and thrift, is conceded ; in breeding animals, as a class, up to a higher standard in almost any direction, there will almost necessarily be produced individuals having the desired characteristics in excess, but the larger portion of the animals of the stock will be up to a better average standard in those characteristics sought, than had these few exceptional and excessive cases not been produced. The want of courage to take bold steps in this line of improvement has been a great hindrance to many a breeder in preventing him from making advancement, for fear that some animals might be produced that were unevenly balanced in their good qualities.

This observation is particularly applicable to the matter of folds or wrinkles. There are very few flocks of sheep where it is not desirable to use a ram that has an amount of wrinkles in excess of what is desirable in the ideal sheep, to bring the flock up to a desirable style, and to develop the greatest wool-bearing capacities.

In breeding for great looseness of skin, our Vermont breeders have the facts of history and the authority of breeders of olden time to sustain them in their arguments as to its desirability.

Petri, the French writer upon Merinos, says a ram should have "a heavy folded skin." He also observes "that the lambs which bring into the world fine, soft hair and a great number of folds, and whose tails are, in appearance, shortened by the large folds around them, bear the indication of great softness and quantity of wool."

Mr. Charles S Fleischmann visited Germany in 1844 and '45, and made very minute investigations of the principles upon which wool raising and sheep breeding was conducted in Silesia, and the results of the experience in that country in their efforts to improve the Spanish Merino sheep in its capacity as a producer of fine wool. Mr. F gives the results of these observations and investigations in the Patent Office Report for 1847. From this article we make a few extracts :

"Twenty years ago, bucks with a smooth, tight skin, which had extremely fine wool, were considered the best; but their fleeces were light in weight, and had a tendency to run into twist. *The German Merino wool grower had to come back to the original form of rams, with a loose skin, many folds, and heavy fleeces*, and since then they have succeeded in uniting, with a great quantity of wool, a high degree of fineness."

"This kind of heavy folded animals, rams and ewes, are now considered the best for breeding and wool bearing"

"According to Petri, who traveled in Spain, with a view of collecting information upon Merino wool culture, the Spanish consider Merino sheep, with folds, as a sign of an improved and thorough breed." "More or less folds upon an animal give proof of the greater or less quantity of wool ; but these folds must be covered with as fine and good a wool as it is on the adjacent parts of the body."

"The Spaniards kill all those lambs which are born with few or no folds, and fine, short hair, or almost naked, because experience has taught them that the offspring of such animals bear a fine wool, *but produce, by degrees, animals with flabby, light* fleeces, which *gradually* loose the folds, and become *thinner and thinner in the fleece,* and are consequently less advantageous to the wool-grower than those sheep which are produced from lambs with plenty of folds and a thick cover of soft, fine hair."   *   *   *   *

Only to the Merino breed belong the close and thick-set fleece, which, in respect to their size, produces the greatest quantity of wool.   The folds are not a *necessary* condition of fineness, *but of quantity,* and are peculiar to the Spanish full-blaoded Merinos.

We believe the experience of Vermont breeders has been the same as those of Germany and of Spain.

Consul Jarvis may be quoted as an example of failure to realize any improvement in the wool-raising capacity of his sheep, for want of a taste for these folds, and a large or desirable amount of oil in the fleece.   After breeding his imported Merinos for nearly or quite fifty years with these views, he was only able to say the pure-blood Merino part of it (his flock) will not vary materially from the original weight.

The large majority of the best breeders of Merinos in Vermont believe that all who are breeding for smooth sheep, bearing dry wool, will sooner or later find they cannot keep their flocks up to a high profitable standard in the line of heavy fleeces.   Such breeders will find after a lapse of years they will be obliged to report their fleeces "will not vary materially from the original weight;" probably many of them obliged to report a decline from "the original weight."

If we admit that our best flocks of Merinos have oil and wrinkles in excess of the wants of the practical wool-grower for his wool-bearing sheep, as a class, we contend that we are not breeding altogether with a view of wool-growing in Vermont, but our most profitable product is blood, that will produce improvements in the wool-bearing capacities of the flocks in localities where it is hard to keep them up to the most profitable standard.   Hence it is for our best interest—as it is for theirs—that we should be able to furnish them with sheep having these qualities in a very marked degree, and greatly in excess of what may, perhaps, be their ideal.   So long as nearly all the sheep in the flocks of the wool-producing regions of our own and other countries lack a sufficiency of wrinkles to make them stylish in the general appearance, and give them the greatest capacity for dense, heavy fleeces, and have not sufficient oil to properly preserve the strength, elasticity, fineness, and felting quantity of the fibre of their fleeces, Vermont breeders should, and probably will, have a demand for all the sheep she can spare from her breeding flocks possessing these characteristics in a high degree, even though over-fastidious persons may claim our sheep have too much oil and too many wrinkles.   The experience of our breeders has taught them it is not safe to allow even these fastidious persons to select from their flocks, if they wish to retain their most wrinkly and oiliest sheep.

We do not wish to be understood as advocating or encouraging the cultivation of that class of large, heavy wrinkles over the body, that bear a very coarse quality of wool or jar hairs, or of that kind of oil, or, more properly speaking, gum, that gathers in lumps or clots instead of circulating freely through the fleece to the ends of the fibre.   Our best breeders are not breeding for that type of sheep, but have made great improvement in breeding sheep with a fineness of fibre upon the wrinkles more nearly corresponding with the quality in the body of the fleece ; and from the advances they have been able to make in this direction, it is reasonable to suppose not many years will elapse before our Merinos will yield as fine wool upon their wrinkles as upon the plain parts of their bodies.   There is already less difference in the quality of the wool from these places on our best sheep than most breeders imagine. Having increased the weight of fleece to the highest point, it would seem to be profitable, with the present size of carcass, and developed as fine quality in the body of the fleece as is desirable, our breeders may turn their intention more to increasing the size of the sheep and its evenness of fleece, with the assurance that they may accomplish as much in these directions as they have in the others, and carry along with them all the great and important improvements they have already made.

We know that wool manufacturers and wool buyers would not be in the fashion did they not decry our Vermont wools, that they would advise our breeders to breed light, dry fleeces, but so long as they readily buy our heaviest wool, and pay very nearly the same price per pound as they will that from our flocks shearing not more than half as much, raising the heavier fleeces must be much the most profitable, and the experience of our breeders for past years is ample to prove that the advice of manufacturers or wool buyers in matters of breeding is far from disinterested or profitable for the wool-grower to follow.

To summarize the aims of Vermont breeders, we believe they have received such encouragement from the vast changes for the better already made in our Merino sheep, to hope that still higher degrees of excellence may be attained.   They are not content with the success they have already achieved, but in an enterprising spirit they prepare, favored by an unequalled climate and location for their purpose, to increase the size of their sheep, improve their forms and constitutional development, and even, if it is possible, to increase their wool-bearing capacities, and the evenness and quality of their fleeces, until the average of their flocks shall be equal to the very best specimens yet produced, and the best shall then be far in advance in all good qualities of any we have yet produced.

LIST OF STOCK RAMS.

EXPLANATION.

We present the following list of stock-rams, not as a perfect one, but by far the largest and most complete that has ever been compiled and published. No part of the work of preparing the Register for publication has cost us more research and labor than this.

The imperfect manner most breeders have kept their records (where, indeed, they have kept any at all), will at this late day render it impossible to make any list of stock rams with the entire pedigree that will be perfect.

While breeders have generally aimed to be particular about the blood of the stock rams they have used, very few have seemed to deem it a matter of importance to trace out or make any record of individual pedigrees. A very small number of breeders have kept any record of purchases or sales, very few have thought a certificate and pedigree at the time of purchase of any consequence, and fewer still have kept any record of their breeding. Consequently in compiling the following list and pedigrees of stock-rams, memories have to be largely consulted and depended upon.

To systemize and reconcile all disagreements and differences in the memories of those we have had to consult, has been no trifling matter, and has consumed no small amount of time, patience or labor. And still more time and research would be necessary than we should be justified in giving to this list before we should probably feel ourselves satisfied with it, or could feel justified in pronouncing it complete. In giving the histories of the rams so far as we could ascertain their change of ownership, we may sometimes have made mistakes, but in the lines of blood and pedigrees, have aimed at great accuracy, and secured it in as great degree as it is possible at this late day, and we believe the conclusions we have arrived at are entitled to the confidence of breeders.

EXPLANATION OF THE NAMES FOR THE LINES OF BLOOD.

The Spanish Merino sheep of Vermont that have been kept pure since they have been introduced. have been mainly if not entirely derived from five sources or flocks. First, that of Consul Jarvis, by direct importation by himself from Spain, composed of representatives from the Paulan, Agueirres, Escurial, Negretti, and

Montarco Cabanas. According to Mr. Jarvis these were bred "separately—that is, each kind by itself—from 1811 to 1816, but in that year I began mixing all together, and have ever since bred so without discrimination." This combination of different bloods we call Jarvis bloods. Second, from the flocks of Andrew Cocks, Long Island, from which Zebulon Frost and Leonard Beedle brought the foundations of the Rich, Beedle and Frost flocks. The letter of Effingham Lawrence, published elsewhere in this work (originally published by Mr. Randall, in 1844), will explain the manner this flock was founded.

The blood of this flock, up to the time that J. Thurman Rich and Tyler Stickney introduced strains of the Jarvis blood by rams from his flock, may be designated Cocks blood.

Third, from the flock of Stephen Atwood, of Connecticut, by Messrs. Cook, Hammond & Hall, and several other breeds, have been introduced descendants from the importation of Col. Humphreys. From what Spanish Cabana Col. Humphreys derived these sheep is still and probably ever will be a mystery. The blood from this flock has everywhere been so long called Atwood, and is so seldom called by any other name, that we see no good reason for giving it any other.

Fourth, From the flocks of Buffum and Joseph I. Bailey, of Rhode Island, were brought by Messrs. Murray, Munger & Bundy (and later, several other breeders), sheep that were descended from importations by Capt. Paul Cuffe, believed to have been selected in Spain by Consul Jarvis. These we have followed custom by calling them Rhode Island blood.

Fifth, from the flock of Jacob N. Blakeslee, of Connecticut, have been introduced by Messrs. Sanford, Bingham and others, sheep descended from the importations of Fairchild, Peck & Woodward (erroneously called the Peck & Atwater importation), and the importation of Abraham Heaton and others. The former were from the Cabaña of the Duke Infantado, and the latter from the Gaudaloupe and Negretti Cabanas. These sheep, from the time of their purchase of Mr. Blakeslee, have taken his name, and we see no reason for changing it. There is very little of this blood left in the flocks of Vermont.

Their more marked characteristics were, in their moderate size, smooth bodies, covered with a fleece unrivalled by that of any other pure Merino family of that day for length of staple, fine, crimpy, elastic fibre, with a free circulating oil that gave great softness to the feel, and brilliancy to the appearance of the whole fleece. They were somewhat inferior to the other families of Merino sheep in constitution. The great reaction from the Merino for Saxon fineness and weak constitution demanded a sheep of a much more decided type for heavy fleece, strong constitution, and more of the distinguishing marks of the Merino as exhibited in heavy folds and wrinkles. For these reasons the Blakeslee blood has been bred out of our flocks until those few where it was introduced have a very small fraction of it left. Of the other bloods, the Atwood in many flocks has been kept pure and unmixed with any other family, which is not the case with any other of the five families.

The Cocks blood was kept pure by the Messrs. Rich until after 1840, when one or two crosses of Jarvis blood were introduced, followed soon after with an introduction of a strain of Atwood blood, followed later with introduction of both these strains of blood, by using to some extent rams that had both Jarvis and Atwood bloods mingled with the Cocks. Notwitstanding the introduction of these bloods, the Rich flock still retains a preponderance of Cocks blood.

In 1835, Tyler Stickney united the Cocks and Jarvis strains of blood by using a Jarvis ram on Cocks' ewes, and subsequently continued to use rams in which these bloods were combined with the Atwood.

Erastus R. Robinson commenced breeding in 1836 with the Cocks blood of sheep, and so bred them until 1845, when a ram was introduced combining the blood of the Atwood, Jarvis and Cocks strains. In the same year and the one following, the Atwood blood undiluted was introduced by using a ram of that blood to a portion of the flock, and soon after more of these bloods were introduced by a purchase of thirty ewes in which these three strains of blood—Atwood, Jarvis and Cocks—were combined.

It will readily be seen that the three flocks of Messrs. Rich, Stickney and Robinson came finally to possess the same three bloods, the Atwood, Jarvis and Cocks, and although the proportions might differ somewhat, in the main and substantially they were the same. The individual tastes of each may have somewhat varied their practice as breeders, and consequently may have affected the characteristics of the individual sheep that were bred in their flocks, but when we call a sheep's blood Rich, Stickney and Robinson, we mean one bred from their flocks, and combining the bloods of the Atwood, Cocks and Jarvis flocks.

Two other breeders contemporary with these three, made something the same crosses or combinations of bloods, Messrs. S. C. Remele and Prosper Elitharp, and although the latter introduced more of the Atwood blood, and gradually changed it to that alone, from 1844 to 1850 all these five flocks contained these same bloods, and Mr. Remele's does still. There is this difference, however : both Mr. Elitharp and Mr. Remele bought and introduced into their flocks pure bred Atwood ewes. Therefore, it is possible that pure Atwood sheep may be pedigreed through Mr. Remele's flock, as we know that large numbers are through Mr Elitharp's.

The Rhode Island blood has come down to us through the flocks of David and German Cutting. For a few years after they commenced their flocks they bred only this blood, but in 1846 they introduced Atwood blood, and for several years subsequently nearly or quite all the rams they used were of this blood. They also introduced this blood by way of ewes purchased, until finally it became by this and more by the rams used, largely composed of this blood, but we believe no special efforts were made to keep this blood pure or distinct from the other. Cutting blood, then, means a combination of the Atwood and Rhode Island flocks.

There were other breeders that made about the same crosses of the different strains or families of blood, as the histories of their flocks will show. It is not deemed necessary to name them all here.

We are obliged to commence our list with a Jarvis ram, dropped about twenty-five years after the Consul made his importations. Although there are individual rams mentioned previous to that time, there is no pedigree given of any, nor any line of ancestry previous to that time that is continuous through individuals, only through flocks. This want of a perfect line of ancestry through individuals from the importations from Spain to 1835, and in the case of the Atwoods to 1844, is one to be deeply regretted, but it is inevitable. No amount of research can now supply it; we are obliged to start from those dates, with representatives from flocks that we have every reasonable assurance had been kept pure. What is true of those with which we have started, is also true of some others that we, from time to time, have found necessary to introduce into this list. We have been obliged to start pedigrees through them from well-known, pure bred flocks, instead of through a line of individual ancestry. There are very few cases where the pedigree can be traced through the ram for many generations. Although this may not materially affect the standing of those that have been bred heretofore, it is evident it will unfavorably affect the standing of those that are bred after this date, if there is not care and attention given to the matter of keeping records of breeding and pedigrees of not only flocks, but of all individual members of them.

STOCK RAMS.

THEIR PEDIGREES AND HISTORIES AS FAR AS ASCERTAINED.

1 CONSUL (Sanford's).

Bred by William Jarvis about 1838. Sold by him to Ward M. Lincoln, Brandon Vt., and purchased of him by W. R. Sanford, Orwell,Vt ,who sold him at the New York State fair,at Poughkeepsie, 1844, after breeding from him for many years. Live weight, about 160 lbs. Jarvis blood.

2 CONSUL (Stickney's).

Bred by Wm. Jarvis, 1855, purchased of him by Tyler Stick- ney, Shoreham, Vt., when a teg, soon after he was weaned. He was used for a number of years by Mr. Stickney and his neighbors, then sold in 1843 to G. A. Austin and John Looker, Orwell, Vt , and by them to J. Thurman Rich, Richville, Vt.*

This ram weighed at maturity about 130 to 140 lbs. His heavi- est fleece was 9 lbs. 2 ozs. washed wool. Jarvis blood.

3 JARVIS.

Bred by Wm. Jarvis, sold to W. R Sanford, afterwards, when nine years old, to Merrill Bingham, Cornwall, upon whose farm he died, after having been used as a stock ram by Mr. B. two years. Jarvis blood.

4 HERO.

Bred by Tyler Stickney, Shoreham, Vt., 1840. Sire, Consul (2), dam, a pure Cocks ewe, bred by Charles Rich, Shoreham. He was sold when two or three years old, to A. L. Bingham, Cornwall, Vt. His heaviest fleece weighed 13 lbs. Stickney blood.

5 FORTUNE.

Bred, 1844, by Tyler Stickney, Shoreham. Sire and dam same as Hero. Sold when a teg to L. C. Remele, Shoreham, by him to Jonathan Willson, Shoreham, and by him to S. W. Jewett, of Wey- bridge, in whose hands he attained great celebrity. He weighed about 120 pounds and gave at his third fleece 13 lbs. 4 ozs. of wool, imperfectly washed in the brook. Stickney blood.

*We have in our possession letters from Messrs. Austin and Looker, besides the statement of Mr. Stickney, certifying to these facts.

6 DON PEDRO.

Bred by Alfred Hull, Wallingford, of Jarvis blood. Sold by Mr. Hull to Wm. Lane, Cornwall, then by Mr. Lane to S. W. Jewett, Weybridge. This ram was a large, fine-formed sheep, would weigh 140 to 150 lbs., with a vigorous constitution. Sheared about 13 lbs. (imperfectly) brook-washed wool, as his heaviest fleece. Jarvis blood.

7 RHODE ISLAND.

Probably bred by J. I. Bailey, Newport, R. I., sold by him to Munger Murray & Bundy, who brought him to Vermont, and after changing hands several times, became the property of L. D. Greg. ory, Weybridge, where he made great improvements in the flocks of the neighborhood. Rhode Island blood.

8 BLACK HAWK.

Bred by M. W. C. Wright, Shoreham, Vt., about 1842. Sire, Fortune (5), dam, an excellent ewe, bred by Wm. Jarvis. She was only medium in size, but had a fine, long-stapled fleece. Mr. Wright sold this ewe to S. W. Jewett, who afterwards bred a ram lamb from her. Her teats had been cut off by some bungling sheep-shearer, and her lambs were raised on cow's milk Mr. Jewett sold this lamb he bred from her to A. Chapman, who re-sold him to Mr. Jewett, the next year. Black Hawk weighed about 100 to 110 lbs., and died the property of Mr. Wright. Jarvis and Cocks blood.

9 OLD BLACK.

Bred by Stephen Atwood, Connecticut, 1841. Sold by him to W. R. Sanford, Orwell, Vt., who afterwards sold an interest in him to Messrs. Hammond and W. R. Remele, of Middlebury, Vt. His heaviest fleece weighed about 14 lbs., his live weight about 135 lbs. Sold when fifteen years old to Capt. J. Sheldon, Fairhaven, Vt. He lived to be nineteen years old. Atwood blood.

10 MATCHLESS.

Bred by Stephen Atwood, Conn., 1841. Sold to Messrs. Hammonds & Hall, with the first purchase of rams and ewes they made of Mr. Atwood. Best fleece, about 12¼ lbs. Live weight, 135 to 140 lbs. Atwood blood.

11 LITTLE RAM.

Bred by Stephen Atwood, Conn., 1841. Purchased at the same time as Matchless, by Messrs. Hammonds & Hall, and used by them as a stock ram for a short time. Atwood blood.

12 ATWOOD.

Bred by Stephen Atwood in 1842. Sold 1844 by Mr. Atwood to Mr. S. L. Bissell, Shoreham, Vt., at the N. Y. State Fair at Poughkeepsie, who, on the way home, sold a half interest in him to

M. W. C. Wright, of Shoreham, and Mr. Wright sold a half of his interest to S. W. Jewett, of Weybridge. Soon after, he was sold to L. C. Remele, Shoreham, and Prosper Elitharp, Bridport. This ram weighed about 100 lbs.; his heaviest fleece about 15 lbs., which, after cleansing, weighed 6 lbs. He made great improvements in the flocks where he was used, and his blood has descended through the pedigrees of many of the most celebrated sheep of Vermont. He died in 1850, the property of Mr. Elitharp. Atwood blood.

13 ELITHARP.

Bred by Prosper Elitharp, Bridport, Vt., 1845. Sire, Atwood (12); dam bred by L. C. Remele, sired by Black Hawk (8); 2d dam bred by Wm. Jarvis. Sold to Erastus R. Robinson, and used by him as a stock ram. Atwood, Jarvis and Cocks blood.

14 COOK RAM.

Bred by Stephen Atwood, Conn. Sold 1841 to David & C. B. Cook, of Charlotte, Vt., and used by them as a stock ram on the ewes purchased at the same time, until 1845. Atwood blood.

15 BLACK BUCK.

Bred by Stephen Atwood, 1845. Sold when a teg to C. B. Cook, Charlotte, and used by him as a stock ram. Atwood blood.

16 WOOSTER.

Bred by W. S. & E. Hammond, Middlebury, Vt., 1849. Sold when a teg to A. J. Wooster, Cornwall, Vt. Sire, Old Black (9); dam bred by S. Atwood. Live weight, about 100 lbs.; his first fleece weighed 12½ lbs; his second 19¼ lbs. Atwood blood.

17 YOUNG MATCHLESS.

Bred by W. S. & E. Hammond, 1850. Sire, Wooster (16); dam, light-colored ewe, bred by Messrs. Hammond, sired by Matchless (10); 2d dam bred by S. Atwood. Live weight, about 140 to 150 lbs; heaviest fleece, 23 lbs. A half interest was sold to W. R. Sanford, Orwell, Vt. Atwood blood.

18 OLD GREASY.

Bred by W. S. & E. Hammond, 1850. Sire, Wooster (16); dam, an Atwood ewe, bred by W. S. & E. Hammond, sired by Old Black (9); 2d dam bred by S. Atwood. A half interest in this ram was sold to W. R. Sanford at the same time as in Young Matchless. His live weight was about 105 to 110 lbs., and his heaviest fleece 22 lbs. This ram died in Massachusetts. Atwood blood.

19 BUTLER.

Bred by W. S. & E. Hammond, 1845. Sire, Little Ram (11); dam bred by S. Atwood. Sold when a teg to Albert Chapman, Weybridge, Vt., and by him to Butler A. Goodrich Brandon, Vt., in whose flock he proved an excellent stock ram. Atwood blood.

20 OLD WRINKLY.

Bred by W. S. & E. Hammond, 1853. Sire, Old Greasy (18); dam bred by W. S. & E. Hammond, sired by Old Greasy (18); 2d dam bred by W. S. & E. Hammond, sired by Matchless (10); 3d dam bred by W. S. & E. Hammond. Live weight, 125 to 130 lbs.; best fleece, 23 lbs. Atwood blood.

21 LONG WOOL.

Bred by W. S. & E. Hammond, 1853. Sire, Old Greasy (18) dam, Lawrence ewe, bred by W. S. & E. Hammond, sired by Young Matchless (17); 2d dam bred by Stephen Atwood. Live weight, about 120 to 125 lbs. He sheared over 20 lbs. of long wool of fine style, well filled with white oil. A most excellent sire of ewes. Atwood blood.

22 LITTLE WRINKLY.

Bred by W. S. & E. Hammond, 1855. Sire, Old Wrinkley (20); dam, twin of little Lawrence ewe, bred by W. S. & E. Hammond, sired by Wooster (16). (This ram was sometimes called the Fine Ram.) He weighed about 100 to 110, and sheared about 19½ lbs. Atwood blood.

23 LONG WOOL 2D.

Bred by Albert Chapman, at Sudbury, Vt., 1854. Sire, Long Wool (21); dam, an Atwood ewe, bred by —— Barker, Leicester, Vt. Sold when a teg to Isaac N. Sawyer, Salisbury, N. H. Atwood blood.

24 LAWRENCE RAM.

Bred by W. S. & E. Hammond, 1856. Sire, Old Wrinkly (20); dam, Lawrence ewe, bred by W. S. & E. Hammond, sired by Long Wool (21). This ram was one of the best in Messrs. Hammonds' flock. His live weight was about 120 to 130 lbs.; his heaviest fleece 24 lbs. The dam of this ram was sold to Almon Lawrence, of Monkton, Vt., a short time before she dropped this ram. Messrs. Hammonds subsequently purchased him of Mr. Lawrence, and used him for a number of years as one of the stock rams of the flock. Sold in his old age to Capt. Joseph Sheldon, Fair Haven, Vt. Atwood blood.

25 ATWOOD RAM (Cutting's).

Bred by S. Atwood. Sold to D. & G. Cutting, 1846. Used by them a number or years as a stock ram.

26 GEO. ATWOOD RAM.

Bred by George Atwood. Conn. (son of S. Atwood), 1846. Sold to D. & G. Cutting, and used by them as a stock ram. Atwood blood.

27 SAXTON RAM.

Bred by N. A. Saxton, Waltham, Vt. 1852. Sire, Wooster (16), dam bred by W. S. & E. Hammond. Sold to D. & G. Cutting, 1853, and used by them and others. Atwood blood.

28 YOUNG WOOSTER.

Bred by D. & G. Cutting, 1853. Sire, Wooster (16), dam by Geo. Atwood ram (26), 2d dam, a Rhode Island ewe. Used one year by D. & G. Cutting, then sold while still a teg to N. Cushing, Woodstock, who afterwards sold him to Geo. Campbell, Westminster West, who used him a number of years. Cutting blood.

29 BLACK HAWK.

Bred by W. S & E. Hammond, 1845. Sire, Old Black (9); dam an Atwood ewe. Atwood blood.

30 OLD POMP.

Bred by E. Bridge, Pomfret, Vt. Sire, Black Hawk (29), dam, a Jarvis ewe. Sold to George Campbell, Westminster West, 1846, when a teg. Used by Mr. Campbell for a number of years. Died at 15 years old. Atwood and Jarvis blood.

31 GREASY (CUTTING's).

Bred by David Cutting, Shoreham, 1852. Sire, Wooster (16); dam by Old Black (9), 2d dam, a Rhode Island ewe, bred by Messrs. Cutting. Fleece, at three years, 23½ lbs. He was sold to A. L. Bingham, Cornwall, Vt., and by him to J. Slocum, Western, Pa. Cutting blood.

32 SWEEPSTAKES.

Bred by W. S. & E. Hammond, 1856. Sire, Little Wrinkly (22); dam, Light Colored Ewe, 3d, by Old Greasy (18); 2d dam, Light Colored Ewe, 2d, by Old Greasy, (18); 3d dam, Light Colored Ewe, 1st, by Old Matchless (10); 4th dam, bred by S. Atwood. Except Old Matchless and the last-named ewe, all bred by W. S. & E. Hammond. Sweepstakes was one of the most celebrated of the stock rams of W. S. & E. Hammond, and made great improvements in the many flocks where he was used His live weight was 135 to 140 lbs., his heaviest fleece 27 lbs. He died at the Cream Hill stock farms, Shoreham, Vt., in 1867. Atwood blood.

33 PEERLESS (OR McFARLAND).

Bred by H. W. Hammond, Middlebury, 1860. Sire, Sweepstakes (32); dam an Atwood ewe, bred by W. S. & E. Hammond. Sold to Col. McFarland, Western, Pa, and afterwards to S. Sweet, Bennington, Vt. Atwood blood.

34 CALIFORNIA.

Bred by H. W. Hammond, 1860. Sire, Sweepstakes (32) ; dam, Beauty ; 2d dam, Old Queen, by Long Wool (21) ; 3d dam, Old Queen's dam, by Old Black (9) ; 4th dam, first choice of old ewes, purchased of S Atwood by W. S. & E. Hammond. Sold to Flint, Bixby & Co., California. Atwood blood.

35 AMERICA.

Bred by E. Hammond. 1859. Sire, Sweepstakes (32) ; dam bred by W. S. & E. H. Sold as a teg to N. A. Saxton, Waltham, who sold a half interest to C. B. Cook, Charlotte, and two years after the other half to P. Elitharp. He was used in these flocks as a stock ram, as well as in Mr. Saxton's, until he died. Atwood blood.

36 CROSS TOM.

Bred by E. Hammond. Sire, Sweepstakes (32); dam an Atwood ewe bred by W. S. & E. Hammond. Sold to A. J. Grinnell, Cold-water, Mich., and other parties at the West, afterwards purchased by J. H. Sprague and F. D. Barton, and brought back to Waltham, Vt., where he died. Atwood blood.

37 THOUSAND DOLLAR RAM.

Bred by E. Hammond. Sire, Sweepstakes (32) ; dam same as the dam of Old Queen, by Old Black (9); 2d dam bred by S. Atwood. Sold to A. Willcox, Fayetteville, Onondaga Co., N, Y. Fleece, 25¼ lbs. Atwood blood.

38 OLD ROBINSON RAM.

Bred by E. R. Robinson, Shoreham, Vt. Sire, Elitharp (13); dam by Atwood (12) ; 2d dam, a Rich ewe, bred by E. R. R., used by his breeder as a stock ram for a number of years, then sold to Tyler Stickney, in 1853, where he became very celebrated as a stock ram, living to the age of thirteen or fourteen years. He weighed about 100 lbs , and sheared about fourteen pounds. Robinson blood.

39 LUTE ROBINSON RAM.

Bred by E. R. Robinson, Shoreham, Vt. Sire, Old Robinson (38) ; dam bred by E R. R. Sold to — Clark, at Ticonderoga, N.Y. re-purchased by E. R.R. and his brother, Lucius Robinson, and used, by them as a stock ram. He was afterwards sold to C. D. Lane, Cornwall, who used him a few years, then sold him to A. H. Clapp & A. H. Avery, Manlius, N. Y. He died in 1863 or 4. This ram weighed about 120 to 125 lbs. Robinson blood.

40 TOTTINGHAM.

Bred by E. A. Birchard, Shoreham, Vt., 1858. Sire, Lute Rob-inson (39) ; dam, a Robinson ewe, sired by Old Robinson (38). A half-interest in this ram was sold to B. B. Tottingham, Shoreham, Vt., who kept him until his death, in 1864.   Live weight, 100 to 110 lbs. Fleece, about 19 to 20 lbs.   Robinson blood.

41 TWENTY-ONE PER CENT.

Bred by W. S. & E. Hammond. Sire, Lawrence Ram (24); dam, Old Tulip, bred by W. S. & E. H., by Long Wool (21) ; 2d dam, bred by W. S. & E. H., by Old Black (9) ; 3d dam, bred by S. Atwood. Sold to D. Cosset, Syracuse, N. Y.   Atwood blood.

42 CROSS RAM.

Bred by W. R. Sanford. Sire, Old Greasy (18) ; dam, an Atwood ewe, bred by W. R. S. Sold to R. J. Jones, Cornwall, Vt. Atwood blood.

43 WRIGHT'S CALIFORNIA.

Bred by Victor Wright, Weybridge, Vt.   Sire, Long Wool (21); dam (bred by W. S. & E. Hammond), by Wooster (16) ; 2d dam bred by S. Atwood. A half interest was sold to W. R. Sanford, in 1858, and finally sold entire to Messrs. Hoyt, in California, in 1861. Atwood blood.

44 DON PEDRO.

Bred by W. S. & E Hammond. Sire, one of E. Hammond's stock rams ; dam, a pure Atwood ewe. Sold to R. P. Hall, and by him to Prosper Elitharp, and used by him as a stock ram   Atwood blood.

45 COSSETT (Gold Mine).

Bred by W. R Sanford. Sire, Cross Ram (42) ; dam bred by W. R. S., of Atwood blood. Used by Mr. S. as a stock ram. At three years old, sheared 24 lbs   Sold to Franklin Knox, Whitewater, Wis. Atwood blood.

46 GREEN MOUNTAIN.

Bred by W. R. Sanford, 1858. Sire, Cross Ram (42) ; dam by Young Matchless (17) ; 2d dam bred by S. Atwood. Sold to Jesse Hinds, Brandon, Vt., and used by him as a stock ram, then sold to Messrs. Buells, Orwell, who used him as a stock ram two years. Atwood blood.

47 STOWELL'S SWEEPTAKES.

Bred by E. S Stowell, Cornwall, Vt., 1860. Sire, Peerless (33); dam by Long Wool (21) ; 2d dam bred by W. S. & E. Hammond. Sold to Carey, Boyer & Twitchell, Wyandotte Co., Ohio. Atwood blood.

48 DEAN'S LITTLE WRINKLEY.

Bred by H. W. Hammond. Sire, Sweepstakes (32) ; dam, an Atwood ewe, bred by W. S. & E. Hammond. Purchased by F. H. Dean, Cornwall, and obtained great celebrity in his hands as a stock ·

ram. Sold when old to Peter Martin, Western New York, where he sired a few lambs. Atwood blood.

49 OLD GRIMES.

Bred by W. S. & E. Hammond, 1858. Sire, Sweepstakes (32); dam, an Atwood ewe, bred by W. S & E. H. Sold when a lamb to Geo. Campbell, Westminster West, Vt. Used by him a number of years as a stock ram, and we believe by Mr. Chamberlain, Red Hook, N. Y. Atwood blood.

50 YOUNG SWEEPSTAKES.

Bred by D. Cutting, 1863. Sire, Sweepstakes (32); dam, an Atwood ewe, bred by N. A. Saxton. Atwood blood.

51 MOUNTAINEER.

Bred by J. T. & V. Rich, Shoreham, Vt., 1863. Sire, Tottingham (40); dam bred by J. T. & V. R., from their Rich stock. Rich blood.

52 OLD DEA. JAMES' RAM.

Bred by E. A. Birchard, 1859. Sire, Tottingham (40); dam, a Robinson ewe, bred by E. A. B. Sold when a yearling to Dea. Samuel James, Weybridge, and G. N. Payne, Bridport. Dea. James sold his interest in 1872 to O. S. Hamilton, Bridport. Died 1876. Robinson blood.

53 E. RICH'S SWEEPSTAKES.

Bred by Elisha Rich, Whiting, Vt. Sire, Sweepstakes (32); dam, a Robinson ewe. Used for a number of years as a stock ram by his breeder and others. Atwood and Robinson blood.

54 SANFORD & TREADWAY RAM.

Bred by W. S. & E. Hammond. Sired by Little Wrinkly (22); dam, an Atwood ewe, bred by W. S. & E. Hammond. Sold to L. Treadway, Shoreham, who afterwards sold an interest in him to E. Sanford, Cornwall. He was sold afterwards to C. D. Lane, Cornwall. Atwood blood.

55 KEARSARGE.

Bred by H. W. Hammond, 1863. Sire. Sweepstakes (32); dam, 1st choice of old ewes bred by W. S. & E. Hammond, when they were divided by E. H. and H. W. H. She was sired by Long Wool (21). Used as a stock ram by his breeder, as well as by E. Hammond & Son. Died 1866. Atwood blood.

56 SANFORD & GIBBS' RAM.

Bred by E. R. Robinson, 1852. Sire, Lute Robinson Ram (39); dam bred by E. R. Robinson. Sold to S. S. Gibbs, Cornwall, and a half interest by Gibbs to E. Sanford, Cornwall. Used extensively as a stock ram in other flocks besides his owners'. Robinson blood.

12

57 COMET.

Bred by W. R. Sanford, 1861. Sire, V. Wright's California (43); dam bred by W. R. S., by Old Greasy (18); 2d dam bred by W. R. S. Live weight at three years about 115 lbs.; third fleece, 24¾ lbs. Used extensively as a stock ram. It is said that the income of this ram at three years old was $3,000. Atwood blood.

58 EUREKA.

Bred by W. R. Sanford, 1861. Sire, Comet (57); dam, an Atwood ewe, bred by W. R. S. This ewe was sold by Mr. S. when in lamb by Comet to F. & H. Root, of Orwell. Her lamb was sold to W. O. Bascom, who sold him when a yearling to S. S. Rockwell, West Cornwall, who named him Eureka. In Mr. Rockwell's hands he obtained great celebrity as a stock ram, improving the flock of his owner, and was extensively patronized by many breeders in his neighborhood. He died in 1860, after earning for his owner $8600, besides his use in Mr. R.'s flock. Atwood blood.

59 DEW DROP.

Bred by E. S. Stowell, 1862. Sire, Stowell's Sweepstakes (47); dam, Sukey, 2d, bred by E. S. S., and out of Old Sukey, bred by and purchased of W. S. & E. Hammond. Dew Drop was sold, 1869, to John Sheldon, Livingston Co., N. Y., after being used as a stock ram by his breeder 3 years. Atwood blood.

60 SANFORD LANE, OR RAPALEE RAM.

Bred by Henry Lane, Cornwall, Vt., 1863. Sire, Cross Ram (42); dam bred by N. A. Saxton. Half-interest sold to E Sanford, West Cornwall, afterwards he was sold to H. Rapalee, Gorham, N. Y. Atwood blood.

61 SILVER MINE.

Bred by E. Hammond, 1861. Sire, Sweepstakes (32); dam bred by W. S. & E. Hammond. Used as a stock ram by E. Hammond. Was sent to Wisconsin, where he made a season, after which he was returned, and died at E. Hammond's. Atwood blood.

62 SANFORD.

Bred by S. Atwood, Conn., 1846. Sold to W. R. Sanford, who sold him to S. L. Bissell, Shoreham, Vt., who, after using him for a stock ram some years, sold him to Amos Walker, Whiting, Vt. Atwood blood.

63 GOLD MINE, OR PERCY.

Bred by E. Hammond, 1862. Sire, Sweepstakes (32); dam bred W. S. & E. Hammond. Sold to D. W. Percy & L. J. Burgess, Rensselaer Co., N. Y. Died, 1868 or 9. Atwood blood.

64 GOLD DROP.

Bred by E. Hammond, 1861. Sire, California (34) ; dam, Old Queen ewe, bred by W. S. & E. Hammond, by Long Wool (21) ; 2d dam, by Old Black (9) ; 3d dam, bred by S. Atwood, and was the first choice of the old ewes purchased of him by W. S. & E. H. Heaviest fleece, 25 lbs. Died in 1865. Atwood blood.

65 WOODSTOCK.

Bred by George Campbell. Sire, Wooster (16) ; dam bred by G. C., by Old Pomp (30); 2d dam a Jarvis and Atwood ewe. Atwood and Jarvis blood.

66 GOLD-FINDER.

Bred by J. Hinds, Brandon, Vt. Sire, Gold Drop (64) ; dam by Green Mountain (46) ; 2d dam bred by W. S. & E. Hammond. Sold to John D. Patterson, California. Atwood blood.

67 OLD ETHAN.

Bred by E. A. Birchard, 1859. Sire, Sanford & Treadway (54); dam, an ewe bred by E R. Robinson. A half-interest was sold in this ram to Rollin Birchard, when he was a year old. His live weight was about 110 lbs. He died the property of E. A. & R. Birchard. Atwood and Robinson blood.

68 YOUNG GRIMES.

Bred by George Campbell, Westminster West, Vt., 1861. Sire, Old Grimes (49) ; dam by Woodstock (65) ; 2d dam by Old Pomp (30). Sold to Harlow Bros., Darien, N. Y. Atwood and Jarvis blood.

69 PRINCE.

Bred by N. A. Saxton. Sire, America (35) ; dam bred by W. S. & E Hammond. Sold A. Barringer Illinois. Atwood blood.

70 GREEN MOUNTIAN.

Bred by E. Hammond, 1864. Sire, Gold Drop (64) ; dam bred by W. S. & E. Hammond. Sired by Long Wool (21) Died, 1868. Atwood blood.

71 KILPATRICK.

Bred by W. R. Sanford, 1864. Sire, Comet, (57) ; dam bred by W. R. S., by California (43). Sold to Jed Hyde, Sudbury, Vt , and by him to L. P. Clark, Addison, Vermont, at whose place he died in 1874. Live weight, 150 pounds. Heaviest fleece, 31 ponnds. Atwood blood.

72 GEN. GRANT.

Bred by W. H. Delong, West Cornwall, Vt. Sire, Sweepstakes (32) ; dam bred by D. E. Robinson. Sold to go to Western N. Y. Atwood and Robinson blood.

73 BONAPARTE.

Bred by H. Robbins & S. C. Parkill, Cornwall, Vt. Sire, Sweepstakes (32), dam, Atwood ewe bred by H. W. Hammond. Sold to Miles Rapalee, Western New York. Atwood blood.

74 OLD GREASY.

Bred by Victor Wright, then of Cornwall, Vt. Sire, California (43); dam, an Atwood ewe, bred by W. S. & E. Hammond. Sold to R. Perrine, West, Pa. Atwood blood.

75 SEVILLE.

Bred by W. R. Sanford, 1862. Sire Comet (57); dam bred by W. R. Sanford. Sire, Cross Ram (42). Sold to R. J. Jones, Cornwall. Died, 1868. Atwood blood.

76 FRANK.

Bred by Edgar Sanford, West Cornwall, Vt., 1858. Sire, Sanford Treadway, (54); dam, a Robinson ewe. Sold to E. R. Pottle, Western New York. Atwood and Robinson blood.

77 BLUCHER.

Bred by W. R. Sanford, 1864. Sire, Comet (57); dam bred by W. R. Sanford. Taken to Streator, Ill., and died there. Atwood blood.

78 WRINKLY.

Bred by E. A. Birchard. Sire, Old Ethan (67); dam bred by E. R. Robinson. Died the property of his breeder. Robinson blood.

79 ELITHARP & BURWELL RAM.

Bred by P. Elitharp, 1868. Sire, Eureka (58); dam, Old Favorite, bred by P. Elitharp. Sold in 1871 to H. C. Burwell. Sold in 1872 to Peet & Severance, and by them taken to California. Atwood blood.

80 IRONSIDES.

Bred by Nelson Richards, Panton, Vt., 1860. Sire, America (35); dam bred by W. S. & E. Hammond, and sold by them to Almon Lawrence, of Hinesburgh, Vt., and by him to Mr. Richards. Ironsides was sold to C. N. Hayward, Bridport, Vt. Atwood blood.

81 YOUNG EUREKA.

Bred by J. J. Crane, Bridport, Vt., 1864. Sire, Eureka (58); dam bred by C. N. Hayward, of Bridport. She was sired by America (35). Used by his breeder and others as a stock ram. Sold to L. J. Wright, who took him to Western New York. Atwood blood.

82 YOUNG IRONSIDES.

Bred by C. N. Hayward, 1863. Sire, Ironsides (80); dam, one of Mr. Hayward's best ewes, bred by P. Elitharp. Sold to C. M. Clark, Whitewater, Wis. Atwood blood.

83 SEA LION.

Bred by Alvin Clark, Shoreham, Vt. Sire, Old Ethan (67); dam bred by E R. Robinson. A half interest in this ram was sold to P. Elitharp. Afterwards he was sold to J. H. Sprague, Waltham, in whose hands he became celebrated as a stock ram. He was very generally known as the Sprague ram. Robinson blood.

84 SILVER MINE 2D (Gleason).

Bred by Rollin Gleason, Benson, Vt., 1863. Sire, Silvermine (61); dam by Sweepstakes (32); 2d dam bred by W. R. Sanford; sired by Wooster (16). Atwood blood.

85 ADDISON.

Bred by J. E. Parker, Whiting, Vt., 1865. Sire, Eureka (58); dam bred by F. H. Dean, Cornwall, Vt.; sired by Sweepstakes (32). Sold to S. B. Lusk, Batavia, N. Y. Atwood blood.

86 PRINCE.

Bred by E. Sanford, Cornwall, Vt., 1863. Sire, Sanford Lane (60); dam bred by N. A. Saxton. Sold to Bingham & Eells, who took him to Ohio; sold finally in Western New York. Atwood blood.

87 COLUMBUS.

Bred by C. Segar, Addison, Vt. Sire, Sea Lion (83); dam, a Robinson ewe, bred by C. S. Sold to L. P. Clark, Addison, Vt., and by him an interest to R. & G. Gage, Addison. Robinson blood.

88 SMALL TOM.

Bred by D. E. Robinson, Shoreham, Vt. Sire, Tottingham (40); dam, a Robinson ewe, bred by D. E. R. A half interest sold to F. & L. E. Moore, who used him as a stock ram. Robinson blood.

89 JENNINGS RAM.

Bred by Erastus R. Robinson. Sire, Old Robinson (38); dam, a pure Robinson ewe. Sold to E Jennings, Cornwall, Vt., and by him to R. J. Jones, who took him to Western Pennsylvania. Robinson blood.

90 BLACK TOP.

Bred by Victor Wright, Middlebury, Vt., 1863. Sire, Gold Drop (64); dam, Queen 3d, bred by E. Hammond. She was sired by Sweepstakes (32), and out of Hammond's Old Queen. Died 1865. Atwood blood.

91 GOLDEN FLEECE.

Bred by E. S. Stowell, Cornwall, Vt., 1862. Sire, Stowell's Sweepstakes (47) ; dam, Sukey 6th by (47) ; 2d dam, old Suke, bred by W. S. & E. Hammond. Golden Fleece was used very extensively as a stock ram by his breeder and others. Made one season at Naples and Honeoye, Western N. Y. Died in 1874, after earning over $20,000 for his owner. Atwood blood.

92 FITCH RAM.

Bred by Ezra Fitch, Bridport, Vt. Sire, America (85) ; dam, an Atwood ewe, bred by P. Elitharp. Atwood blood.

93 KING SOLOMON.

Bred by E. S. Stowell, 1865. Sire, Golden Fleece (91) ; dam, Queen of Sheba's dam, bred by W. S. & E. Hammond. K. S. sold to George Hammond and taken to California. Atwood blood.

94 MAJOR.

Bred by R. Gleason, Benson, Vt., 1862. Sire, Saxton's Prince (69) ; dam, Old Favorite, by Wooster (16). Major was sold in 1865 to Dickinson & Herrick, Ohio. Atwood blood.

95 CHUNK.

Bred by C. N. Hayward, Bridport, Vt., 1864. Sire, Young Ironsides (82) ; dam, an Atwood ewe, bred by C. N. H. Sold in his old age, and, it is believed, taken to Utah. Atwood blood.

96 GREEN MOUNTAIN, JR.

Bred by H. W. Hammond, 1866. Sire, Green Mountain (70), dam, an Atwood ewe, bred by H. W. H. Sold F. H. Dean, and by him to Martin Bros., W. N. Y. Atwood blood.

97 TORRENT.

Bred by F. H. Dean, Cornwall, Vt., 1867. Sire, Little Wrinkly (48) ; dam by Golden Fleece (91) ; 2d dam bred by W. S. & E. Hammond. Torrent was sold in 1869 to Peter & George F. Martin, Rush, New York. Atwood blood.

98 HALL RAM.

Bred by R. P. Hall, Cornwall, Vt. Sire, Silvermine (61) ; dam, an Atwood Ewe, bred by R. P. H. Atwood blood.

99 FORTUNE.

Bred by J. T. & V. Rich, 1866. Sire, Hall Ram (98) ; dam bred by J. T. & V. R., from their old flock. Sold to L. J. Wright, Weybridge, and by him to David Hatmaker, Yates Co., New York. Atwood and Rich blood.

100 MAJOR.

Bred by Peter Martin, Rush, N. Y. Sire, Green Mountain, Jr. (96) ; dam, an Atwood ewe, bred by F. H. Dean ; sired by Little Wrinkly (48). Sold to Aaron Barber, Livingston Co., New York, and by him to parties in Michigan. Atwood blood.

101 KEYSTONE.

Bred by Peter Martin, 1870. Sire, Torrent (97) ; dam bred by C. N. Hayward ; sired by Chunk (95). Sold to D. P. Dewey, Grand Blanc, Mich. Atwood blood.

102 OLD WRINKLY.

Bred by Peter Martin, 1870. Sire, Torrent (97) ; dam bred by C. N. Hayward ; sired by Chunk (95). Sold to S. S. Lusk. Died in 1874. Atwood blood.

103 BURWELL'S COMET.

Bred by G. D. Miner, Bridport, 1864. Sire, Comet (57) ; dam an Atwood ewe, bred by Victor Wright. Sold to H. C. Burwell, Bridport, and by him to Ira Fletcher, and by him to parties in Niagara Co., New York. Atwood blood.

104 BULL-DOG.

Bred by E. Hammond & Son, 1866. Sire, Green Mountain (70) ; dam bred by E. Hammond, called Old Doll. Bull-Dog was sold to O. & E. S. Hall, East Randolph, Vt., and died their property in May, 1874. Atwood blood.

105 TRUMP.

Bred by Peter Martin, 1871. Sire, Torrent (97) ; dam bred by C. N. Hayward and sired by Chunk (95). Sold to John S. Goe, Pa. Atwood blood.

106 TRIUMPH.

Bred by Peter Martin, 1872. Sire, Torrent (97) ; dam by Little Wrinkly (48); 2d dam an Atwood ewe, bred by F. H. Dean. Atwood blood.

107 LITTLE GOLDEN FLEECE.

Bred by Jerome Holden, Westminster, Vt., 1869. Sire, Golden Fleece (91) ; dam bred by E. S. Stowell. Atwood blood.

108 GREASY.

Bred by Peter Martin, 1872. Sire, Green Mountain, Jr. (96) ; dam by Little Wrinkly (48) ; 2d dam, an Atwood ewe, bred by F. H. Dean. Sold to C. Collins, Grand Blanc, Mich. Atwood blood.

109 RED LEG.

Bred by E. S. Stowell, 1866. Sire, Golden Fleece (91) ; dam an Atwood ewe, bred by E. S. S.; sired by Stowell's Sweepstakes

(47); 2d dam, Duchess. an Atwood ewe, bred by E. S. S.; 3d dam
bred by W. R. Remele. Red Leg was sold to Mr. Hoyt, of California,
in 1870.   Atwood blood.

110 PANIC.

Bred by E. S. Stowell, 1868.   Sire, Red Leg (109); dam, Muffle 3d, by Stowell's Sweepstakes; 2d dam, Muffle, by Long Wool
(21); 3d dam bred by W. S. & E. Hammond.   Panic was sold to
Mr. Hoyt, California.   Atwood blood.

111 TORRENT, JR.

Bred by G. F. Martin, Rush, N. Y., 1871.   Sire, Torrent (97);
dam, an Atwood ewe, bred by F. H. Dean; sired by Little Wrinkly
(48).   Atwood blood.

112 LONG WOOL.

Bred by Victor Wright, 1863.   Sire, V. Wright's Old Greasy
(74); dam, an Atwood ewe, bred by V. Wright.   Atwood blood.

113 COMPACT.

Bred by G. F. Martin, 1873.   Sire, Trump (105); dam by Lit-
tle Wrinkly (48); 2d dam, an Atwood ewe, bred by F. H. Dean.
Sold to parties in Genessee Co., Michigan, and re-purchased by G. F.
M.   Atwood blood.

114 MAJOR HAMMOND.

Bred by F. VanVliet, Shelburn, Vt.   Sire, Little Wrinkly (48);
dam, an Atwood ewe, bred by E. Hammond.   Taken to Western
New York, and a half interest sold to Jephtha Potter, Yates Co.   Atwood blood.

115 BULL-DOG.

Bred by Peter Martin, 1871.   Sire, Green Mountain, Jr. (96);
dam by Little Wrinkly (48); 2d dam bred by F. H. Dean.   Sold to
John H. Thompson, Grand Blanc, Mich.

116 CURLY.

Bred by A. C. Bennett, Livonia, N. Y.   Sire, a ram called
Sweepstakes, bred by W. H. B Rogers, Honeoye Falls, N. Y.,
sired by Sweepstakes (32).   Atwood blood.

117 GREEN MOUNTAIN 3d.

Bred by G. F. Martin, 1872.   Sire, Green Mountain, Jr. (96);
dam, sired by Little Wrinkly (48); 2d dam bred by F. H. Dean.
Atwood blood.

118 DAVID.

Bred by E. S. Stowell, 1871.   Sire, Panic (110); dam, No. 44,
bred by E. Hammond.   (Queen of Sheba's dam.)   Atwood blood.

119 DON PEDRO.

Bred by V. Wright. Sire, Long Wool (112); dam, an Atwood ewe, bred by W. R Sanford. Sold to Dea. Samuel James, and by him to Capt. J. Sheldon, Fairhaven, Vt. Atwood blood.

120 LANE RAM.

Bred by Henry Lane, Cornwall, Vt., 1863. Sire, Sanford Lane (60); dam, an Atwood ewe, bred by C. B. Cook, Charlotte, Vt. Sold to C. D. Lane, Cornwall, and J. B. Cherbino, Weybridge, who sold him to Martin Slusher, Orleans Co., New York. Atwood blood.

121 MATCHLESS.

Bred by George H. Hall, Shoreham, Vt. Sire, Seville (75); dam, an Atwood ewe, descended from R. P. Hall and W. R. Sanford's flocks. After being used a number of years as a stock ram, sold and went to Ohio. Atwood blood.

122 DIAMOND DUST.

Bred by E. S. Stowell, 1873. Sire, David (118); dam by Golden Fleece (91); 2d dam, Broken Leg 3d; 3d dam bred by W. S. & E. Hammond, sired by Lawrence Ram (24). Atwood blood.

123 VERMONT.

Bred by Tyler Stickney. Sire, Sanford & Gibbs's ram (56); dam bred by T. Stickney. Stickney blood.

124 BLACK SPOT.

Bred by E. S. Stowell, 1863. Sire, Golden Fleece (91); dam, Apple Blossom, by Stowell's Sweepstakes (47); 2d dam, Old Granny, bred by W. S. & E. Hammond. Atwood blood.

125 HUNCH BACK.

Bred by E. Hammond & Son, 1866. Sire, Green Mountain (70); dam, an Atwood ewe, bred by E. Hammond. Atwood blood.

126 GEN. FREMONT.

Bred by T. Stickney, 1865. Sire, Vermont (123); dam bred by T. S., sired by Old Robinson Ram (38); 2d dam bred by T. Stickney. One of the heaviest shearing rams of his day. His first fleece weighed 17 lbs.; second, 28½ lbs; third, 34½ lbs.; fourth, 30¾ lbs.; fifth, 29 lbs.; sixth, 26 lbs.; seventh, 25 lbs.; eighth, 25 lbs.; ninth, 27½ lbs.; total for the nine years, 243 lbs. His live weight was about 160 lbs. He died in 1875, the property of T. Stickney & Son. It will be seen that the nine fleeces above averaged 27 lbs. 1 oz. for the nine years. Stickney blood.

127 LAME RAM.

Bred by E. Hammond & Son, 1868. Sire, Bull-Dog (104); dam, one of the Queen family of ewes, bred by E. Hammond. Sold J. M. Kirkpatrick, Ohio. Atwood blood.

14

128 GREASY (Barton's).

Bred by F. D. Barton, Waltham, Vt., 1866. Sire, Green Mountain (70); dam bred by W. R. Sanford, sired by Comet (57). Greasy weighed about 135 to 140 lbs.; his heaviest fleece weighed 31 lbs. He died in 1874, the property of his breeder. Atwood blood.

129 KING.

Bred by F. D. Barton, 1867. Sire, Green Mountain (70); dam, a Queen ewe, bred by E Hammond, sired by Gold Drop (64), and out of Hammond's Old Queen ewe, by Long Wool (21). Atwood blood.

130 VERMONT, or TOWLE'S LITTLE RAM.

Bred by Hall & Towle, Cornwall, Vt. Sire, Don Pedro (119); dam, an Atwood ewe, bred by R. P. Hall. Atwood blood.

131 OLD CAP.

Bred by T. Stickney. Sire, Gen. Fremont (126); dam, bred by T. Stickney. Stickney blood.

132 FEARNAUGHT.

Bred by E. S. Stowell, 1872. Sire, Panic (110); dam, an Atwood ewe, bred by E. S. Stowell. Atwood blood.

133 JOSEPH.

Bred by Joseph Marsh, Hinesburgh, Vt. Sire, Lawrence Ram (24); dam, an Atwood ewe, bred by W. S. & E. Hammond. Sold to A. A. Farnsworth, Brooksville, Vt. Atwood blood.

134 DOTY RAM.

Bred by D. F. Doty, Bridport, Vt. Sire, Young Eureka (81); dam, bred by D. F. Doty, sired by America (35); 2d dam, bred by P. Elitharp. Sold to Childs Bros. and taken to Colorado. Atwood blood.

135 SANFORD.

Bred by W. R. Sanford, 1862. Sire, Cosset (45); dam bred by W. R. S., sired by Victor Wright's California (42). Half interest sold B. L. & J. W. Buell, Orwell, and finally to Columbus Delano, Ohio. Died 1872. Atwood blood.

136 WASHOE.

Bred by S. W. Remele, Middlebury, Vt., 1859. Sire, Sweepstakes (32); dam, an Atwood ewe, bred by S. W. R. Sold to H. Hemmenway, Whitewater, Wis. Atwood blood.

137 WRINKLY.

Bred by F. D. Barton, 1868. Sire, Kearsarge (55); dam bred by W. R. Sanford, sired by Comet (57). Sold to C. D. McConnell, Ripon, Wisconsin. Atwood blood.

138 BOTTUM'S LAWRENCE.

Bred by N. & N. Bottum, Shaftsbury, Vt. Sire, Lawrence Ram (24) ; dam, an Atwood ewe, bred by W. S. & E. Hammond. Atwood blood.

139 OLD BLACK.

Bred by David Cutting, Shoreham, Vt. Sire, Greasy (31); dam bred by D. Cutting. Sold to Col. Bela Howe, Hiram Rich and F. D. Douglass, Shoreham, Vt. Cutting blood.

140 MONITOR.

Bred by F. D. Douglass, Shoreham, Vt., 1861. Sire, Old Black (139) ; dam bred by E. R. Robinson, sired by Old Robinson (38). Monitor's live weight was 101 lbs.; his heaviest fleece 22 lbs., 4 oz. Cutting and Robinson blood.

141 MONITOR (Cutting's).

Bred by D. & G. Cutting, 1860. Sire, Greasy (31) ; dam bred by D. & G. C., sired by Geo. Atwood (26) ; 2d dam, Atwood (25). Cutting blood.

142 MONITOR.

Bred by E. Hammond, 1869. Sire, Sweepstakes (32) ; dam, an Atwood ewe, bred by W. S. & E. Hammond. Sold to S. L. Bissell, Shoreham, Vt.,whose used him for a number of years as a stock ram. Atwood blood.

143 ADDISON CHIEF.

Bred by D. & G. Cutting, 1861. Sire, Monitor (141) ; dam, bred by D. & G. C. Cutting blood.

144 PECK RAM.

Bred by Hiram Peck, Cornwall, Vt., 1857. Sire, Jennings Ram (89) ; dam, a Robinson ewe, bred by E. Hamilton, Cornwall, Vt., sired by one of E. R. Robinson's stock rams ; 2d dam, bred by E. R. Robinson. Mr. Peck moved to Schroon, N. Y., and took his ram with him. While Mr. Peck owned him he was kept a part of the time by Mr. Merrill Bingham and E. G. Farnham. He was finally purchased of Mr. Peck by N. T. Sprague, Brandon, Vt. Robinson blood.

145 YORK STATE.

Bred by T. Stickney, 1856. Sire, Robinson (38) ; dam, bred by T. Stickney. Sold to E. N. Bissell, and by him in 1860 to Calvin Armstrong, Wayne Co., Ohio. Stickney blood.

146 E. N. BISSELL'S No. 1.

Bred by E. N. Bissell, Shoreham. Sire, Monitor (142) ; dam, bred by George Atwood, Conn. Sold to John Cridler, Dansville, N. Y., 1867. Atwood blood.

147 FIFTY-TWO.

Bred by E N. Bissell, 1865. Sire, Sweepstates (32) ; dam. an Atwood ewe, bred by E. Hammond. Sold to George Clark, Ticonderoga, N. Y. Atwood blood.

148 ADDISON CHIEF.

Bred by S. L. Bissell, 1866. Sire, Golden Fleece (91) ; dam, an Atwood ewe, bred by George Atwood, Connecticut. Atwood blood.

149 CHUNK.

Bred by Peter Martin, 1870. Sire, Torrent (97) ; dam, by Curly (116) ; 2d dam, bred by Henry Lane, Cornwall, Vt. Atwood blood.

150 WRINKLY (Fuller's).

Bred by H. W. Hammond, 1863. Sire, Sweepstakes (32) ; dam bred by Hammonds. Sold to A. E. Fuller & B. W. Couch, Pomfret, Vt., who used him as a stock ram three years, and then sold him to J. W. Judkins, Missouri. Atwood blood.

151 BENEDICT & BOTTUM.

Bred by C. I. Benedict, Arlington, Vt., 1858. Sire, Sweepstakes (32) ; dam an Atwood ewe, bred by W. S. & E. H. Sold Benedict & Bottum, Shaftsbury, Vt. Atwood blood.

152 BENEDICT.

Bred by C. I. Benedict, Arlington, Vt., 1861. Sire, Benedict & Bottum (151) ; dam, an Atwood ewe, bred by W. S. & E. Hammond. Atwood blood.

153 PRINCE OF GOLD-DROPS.

Bred by E. Hammond, 1863. Sold in the ewe to N. & N. Bottum, Shaftsbury, Vt. Sire, Gold Drop (64) ; dam, by Sweepstakes (32), called by Messrs. Bottum their $800 ewe. This ram died in 1866 or '67. Atwood blood.

154 COMET, Jr.

Bred by W. R. Sanford, 1864. Sire, Comet (57) ; dam, an Atwood ewe, bred by W. R Sanford. Sold to A. E. Fuller, Pomfret, Vt., and by him to J. W. Judkins, of Missouri. Atwood blood.

155 PONY GRIMES.

Bred by Harlow Bros., Darien, Western, N. Y., 1865. Sire, Young Grimes (68) ; dam, by Young Grimes (68) ; 2d dam bred by Geo. Campbell. Atwood, Jarvis and Cutting blood.

156 BACON'S FARNSWORTH RAM.

Bred by A. A. Farnsworth, Brooksville, Vt. Sire, Joseph (133) ; dam, an Atwood ewe, bred by A. A. F. Sold O. O. Bacon and others, Waltham, Vt. Atwood blood.

157 EMPIRE.

Bred by G. F. Martin, 1874. Sire, Torrent (97); dam an Atwood ewe, bred by F. H. Dean, sired by Little Wrinkly (48). Sold to S. S. Lusk. Atwood blood.

158 LITTLE MONITOR.

Bred by Peter Martin, 1874. Sire, Torrent (97); dam, by Maj. Hammond (114); 2d dam, an Atwood ewe, bred by C. N. Hayward. Sold S. S. Lusk. Atwood blood.

159 VICTOR.

Bred by E. Hammond & Son. Sire, Green Mountain (70); dam, an Atwood ewe, bred by E. Hammond. Atwood blood.

160 TARIFF.

Bred by H. W. Hammond, 1866. Sire, Green Mountain (70); dam, an Atwood ewe, bred by H. W. H. Sold N. E. Wheeler, and by him to E. Townsend, Pavilion Centre, N. Y. Atwood blood.

161 OLD GENESEE.

Bred by Ira Moore, Woodstock, Vt., 1864. Sire, Eureka (58); dam, a Hammond ewe. Sold E. Townsend, Pavilion Centre, N. Y., and N. E. Wheeler, Middlebury, Vt. Atwood blood.

162 GENESEE.

Bred by E. Townsend, Pavilion Centre, N. Y., 1874. Sire, Addison (85); dam, by Old Genesee (161); 2d dam, bred by F. H. Dean, sired by Little Wrinkly (48). Atwood blood.

163 WEBSTER & HALL.

Bred by R. P. Hall, Cornwall, Vt. Sire, Hall Ram (98); dam, an Atwood ewe, bred by R. P. Hall. A half-interest sold to Webster. Atwood blood.

164 TOWLE RAM.

Bred by John Towle, Cornwall, Vt. Sire, Webster & H. (163); dam, an Atwood ewe, bred by R. P. Hall Sold A. H. Hubbard, Whiting, Vt, by him to H. A Cutts, Orwell, and by him to O. H. Bascom, Orwell. Atwood blood.

165 HUBBARD RAM.

Bred by A. H. Hubbard, Whiting, Vt., 1871. Sire, Towle Ram (164); dam, an Atwood ewe, bred by A. H. H., sired by Vermont (130); 2d dam, bred by Hall & Towle, Cornwall, Vt. Sold to George Hammond, and by him to Charles Sanford, Orwell, and by him to O. H. & W. O. Bascom, same place. Atwood blood.

166 YOUNG GOLD DROP.

Bred by Edwin Hammond. Sire, Gold Drop (64) ; dam, an Atwood ewe bred by E. Hammond, sired by Sweepstakes (32). Sold Baker & Harrigan, Comstock's Landing, N. Y. Atwood blood.

167 CHUB.

Bred by R. J. Jones. Sire, Seville (75) ; dam, an Atwood ewe, bred by R. J. J. Atwood blood.

168 NEVADA.

Bred by R. J. Jones. Sire, Chub (167) ; dam, an Atwood ewe, bred by R. J. J. Sold to Peet & Severance, and taken by them to Nevada. Atwood blood.

169 ALL-RIGHT.

Bred by R. J. Jones, 1871. Sire, Nevada (168); dam by Golden Fleece (91) ; 2d dam, an Atwood ewe, bred by R. J. J. Atwood blood.

170 CHAMPION.

Bred by W. R. Sanford & Son, 1867. Sire, Blucher (77) ; dam, an Atwood ewe, bred by W. R. S. Champion was sold to A. E. Fuller, Pomfret, Vt., and re-sold to W. R. Sanford & Son, who used him as a stock ram, and at whose place he died. Atwood blood.

171 GIBBS RAM.

Bred by S. S. Gibbs. Sire, Sanford & Gibbs (56); dam, a Robinson ewe, bred by S. S. Gibbs. Robinson blood.

172 PHIL SHERIDAN.

Bred by Oliver Severance, Middlebury, Vt., 1864. Sire, Golden Fleece (91) ; dam, an Atwood ewe, bred by Victor Wright. Sold to M. A. Munroe, Middlebury, Vt. Died 1878. Atwood blood.

173 VICTOR WRIGHT'S WRINKLY.

Bred by Victor Wright, 1863. Sire, Gold Drop (64); dam, an Atwood ewe, bred by E. Hammond. Died 1868. Atwood blood.

174 CHAMPION, Jr.

Bred by A. E. Fuller, Pomfret, Vt., 1869. Sire, Champion (170) ; dam, an Atwood ewe, bred by A. E. F.; 2d dam, an Atwood ewe, bred by W. R. Sanford. Sold 1872, to M. P. Tone, Delaware, Ohio. Atwood blood.

175 HAMILTON & CRANE RAM.

Bred by Byron W. Crane, Bridport, Vt. Sire, Don Pedro (119) ; dam, an Atwood ewe, bred by E. Hammond. Sold by him to one Johnson, and by him to Crane. A half interest in this ram was sold to A. C. Hamilton, Bridport, Vt., and finally entire to Houser & Gilmore, Marion, Ohio. Atwood blood.

176 BONAPARTE.

Bred by B. J. Myrick, Bridport, Vt., 1870. Sire, Doty Ram (134); dam, a Robinson ewe, bred by E. A. Birchard. Sold to Cherbino & Williamson, Middlebury, Vt., and by them to Mills & Condit, Ohio. Atwood and Robinson blood.

177 SILVER HORNS.

Bred by H. C. Burwell, Bridport, Vt., 1872. Sire, Bonaparte (176); dam, an Atwood ewe, bred by H. C. Burwell, sired by Ironsides (80); 2d dam, an Atwood ewe, bred by C. N. Hayward, sired by America (35); 3d dam, bred by Prosper Elitharp. Sold when a yearling to Cherbino & Williamson. Died 1876. Robinson and Atwood blood.

178 ROUGH AND READY.

Bred by T. Stickney & Son, 1865. Sire, Sanford & Gibbs (56); dam, an ewe bred by T. Stickney. Sold N. Dyer, Colorado. Stickney blood.

179 MAXIMILIAN.

Bred by J. Holden. Sire, Bull-Dog (104); dam, Atwood ewe, bred by N. A. Saxton, and called by him Queen. Atwood blood.

180 FASHION.

Bred by George H. Hall. Sire, Matchless (121); dam, an Atwood ewe, bred by R. J. Jones. Used as a stock ram a number of years, and sold to E. A. Birchard, Shoreham, Vt. Atwood blood.

181 MAJOR.

Bred by Geo. H. Hall. Sire, Fashion (180); dam, an Atwood ewe, bred by R. J. Jones. Used as a stock ram a number of years, and sold to Mills & Green, Ohio. Atwood blood.

182 CHUNKHEAD.

Bred by L. P. Clark, Addison, Vt., 1865. Sire, Green Mountain (70); dam, an Atwood ewe, bred by Victor Wright. Died 1878, the property of L. P. C. Atwood blood.

183 PATRICK HENRY.

Bred by L. P. Clark, 1873. Sire, Kilpatrick (71); dam, an Atwood ewe, bred by L. P. C., sired by Gold Drop (64); 2d dam, by Sweepstakes (32); 3d dam, an Atwood ewe, bred by Victor Wright. Patrick Henry's live weight was 147 lbs., after his last fleece. His first fleece weighed 19½ lbs.; second, 27¼ lbs.; third, 37 lbs. (cleansed 9 lbs. 10 oz.); fourth, 32½; fifth, 31½; average five fleeces, 29½ lbs. This ram made seasons in Ohio and New York. He died the property of L. P. C., in 1878. Atwood blood.

184 PERKINS' BACON RAM.

Bred by O. C. Bacon, Waltham, Vt. Sire, Bacon's Farnsworth (156); dam, an Atwood ewe, bred by O. C. Bacon, sired by Don Pedro (119); 2d dam by V. Wright's Old Greasy (74); 3d dam bred by R. P. Hall. Sold to A. E. Perkins, Pomfret, Vt. Died 1869. Atwood blood.

185 USURPER.

Bred by E. S. Stowell, 1864. Sire, Golden Fleece (91); dam, Old Hammond ewe, bred by W. S. & E. Hammond. Sold Wood Bros., Michigan. Atwood blood.

186 DEXTER.

Bred by E. S. Stowell, 1864. Sire, Golden Fleece (91); dam, Griselda, by Comet (57); 2d dam, an Atwood ewe, bred by W. R. Sanford. Atwood blood.

187 NUGGET.

Bred by E. S. Stowell, 1864. Sire, Golden Fleece (91); dam, Peachblossom, by Comet (57), bred by W. R. Sanford. Atwood blood.

188 BRIGHAM.

Bred by E. S. Stowell, 1862. Sire, Stowell's Sweepstakes (47); dam, Dancer, bred by W. S. & E. Hammond. Atwood blood.

189 SEARCHER.

Bred by E. S. Stowell, 1865. Sire, Golden Fleece (91); dam, bred by W. S. & E. Hammond. Sold to A. Huff, Hillsdale, Mich. Atwood blood.

190 COL STOWELL.

Bred by E. S. Stowell, 1870. Sire, King Solomon (93); dam, Beauty, 2d, by Golden Fleece (91); 2d dam, Beauty, bred by W. R. Remele. Atwood blood.

191 GOLDEN HORN.

Bred by E. S. Stowell, 1872. Sire, Panic (110); dam, Princess, 3d, by Golden Fleece (91); 2d dam, bred by W. S. & E. Hammond. Sold C. H. Ripley, Colorado. Atwood blood.

192 BLACKSTONE.

Bred by E. S. Stowell, 1873. Sire, David (118); dam, Daffodilla, by S. Sweepstakes (47); 2d dam, bred by W. R. Remele. Atwood blood.

193 MUFFLER.

Bred by E. S. Stowell, 1874. Sire, David (118); dam, Muffle, 4th, by Golden Fleece (91); 2d dam, Muffle, 3d, by S. Sweepstakes (47); 3d dam, by Long Wool (21). Atwood blood.

194 IRON-CLAD.

Bred by E. S. Stowell, 1873. Sire, David (118) ; dam, Belle, 2d, by Golden Fleece (91) ; 2d dam, Belle by Comet (57) ; 3d dam, bred by W. R. Sanford. Atwood blood.

195 WOOLLY.

Bred by E. S. Stowell, 1874. Sire, Diamond Dust (122) ; dam, Light Colored Ewe, 3d, by Golden Fleece (91) ; 2d dam, Light Colored Ewe, 2nd, by S. Sweepstakes (47) ; 3d dam, Light Colored Ewe, bred by W. S & E. Hammond. Sired by Long Wool (21). Woolly was sold to go to California. Atwood blood.

196 LONG WOOL.

Bred by E. S. Stowell, 1875. Sire, David (118) ; dam, Sukey, 9th, by Golden Fleece (91) ; 2d dam, Sukey, 7th, by S. Sweepstakes (47) ; 3d dam, Sukey, 2d, by Long Wool (21) ; 4th dam, Old Suke, bred by W. S. & E. Hammond. Atwood blood.

197 GOLIATH.

Bred by E. S. Stowell, 1875. Sire, Iron-Clad (194) ; dam Big Ewe, 8th, by Panic (110); 2d dam, Big Ewe, 4th, by Golden Fleece (91); 3d dam, Big Ewe, 2d, by S. Sweepstakes (47) ; 4th dam, by W. S. & E. Hammond. Atwood blood.

198 AND 199 CASTOR AND POLLUX. (TWINS).

Bred by E. S. Stowell, 1875. Sire, Diamond Dust (122) ; dam, Leonesa, 8th, by Golden Fleece (91); 2d dam, Leonesa, by S. Sweepstakes (47) ; 3d dam, bred by W. S. & E. Hammond. Atwood blood.

200 COLUMELLA.

Bred by E. S. Stowell, 1876. Sire, Goliath (197) ; dam, Light Colored Ewe 15th, by Panic (110). Atwood blood.

201 JASON.

Bred by E. S. Stowell, 1876. Sire, Ironclad (194) ; dam, Beauty 17th, by Red Leg (109). Atwood blood.

202 BIRCHARD & TOTTINGHAM.

Bred by E A. Birchard, Shoreham, Vt. Sire, Tottingham (40); dam, a Robinson ewe. A half interest sold to G. D. Bush, Orwell, Vt., who sold his interest to B. B. Tottingham and H. W. Jones, Shoreham, Vt. Robinson blood.

203 ROCKET.

Bred by Wm. M. Holmes, Greenwich, N. Y., 1865. Sire, Gold Mine (63) ; dam bred by N. & N. Bottum, Shaftsbury, Vt., from pure Hammond stock. Died 1869 or '70. Atwood blood.

15

204 MADOC.

Bred by Wm. M. Holmes, 1863. Sire, Gold Mine (63); dam, Light Colored Ewe 3d. dam of Sweepstakes (32). See Randall's Practical Shepherd, p. 121. Atwood blood.

205 LONG WOOL.

Bred by L. P. Clark, Addison, Vt. Sire, Kilpatrick (71); dam by Sweepstakes (32). Sold to E. C. Eells, and by him to John Moore, Short Creek, Ohio. Atwood blood.

206 YOUNG CHUNK.

Bred by C. N. Hayward, 1866. Sire, Chunk (95); dam, an Atwood ewe, bred by C. N. H. A half interest was sold in this ram to R. Lane, Cornwall, who kept him as a stock ram for a number of years. Atwood blood.

207 MONITOR.

Bred by Rector Gage, Addison, Vt., 1874. Sire, Kilpatrick (71); dam, an Atwood ewe, bred by R. G. Sold C. R. Jones, Hubbardton, Vt., and by him to parties in Western New York. Atwood blood.

208 LONGFELLOW.

Bred by R. Gage, 1875. Sire, Patrick Henry (183); dam, an Atwood ewe, bred by R. G., sired by Little Wrinkly (48). Sold to C. R. Jones, Hubbardton, Vt., and by him to S. B. Lusk, Western New York. Atwood blood.

209 VIGOR.

Bred by L. P. Clark. Sire, Chunkhead (182); dam by Sweepstakes (32); 2d dam, an Atwood ewe, bred by N. A. Saxton. Sold to F. S. Higby, who sold him to parties in Ohio. Atwood blood.

210 GENERAL.

Bred by L. P. Clark, 1874. Sire, Pat Henry (183); dam, Old Favorite, by Green Mountain (70); 2d dam by Sweepstakes (32); 3d dam by Black Top (90); 4th dam bred by N. A. Saxton. A half interest was sold January, 1878, to E. Townsend, Pavilion Centre, N. Y. Atwood blood.

211 MONARCH.

Bred by L. P. Clark, 1873. Sire, Pat Henry (183); dam, L. P. Clark's 33, by Little Wrinkly (48); 2d dam by Little Wrinkly (48). Atwood blood.

212 STOGA.

Bred by Geo. H. Hall, 1874. Sire, Major (181); dam by Fashion (180); 2d dam, an Atwood ewe, bred by R. J. Jones. Sold to E. N. Bissell, Shoreham; by him to E. C. Eells, and by him to Alexander T. Henderson and Oliver Watkins, St. Clairsville, Belmont Co., Ohio. Atwood blood.

213 WOOLLY.

Bred by H. G. Hibbard, Orwell, Vt. Sire, Hubbard (165); dam, an Atwood ewe, bred by H. G. H., sired by Gold Drop (64). Atwood blood.

214 HIBBARD.

Bred by H. G. Hibbard, 1871. Sire, Woolly (213); dam, an Atwood ewe, bred by H. G. Hibbard. Sold to E. N. Bissell, and a half interest by him to Humbert Bros., Caldwell Prairie, Illinois. Atwood blood.

215 FREMONT, JR.

Bred by J. Q. Stickney, Whiting, Vt. Sire, Fremont (126); dam, an Atwood and Robinson ewe, bred by Seneca Root, Hubbardton, Vt. Stickney blood.

216 T. S. & SON'S 140.

Bred by T. Stickney & Son, Shoreham, Vt. Sire, Tottingham (40); dams all bred by T. S. & Sons, 1st by Fremont (126), 2d by Stickney's Matchless, 3d dam by Old Robinson Ram (38). Stickney blood.

217 T. S. & SON'S 150.

Bred by T. Stickney & Son. Sire, Silver-Horn (177); dam by Fremont (126); 2d dam by Stickney & Son's Old Cap; 3d dam by Old Robinson (38). Stickney blood.

218 T. S. & SON'S 217.

Bred by T. S. & Son. Sire, Fremont, Jr. (215); dam by Rough and Ready (178); 2d dam, T. Stickney's Matchless; 3d dam by Old Robinson (38). Stickney blood.

219 SILVER-RING.

Bred by James Hill, Bridport, Vt., 1864. Sire, Fitch Ram (92); dam, an Atwood ewe, bred by P. Elitharp. Sold to H. C. Burwell, at whose place he died. Atwood blood.

220 BLACK HAWK.

Bred by D. E. Hill, Bridport, Vt. Sire, one of Victor Wright's stock rams that was sired by one of E. Hammond & Son's stock rams; dam, an Atwood ewe, bred by D. E. Hill; sired by Dean's Wrinkly (48); 2d dam, an Atwood ewe, bred H. W. Hammond. Atwood blood.

221 BISMARCK.

Bred by H. C. Burwell, 1873. Sire, Silver-Horns (177); dam, Burwell's Old Queen, by Elitharp & Burwell Ram (79); 2d dam, Beauty, 2d, by Sea Lion (83); 3d dam, Beauty, by V. Wright's Old Greasy (74); 3d dam, an Atwood ewe, bred by E. S. Stowell. Atwood and Robinson blood.

222 STUB.

Bred by H. C. Burwell, 1873. Sire, Silver-Horns (177); dam, Lady Elitharp, by Elitharp & Burwell (79); 2d dam, 1st choice of Elitharp ewes. Half-interest sold to F. S. Higbee, Ohio. Robinson and Atwood blood.

223 EUREKA 3D.

Bred by J. J. Crane, Bridport, Vt., 1873. Sire, Silver-Horns (177); dam, an Atwood ewe, bred by J. J. C., by Rockwell's Eureka (58); 2d dam, an Atwood ewe, bred by P. Elitharp. Owned by J. J. Crane & H. C. Burwell. Atwood and Robinson blood.

224 TIP.

Bred by H. B. Burwell, 1875. Sire, Bismarck (221); dam by Silver-Ring (219); 2d dam, Beauty 2d, by Sea Lion (83); 3d dam, Old Beauty, by one of V. Wright's stock rams. 4th dam, bred by E. S. Stowell. Robinson and Atwood blood.

225 BLACK-TOP.

Bred by H. C. Burwell, 1875. Sire, Bismarck (221); dam, bred by D. E. Hill. Sire, Little Wrinkly (48); 2d dam bred by E. Hammond. Robinson and Atwood blood.

226 WRINKLY.

Bred by O. C. Bacon, Waltham, Vt. Sire, Farnsworth (156); dam, an Atwood ewe, bred by R. P. Hall. Sold to A. D. Hayward, Weybridge, Vt. Atwood blood.

227 GREEN MOUNTAIN.

Bred by N. & N. Bottum, Shaftsbury, Vt. Sire, Green Mountain (70); dam, an Atwood ewe, bred by N. & N. B. Atwood blood.

228 PEET & MEAD RAM.

Bred by W. R. Sanford, 1865. Sire, Comet (57), dam, an Atwood ewe, bred by W. R. S. Sold to Messrs. Peet & Mead, Cornwall, Vt., and by them to Horace Mead. Atwood blood.

229 RED-LEG, 2D.

Bred by R. Gleason. Sire, Red-Leg (109); dam, by Sweepstakes (32). Sold to William N. Batchelder, Fort Collins, Colorado. Atwood blood.

230 TEMPEST.

Bred by F. H. Dean, West Cornwall, Vt. Sire, Little Wrinkly (48); dam, an Atwood ewe, bred by F. H. D. Atwood blood.

231 CAPTAIN SAM.

Bred by E. S. Stowell, 1869. Sire, Red-Leg (109), dam, Broken-Leg. 3d, by Golden Fleece (91); 2d dam, Old Broken-Leg, bred by W. S. & E. Hammond. Atwood blood.

232 LITTLE PHIL.

Bred by Peter Martin, Rush, N. Y., 1875. Sire, Triumph (106); dam by Green Mountain, Jr. (96); 2d dam bred by F. H. Dean, by Little Wrinkly (48). Sold A. M. Willett, Muir, Ionia Co , Michigan. Atwood blood.

233 FORD RAM.

Bred by Chas. R. Ford, Cornwall, Vt. Sire, Dew Drop (59); dam, an Atwood ewe, bred by E. S. Stowell. Atwood blood.

234 JIM.

Bred by E. R. Clay, Middlebury, Vt., 1865. Sire, Gold Drop (64); dam, an Atwood ewe, bred by Henry Lane. Sold Cherbino & Williamson, who afterwards sold an interest to C. D. Lane. Atwood blood.

235 DRAKE RAM.

Bred by Milo J. Ellsworth, Cornwall, Vt., 1864. Sire, Gibbs Ram (171); dam bred by Charles Benedict, Cornwall, Vt., of which sire and dam were bred by P. Elitharp. Sold to A. L. Drake, Cornwall, and by him to Cherbino & Williamson; by them a half interest to C. D. Lane, and finally to Messrs. Hiatt, Morgan Co., Ohio, who named him Old Grant. Robinson and Atwood blood.

236 DEAN.

Bred by H. W. Jones, Shoreham, Vt. Sire, Little Wrinkly (48); dam, a Robinson ewe. Sold to F. & L. E. Moore, Shoreham, Vt., and by them to Wm. Ball, Hamburg, Mich. Atwood and Robinson blood.

237 GREASY.

Bred by Jerome Holden, Westminster West, Vt. Sire, Golden Fleece (91); dam, an Atwood ewe, bred by W. R. Remele, sired by Old Greasy (18). Atwood blood.

238 ROCKY MOUNTAIN.

Bred by Jerome Holden. Sire, Victor, bred by E. Hammond; dam, an Atwood ewe, bred by E. Hammond, called Strongbelly; she by Gold Drop (64). Atwood blood.

239 LEGAL TENDER.

Bred by Peter Martin, 1877. Sire, Little Phil (232); dam, an Atwood ewe, bred by Peter Martin, 1877, sire, Little Phil (232); dam, an Atwood ewe, bred by Peter Martin (his No. 89), sired by Torrent (97); 2d dam bred by C. N. Hayward, sired by Chunk (95). Atwood blood.

240 STOWELL RAM (Fuller's).

Bred by E. S. Stowell, 1872. Sire, King Solomon (93); dam, an Atwood ewe, bred by E. S. S. Sold A. E. Fuller, Pomfret, Vt., and by him to J. H. Hazen, West Hartford, Vt. Atwood blood.

241 JONATHAN.

Bred by E. S. Stowell, 1875. Sire, Diamond Dust (122); dam, Light Colored Ewe 14, by Golden Horn (191); 2d dam, Light Colored Ewe 3d, by Golden Fleece (91); 3d dam, Light Colored Ewe 1. Atwood blood.

242 VICTOR EMANUEL.

Bred by Lyman Clark, Jr., Addison, Vt., 1876. Sire, Patrick Henry (183); dam, an Atwood ewe, bred by L. Clark, Jr. Atwood blood.

243 PERFECT.

Bred by R. Gage, Addison, Vt., 1874. Sire, Patrick Henry (183); dam, an Atwood ewe, bred by Walter Sprague, Vergennes, Vt., from stock bred from N. A. Saxton and A. A. Farnsworth. Sold to H. E. Taylor, Cornwall, Vt., and by him to John Hourse, Romeo, Michigan. Atwood blood.

244 MINGO.

Bred by J. P. Clark, 1875. Sire, Pat Henry (183); dam by Kilpatrick (71); 2d dam by Chunkhead (182); 3d dam, Old Favorite, by Green Mountain (70); 4th dam by Sweepstakes (32); 5th dam by Black Top (90); 6th dam bred by N. A. Saxton. Sold to E. C. Eells, Middlebury, Vt., and by him to Uriah Cahill, J. & B. Cahill, Wm. H. Hanks and John McMillan, Union Co., Ohio. Atwood blood.

245 PIPER RAM.

Bred by Sylvester Piper, Middlebury Vt. Sire. Peet & Mead Ram (228); dam, a Robinson ewe, bred by E. G. Farnham, Cornwall, Vt. Atwood and Robinson blood.

246 BACON'S GREASY.

Bred by O. C. Bacon, Waltham, Vt. Sire, Wrinkly (226]; dam, an Atwood ewe, bred by O. C. Bacon; 2d dam bred by R. P. Hall. Sold to W. W. Ward, Waltham, Vt. Atwood blood.

247 BUCK MOUNTAIN.

Bred by J. C. Cherbino, Weybridge, Vt. Sire, Bacon's Greasy (246); dam, an Atwood ewe, sired by Lane Ram (120); 2d dam, bred by A. A. Farnsworth. Sold C. D. Lane. Atwood blood.

248 PEERLESS.

Bred by E. Sanford. Sire, Little Wrinkly (48); dam bred by E. S. Sold to R. Lane, Cornwall, Vt., and by him to Woods Bros., Mich. Atwood and Robinson blood.

249 FORTUNE.

Bred by W. H. Delong, West Cornwall, Vt. Sire, Little Wrinkly (48); dam, a Robinson ewe. Atwood and Robinson blood.

250 ECHO.

Bred by W. H. & R. Cook, Shoreham, 1868. Sire, Tottingham (40) ; dam bred by J. T. & V. Rich. Sold Charles C. Forbes, Shoreham. Robinson blood.

251 R. GAGE'S 76.

Bred by R. Gage, Addison, Vt. Sire Young Long Wool, by Clark's Long Wool; dam, by Little Wrinkly (48). Sold E. C. Eells, Middlebury, Vt. Atwood blood.

252 HAMBURGH.

Bred by W. R. Sanford, 1861. Sire, V. Wright's California (43) ; dam, an Atwood ewe, bred by W. R S. Sold to A. E. Perkins, Pomfret, Vt., by him to D. & T. Hazen, Newbury, Vt., and by them to Geo. Campbell, who took him to Hamburgh in 1863, and sold him there to Count Shur, Thoss, Silesia. Atwood blood.

253 CHARLIE.

Bred by W. W. Winchester, Bridport, Vt. Sire, Young Chunk (206) ; dam, an Atwood ewe, bred by O. Severance, Middlebury, Vt. Sired by Gold-Drop (64). Sold to S. O. Jones, Shoreham, Vt., and by him to B. B. Tottingham, of Shoreham. Atwood blood.

254 ELDORADO.

Bred by W. R. Sanford. Sire, Comet (57) ; dam, an Atwood ewe, bred by W. R. S. Sired by Old Greasy (18). Sold and taken to Wisconsin. Atwood blood.

255 GREEN MOUNTAIN, JR.

Bred by W. R. Sanford. Sire Green Mountain (70) ; dam, bred by W. R S. Sire, Gold-Drop (64) ; 2d dam bred by same, sire, Comet (57). Sold A. E. Fuller, Pomfret, Vt. Atwood blood.

256 STOWELL RAM (Perkins's).

Bred by E. S. Stowell, 1869. Sire, King Solomon (93) ; dam an Atwood ewe, bred by E. S. S. Sold to A. E. Perkins, Pomfret, Vt. Died, 1875. Atwood blood.

257 CASIUS.

Bred by A. Chapman, Middlebury, Vt., 1874. Sire, Silver-Horns (177) ; dam, an Atwood ewe, bred by E. S. Rowley, of Charlotte, Vt., from pure Hammond stock. Sold half-interest to C. P. Crane, Bridport, and finally to F. H. Dean, West Cornwall, Vt., who has since used him as a stock ram. Robinson and Atwood blood.

258 GRANT.

Bred by B. F. Field, Cornwall, Vt. Sire, Eureka (58) ; dam, a Robinson ewe. Atwood and Robinson blood.

259 SWEEPSTAKES (Bascom's).

Bred by H. A. Bascom, Shoreham, Vt. Sire, Sweepstakes (32); dam, ewe bred by H. A. B. Atwood, Jarvis, Blakeslee, Rich and Cutting blood.

260 YOUNG KEARSARGE.

Bred by H. W. Hammond. Sire, Kearsarge (55); dam an Atwood ewe bred by H. W. H. Sold Dan Giddings, Essex, Vt. Atwood blood.

261 VANDERBILT.

Bred by Victor Wright, 1866. Sire, Black Top (90); dam, an Atwood ewe, bred by V. Wright. Sold H. Thorp, Charlotte, Vt., and by him to P. C. Abbey, Essex, Vt. Atwood blood.

262 CALIFORNIA

Bred by P. C. Abbey. Sire, Vanderbilt (261); dam by Young Kearsarge (260); 2d dam bred by Victor Wright. Atwood blood.

263 GREEN MOUNTAIN, 3D.

Bred by H. W. Hammond. Sire, Green Mountain (96); dam, an Atwood ewe, bred by H. W. H. Sold in the ewe to Henry Thorp, who sold him to P. C. Abbey, Essex, Vt. Atwood blood.

264 OLD PETER.

Bred by H. A. Bascom. Sire, Grant (258); dam bred by H. A. B. Atwood, Jarvis, Blakeslee, Rich and Cutting blood.

265 HOLMES & JONES RAM.

Bred by W. R Sanford, 1866. Sire, Comet (57); dam an Atwood ewe, bred by W. R. S. Sold to F. Holmes, Sudbury, and C. R. Jones, Hubbardton. Atwood blood.

266 STILES RAM.

Bred by J. B. Stiles, then of Hubbardton, Vt. Sire, Silver Mine, 2d (84); dam, a Cutting ewe. Atwood and Cutting blood.

267 STILES RAM, NO. 2.

Bred by J. B. Stiles. Sire, H. Root's Robinson Ram, by T. Stickney's Fremont (126), out of a Robinson ewe; dam of 267, a Cutting and Atwood ewe, bred by J. B. S., sired by Stiles Ram (266). Sold to C. K. Williams, Whiting, Vt., and by him to H. C. Brown, Whiting. Robinson and Cutting blood.

268 GALE RAM.

Bred by J. Towle, Cornwall, Vt. Sire, Gleason (84); dam, bred by R. P. Hall. Sold —— Gale, Sudbury, Vt. Atwood blood.

269 HUBBARD'S RICH RAM.

Bred by J. T. & V. Rich. Sire, Gold Drop (64) ; dam, bred by J. T. & V. Rich. Sold to A. H. Hubbard, Whiting, Vt. Atwood and Rich blood.

270 KILPATRICK, 2D.

Bred by O. H. & W. O. Bascom, Orwell, Vt. Sire, Kilpatrick (71) ; dam, an Atwood ewe, bred by O. H. & W. O. B. Atwood blood.

271 KILPATRICK, JR.

Bred by S. G. Holyoke, St. Albans, Vt. Sire, Kilpatrick (71) ; dam, an Atwood ewe, bred by S. G. H. Atwood blood.

272 BLUCHER, 2D.

Bred by W R. Sanford. Sire, Blucher (77) ; dam, an Atwood ewe, bred by W. R. Sanford. Atwood blood.

273 ROMEO (O H. & W. O. B.'s 101).

Bred by A. H. & W. O. Bascom, 1877. Sire, Hubbard (165); dam, Beauty, by Blucher 2d (272); 2d dam, an Atwood ewe, bred by O. H. & W. O. B. Atwood blood.

274 DICKENS (O. H. & W. O. B.'s 102).

Bred by O. H. & W. O. Bascom, 1877. Sire, Hubbard (165) ; dam, Wrinkly by Kilpatrick, 2d (270). Atwood blood.

275 DUKE.

Bred by H. W. Jones, Shoreham, Vt. Sire, Birchard & Tottingham (202) ; dam, a Robinson ewe. Sold F. & L. E. Moore, Shoreham, Vt. Robinson blood.

276 DON PEDRO.

Bred by F. & L. E. Moore. Sire, Duke (275) ; dam, a pure Robinson ewe. Sold to Wm. Ball, Michigan. Robinson blood.

277 SNOW-FLAKE.

Bred by James Forbes, Jr., 1872. Sire, Charlie (253) ; dam bred by J. F., Jr. She was sired by a Stickney ram ; her dam was an Atwood ewe. Sold to Dennison Blackmer and F. H. Farrington, Brandon, Vt. Atwood and Stickney blood.

278 ADDISON CHIEF.

Bred by F. & L. E. Moore, Shoreham, Vt. Sire. Duke (275) ; dam bred by L. Catlin. of Cutting and Robinson blood. Sold to Wm. Ball, Hamburg, Mich. Robinson and Cutting blood.

16

279 LITTLE RAM.

Bred by N. A. Saxton, 1863. Sire, Prince (69) ; dam, an Atwood ewe, bred by N. A. Saxton. Sold to J. E. Ainsworth and U. D. Twitchell, Middlebury, Vt, Atwood blood.

280 RUTHERFORD B.

Bred by F. & L. E. Moore, 1874. Sire, Don Pedro (276) ; dam, bred by F. & L. E. M. Robinson blood.

281 HUGO.

Bred by Wm. Ball, Hamburgh, Michigan, 1877. Sire, Rutherford B. (280) ; dam, bred by F. & L. E. Moore. Robinson blood.

282 STICKNEY RAM (Bissell's).

Bred by T. Stickney & Son. Sire, Fremont (126) ; dam bred by T. Stickney & Son. Sold to E. N. Bissell ; by him to F. H. Farrington, Brandon, Vt., and by him to parties in Eastern Vermont. Stickney blood. ,

283 YOUNG BLACK TOP.

Bred by Victor Wright. Sire, Black Top (90) ; dam, an Atwood ewe, bred by V. Wright. Sold Rollin J. Smith, Sudbury, Vt. Atwood blood.

284 WRINKLY JIM.

Bred by Wm. Ball. Sire, Rutherford B. (280) ; dam, a Robinson ewe, bred by B. B. Tottingham. Atwood and Robinson blood.

285 MAXMIILIAN.

Bred by J. Q. Stickney, Whiting, Vt., 1875. Sire, Fremont, Jr. (215) ; dam, an ewe bred by J. Q. S., a direct descendant of the Stickney flock. Sold to E. W. Hardy, Osceola, Mich. Stickney blood.

286 COMMODORE.

Bred by W. Ball. Sire, Maximilian (285) ; dam bred by W. B., by Addison Chief (278) ; 2d dam bred by F & L E. Moore. Robinson blood.

287 KEELER RAM.

Bred by James Piper, Middlebury, Vt. Sire, Piper Ram (245) ; dam, an Atwood ewe, descended from E. Hammond's flock through that of Augustus Taylor, Hancock, Vt. Sold M. Keeler, Cornwall, Vt., and by him to parties West. Atwood & Robinson blood.

288 BINGHAM'S MOORE RAM.

Bred by F. & L. E. Moore. Sire, Birchard & Tottingham, (202) ; dam, a Robinson ewe. Sold to W. H. Bingham, West Cornwall, Vt. Robinson blood.

289 LITTLE CORPORAL.

Bred by E. Barnum, Shoreham, Vt. Sire, Old Ethan (67) ; dam raised by E. Barnum. Sold to L. E. Moore, and by him to Wm. Ball, Hamburgh, Mich. Atwood and Robinson blood.

290 BUELL'S TOWLE RAM.

Bred by J. Towle, 1872. Sire, Gale Ram (268) ; dam, an Atwood ewe, bred by J. Towle. Sold to B. L & J. W. Buell, Orwell, Vt. Atwood blood.

291 HORACE GREELEY.

Bred by L. P. Barnes, Whiting, Vt. Sire, Towle (164) ; dam bred by L. P. B., sired by Dew Drop (59) ; 2d dam by Webster & Hall (163); 3d dam, a Robinson ewe. Atwood and Robinson blood.

292 WRINKLY.

Bred by F. Hooker, Cornwall, Vt., 1869. Sire, Little Wrinkly (48) ; dam, an Atwood ewe, bred by F. Hooker, sired by Little Ram (279) ; 2d dam, an Atwood ewe, bred by S. W. Remele, Middlebury, Vt. Atwood blood.

293 BUELL'S SANFORD.

Bred by B. L. & J. W. Buell, Sudbury, Vt. Sire, Sanford (135); dam, an Atwood ewe, bred by W. R. Sanford. Atwood blood.

294 GRAVES'S RAM.

Bred by H. M. Graves, Salisbury, Vt. Sire, Tottingham (40) ; dam bred by L. C. Remele, of Atwood, Jarvis and Rich blood. Sold to Julius G. Barker, Leicester, Vt. Robinson blood.

295 AMERICA.

Bred by Geo. D. Bryant, Shoreham, Vt., 1868. Sire, Birchard & Tottingham (202) ; dam bred by the Robinsons. Sold E G. Farnham, Shoreham, Vt., and by him to Peck & Parker, who took him to California. Robinson blood.

296 HENRY.

Bred by J. Towle, Cornwall, Vt. Sire, Vermont (130) ; dam an Atwood ewe, bred by R. P. Hall. Sold to Henry Manchester, Cornwall, Vt. Atwood blood

297 GREASY.

Bred by J. O. Hamilton, Bridport, Vt., 1872. Sire, Dea. James's Ram (52) ; dam bred by J. O. H. Sold C. H. James, Cornwall, Vt., and by him to —— Murdock, Crown Point, N. Y. Robinson and Cutting blood.

298 CORLIS'S STICKNEY RAM.

Bred by T. Stickney & Son, 1876. Sire, Fremont (126) ; dam, bred by T. Stickney. Sold E. S. Corlis, Nepeuskun, Wis. Stickney blood.

299 KILLKYSER.

Bred by E. Townsend, Pavilion Centre, N. Y. Sire, Prince (86); dam bred by H. A. Bascom, Shoreham, Vt., by Sweepstakes (32); 2d dam by Young Wooster (28); 3d dam bred by H. A. B. Atwood, Cutting and Blakeslee blood.

300 PRINCE ALEXIS.

Bred by Walter Cole, Batavia, N. Y. Sire, Killkeyser (299); dam bred by S. S. Rockwell, Cornwall, Vt. Cutting, Cock and Blakeslee blood.

301 WELLMAN'S PRINCE.

Bred by C. Hatch, Warsaw, N. Y. Sire, Prince Alexis (300); dam bred by H. A. Bascom, East Shoreham, Vt. Cutting, Cock and Blakeslee blood.

302 WALWORTH.

Bred by E. Townsend, 1874. Sire, Wellman's Prince (301); dam and blood same as 299. Sold to O. Cook, Whitewater, Wis. Cutting and Blakeslee blood.

303 RATLER.

Bred by V. V. Blackmer, Orwell, Vt., 1863. Sire, Comet (57); dam sired by a Sanford ram, by Cosset (45); 2d dam bred by E. Hammond. Sold to A. F. Knox, Whitewater, Wis. Atwood blood.

304 GREEN MOUNTAIN.

Bred by Prosper Elitharp, Bridport, Vt., 1864. Sire, Golden Fleece (91); dam, an Atwood ewe, bred by P. Elitharp. Sold to H. Hemmenway, Whitewater, Wis. Atwood blood.

305 GREEN MOUNTAIN BOY.

Bred by Henry W. Walker, Shoreham, Vt. Sire, a ram bred by H. W. W., sired by Little Wrinkly (22); dam, a Rich ewe, bred from J. T. Rich flock. The dam of Green M. B. was sired by Old Robinson (38); 2d dam by D. & G. Cutting's Atwood Ram (25); 3d dam bred by H. W. W., from the Rich flock. Sold to H. Williams, LaGrange, Wis. Atwood and Rich blood.

306 ELITHARP, NO. 3.

Bred by O. Cook, Whitewater, Wis. Sire, Green Mountain (304); dam bred by T. Stickney & Son, Shoreham, Vt., and sold to A. F. Knox, Wisconsin. Atwood and Stickney blood.

307 ELITHARP, NO. 5.

Bred by O. Cook. Sire, Elitharp, No. 3 (306); dam bred by L. C. Remele, Shoreham, Vt. Atwood, Jarvis and Cock blood.

308 WRINKLY.

Bred by R. Lane, Cornwall, Vt., 1874. Sire, Peerless (248); dam bred by S. S. Rockwell, sired by Eureka (58). Sold O. Cook, Whitewater, Wis. Atwood and Robinson blood.

309 SILVER MINE, JR.

Bred by E. Hammond. Sire, Silvermine (61); dam, an Atwood ewe, bred by W. S. & E. Hammond. Sold G. J. Hollenbeck, Hoosick, N. Y. Atwood blood.

310 LOYAL.

Bred by L. C. Remele, Shoreham, Vt. Sire, Henry (296); dam bred by L. C. Remele. Atwood and Robinson blood.

311 GEORGE.

Bred by E. Hammond. Sire, Sweepstakes (32); dam, an Atwood ewe, bred by W. S & E. Hammond. Sold to R. Perrine, Western Pennsylvania. Atwood blood.

312 PERRINE'S LITTLE WRINKLY.

Bred by R. Perrine, Western Pennsylvania, 1867. Sire, V. Wright's Old Greasy (74); dam, an Atwood ewe, bred by J. Meholin, sired by a Hammond ram from an ewe bred by R. P. Hall, Cornwall, Vt. Atwood blood.

313 BULL-DOG, 2D.

Bred by H. W. Hammond, 1869. Sire, Bull-Dog (104); dam, an Atwood ewe, bred by H. W. H. Atwood blood.

314 LITTLE HAMMOND.

Bred by H. W. Hammond, 1872. Sire, Bull-Dog 2d (313); dam, an Atwood ewe, bred by H. W. H. Sold R. Perrine, Western Pennsylvania. Atwood blood.

315 YOUNG HAMMOND.

Bred by R. Perrine. Sire, Little Hammond (314); dam by Perrine's Little Wrinkly (312); 2d dam by Old Greasy (74); 3d dam by George (311); 4th dam bred by E. Hammond. Sold to McCully, Connell & Carson. Atwood blood.

316 ADVANCE.

Bred by H. C. Burwell, Bridport. Vt., 1876. Sire, Bismarck (221); dam, Princess, bred by H. C. Burwell, by Elitharp & Burwell (79); 2d dam, Atwood ewe Josey, bred by H. C. Burwell; 3d dam bred by C. N. Hayward, by his Ironsides (80). Sold to B. W. Cope, Smithfield, Ohio, and by him to C. H. Beall, Washington Co., Pennsylvania. Atwood and Robinson blood.

317 B. W. COPE'S NO 1.

Bred by B. W. Cope, 1877. Sire, Bismarck (221); dam, Princess, same as in 316. Atwood and Robinson blood.

318 B. W. COPE'S 85.

Bred by B. W. Cope, 1877. Sire, Silver Horns (177); dam bred by Chas. D. Lane. Sold to J. S. & T. J. Close, St. Clairsville, Ohio. Atwood and Robinson blood.

319 B. W. COPE'S 86.

Bred by B. W. Cope, 1877. Sire, General (210); dam, Young Greasy, bred by L. P. Clark, Addison, Vt. Sire, Kilpatrick (71); 2d dam by Chunkhead (182); 3d dam by Sweepstakes (32). Sold to Samuel Grubb, Mt. Pleasant, Ohio. Atwood blood.

320 YOUNG BONAPARTE.

Bred by Cherbino & Williamson, Middlebury, Vt., 1875. Sire, Bonaparte (176); dam, an Atwood ewe. Sold B. W. Cope, Smithfield, Ohio, and by him to Shep Ong, Smithfield. Robinson and Atwood blood.

321 B. W. COPE'S 88.

Bred by B. W. C., 1877. Sire Young Hammond (315); dam bred by H. C. Burwell. Sold R. Van Voorhis, Monongahela, Pa. Atwood and Robinson blood.

322 B. W. COPE'S 89.

Bred by B. W. C., 1877. Sire, Advance (316); dam by Young Chunk (206); 2d dam, a descendant of N. A. Saxton's flock. Sold R. Van Voorhis. Atwood and Robinson blood.

323 ROBINSON RICH.

Bred by J. T. & V. Rich, 1855. Sire, Old Robinson (38); dam bred by J. T. & V. R. Sold to H. Hemenway, Wisconsin. Robinson blood.

324 CLARK'S COOK RAM.

Bred by O. Cook, Whitewater, Wis., 1872. Sire, Elitharp, No. 3 (306); dam by Green Mountain Boy (305); 2d dam bred by L. C. Remele, Shoreham, Vt. Sold C. M. Clark, Whitewater. Atwood and Robinson blood.

325 ROYAL PRINCE.

Bred by S. B. Lusk, Batavia, N Y., 1873. Sire, Addison (85); dam, bred by E. Townsend, sired by Prince (86); 2d dam by the Goodhue Ram, bred by N. A. Saxton; 3d dam, a Robinson ewe, bred by E. G. Farnham, Shoreham, Vt. Sold Geo. Lawrence, Jr., Waukesha, Wis. Atwood and Robinson blood.

326 PRINCE.

Bred by S S.Lusk, Victor, N.Y., 1872. Sire, Old Wrinkly (102); dam by Pony Grimes (155) ; 2d dam bred by E. B. Pottle, Naples, N. Y. Sired by Frank (76) ; 3d dam, bred by W. R. or Edgar Sanford. Sold George Lawrence, Waukesha, Wis. Atwood and Robinson blood.

327 ELLSWORTH'S EUREKA.

Bred by A. F. Ellsworth, Whiting, Vt., 1864. Sire, Eureka (58) ; dam bred by A. F. E. Sired by an Atwood ram, bred by C. I. Benedict ; 2d dam, Atwood ewe, bred by F. H. Dean. Atwood blood.

328 ELLSWORTH'S REMELE RAM.

Bred by L. C. Remele, Shoreham, Vt.,1873. Sire, Loyal (310); dam bred by L. C. Remele. Sold to A. F. Ellsworth. After using him as a stock ram a number of years, Mr. E. sold him to W. H. Bingham, Cornwall, Vt. Atwood and Robinson blood.

329 A. F. ELLSWORTH'S 58.

Bred by A.F.E., 1874. Sire, Ellworth's Remele ram (328) ; dam, an Atwood ewe, bred by A. F. Ellsworth. Sold C. D. Mason, and taken to Colorado. Atwood and Robinson blood.

330 A. F. ELLSWORTH'S 59.

Bred by A. F. E., 1876. Sire, Ellsworth's Remele Ram (328) ; dam, an Atwood ewe, by Benedict Ram ; 2d dam, bred by F. H. Dean. Atwood blood.

331 DON PEDRO.

Bred by L. P. Barnes, Whiting, Vt. Sire, Wrinkly (292) ; dam, by Hubbard (165) ; 2d dam by Towle ram (164) ; 3d dam, bred by L. P. B. Atwood and Robinson blood.

332 LONG-WOOL.

Bred by F. Hooker. Sire, Wrinkly (292) ; dam, an Atwood ewe, bred by F. H.; 2d dam bred by S. W. Remele. Atwood blood.

333 GREASY.

Bred by F. Hooker. Sire, Wrinkly (292) ; dam, an Atwood ewe, sired by one of R. P. Hall's stock rams ; 2d dam bred by S. W. Remele, sired by Sweepstakes (32). Atwood blood.

334 GOLD-DUST.

Bred by John Towle, Cornwall, Vt., 1863. Sire, Webster & Hall (163) ; dam, an Atwood ewe, bred by Hall & Towle. Sold A. J. & D. W. Childs, Weybridge, Vt. Atwood blood.

335 KING CHARLES.

Bred by D. W. Childs, Green Horn, Col., 1871. Sire, Gold-Dust (334); dam, a descendant of E. Hammond's flock through that of M. T. Shackett, Middlebury, Vt. Atwood blood.

336 ALEXANDER.

Bred by D. W. Childs, 1871. Sire, G. D. (334); dam bred by M. T. Shackett, of pure Hammond blood, and from one of M. T. Shackett's Queen Ewes. Atwood blood.

337 WONDER.

Bred by D. W. Childs, 1877. Sire, King Charles (335); dam, an Atwood ewe, bred by F. H. Dean, sired by Little Wrinkly (48). Atwood blood.

338 DON.

Bred by George H. Hall, Shoreham, Vt., 1875. Sire, Major (181); dam, an Atwood ewe, bred by G. H. Hall. Atwood blood.

339 BEN.

Bred by Geo. H. Hall 1875. Sire, Major (181); dam, an Atwood ewe, bred by G. H. H. Atwood blood.

340 BIG RAM.

Bred by N. A. Saxton, 1863. Sire, Prince (69); dam, an Atwood ewe, bred by N. A. S. Sold J. E. Ainsworth & U. D. Twitchell, Middlebury, Vt. Atwood blood.

341 LITTLE WRINKLY.

Bred by Victor Wright. Sire, Wrinkly (173); dam, an Atwood ewe, bred by Victor Wright. Sold to Rollin J. Smith, Sudbury, Vt. Atwood blood.

342 YOUNG SAM.

Bred by Cherbino & Williamson, Middlebury, Vt. Sire, Silver-Horns (177); dam, an Atwood ewe, bred by J. C. Cherbino. Sired by Lane Ram (120); 2d dam, an Atwood ewe bred by A. A. Farnsworth. Atwood and Robinson blood.

343 HARRY.

Bred by F. Hooker, Cornwall, Vt. Sire, Wrinkly (292); dam, an Atwood ewe, bred by F. H. Sold to H. F. Dean, and by him to A. A. Farnsworth, Brooksville, Vt. Atwood blood.

344 CLAY'S GREEN MOUNTAIN.

Bred by E. R. Clay, Middlebury, Vt., 1866. Sire, Green Mountain (70); dam, an Atwood ewe, bred by Henry Lane, Cornwall, Vt. Atwood blood.

345 WELLINGTON.

Bred by J. G. Wellington, Middlebury, Vt., 1868. Sire, Clay's Green Mountain (344); dam. an Atwood ewe. Sold Moses Sheldon, Salisbury, Vt. Atwood blood.

346 DUNMORE.

Bred by J. O. Hamilton, Bridport, Vt., 1874. Sire, Dea. James's ram (52); dam bred by J. O. H., of Cutting blood. Sold Moses Sheldon. Robinson and Cutting blood.

347 BLACK DIAMOND.

Bred by Victor Wright, 1865. Sire, Long Wool (112); dam by Don Pedro (119). Sold H. Thorp, and by him to E. S. Rowley, Shelburn, and by him to J. H. Sherman, Charlotte, Vt. Atwood blood.

348 SHERMAN.

Bred by E. S. Rowley, 1869. Sire, Black Diamond (347); dam, an Atwood ewe, bred by E. S. R. Sold to J. H. Sherman, Charlotte, Vt.; by him to A. Chapman, and by him to U. D. Twitchell, Middlebury. Atwood blood.

349 A. CHAPMAN'S 13.

Bred by A. Chapman, 1876. Sire, Allright (169); dam, an Atwood ewe, bred by E. S. Rowley. Atwood blood.

350 CROMWELL.

Bred by Rollin J. Smith, Sudbury, Vt. Sire, Little Wrinkly (341); dam, bred by R. J. Smith, sired by Young Black Top (283). Atwood blood.

351 A. CHAPMAN'S 22.

Bred by A. Chapman, 1877. Sire, General (210); dam, an Atwood ewe, bred by S. W. Remele. Atwood blood.

352 A. CHAPMAN'S 26.

Bred by A. Chapman, 1878. Sire General (210); dam, an Atwood ewe, bred by P. Elitharp. Atwood blood.

353 A. CHAPMAN'S 27.

Bred by A. Chapman, 1878. Sire, Bismarck (221); dam, an Atwood ewe, bred by S. W. Remele. Atwood blood.

354 BROOKSVILLE.

Bred by A. A. Farnsworth, 1874. Sire Harry (343); dam, an Atwood ewe, bred by A. A. Farnsworth. Sold A. Chapman. Atwood blood.

355 OLD STOGA.

Bred by C. P. Crane, Bridport, Vt. Sire Young Chunk (206); dam, an Atwood ewe, bred by Chilon Crane. Atwood blood. 17

356 YOUNG STOGA.

Bred by C. P. Crane. Sire, Stoga (355); dam by Sandford Lane (60); 2d dam, by America (35); 3d dam, bred by P. Elitharp. Sold to Geo. Hammond, by him sold to Flint Bros., California. Atwood blood.

357 IRONSIDES, 3D.

Bred by J. J. Crane, 1873. Sire, Silver Horns (177); dam by Young Eureka (81); 2d dam, an Atwood ewe, bred by O. C. Bacon, Waltham, Vt. Sold to C. P. Crane, and used by him as a stock-ram. Atwood and Robinson blood.

358 C. P. CRANE'S 55.

Bred by C. P. Crane, 1877. Sire, Ironsides, 3d (357); dam, bred by C. D. Lane. Sold L. J. Wright, Middlebury, Vt., and named by him, "My Choice." Atwood and Robinson blood.

359 CORLIS' STICKNEY RAM.

Bred by S. Stickney & Son. Sire and dam bred by S. Stickney & Son. Sold to H. Corlis, Ripon, Wis. Stickney blood.

360 W. H. DELONG'S 100.

Bred by W. H. Delong, West Cornwall, 1875. Sire Allright (169); dam, an Atwood ewe, sired by Little Wrinkly (48); 2d dam bred by F. H. Dean. Atwood blood.

361 DELONG'S REMELE RAM.

Bred by L. C. Remele, 1875. Sire, Fremont, Jr. (215); dam, bred by L. C. Remele. Sold W. H. Delong. Stickney blood.

362 FRENCH & MASON RAM.

Bred by A. W. Baldwin, Whiting, Vt. Sire, Golden Fleece (81); dam, a Robinson ewe. Sold French & Mason, New Haven, Vt. Atwood and Robinson blood.

363 DOWD'S WRINKLY.

Bred by S. & S. D. Dowd, New Haven, Vt. Sire, Little Wrinkly (48); dam, an Atwood ewe, bred by S. Dowd. Atwood blood.

364 PRINCE.

Bred by R. Gleason, 1865. Sire, one of Hall & Towle's Stock Rams; dam, bred by R. P. Hall, sired by Longwool (21). Sold to Charles E. Stewart, Kalamazoo, Mich. Atwood blood.

365 CURLEY.

Bred by R. Gleason, 1865. Sire. Gleason (84); dam, bred by R. G. Sire, Sweepstakes (32); 2d dam, bred by W. R. Sanford. Sold to C. E. Stewart. Atwood blood.

366 CONSTITUTION.

Bred by H. C. Burwell, 1875. Sire, Eureka 3d (223); dam, Chunk ewe, by Fitch ram (92). Sold to Cherbino & Williamson, and by them to W. W. Holmes, Short Creek, Ohio. Atwood and Robinson blood.

367 HOLYOKE'S & HUNT'S ATWOOD RAM.

Bred by S. Atwood, Conn. Sire, Old Black (9); dam, bred by S. Atwood. Sold to S. G. Holyoke & Hunt, Franklin Co., Vt. Atwood blood.

368 SAM.

Bred by Cherbino & Williamson, Middlebury, Vt. Sire, Silver Horns (177); dam, an Atwood ewe, bred by J. C. Cherbino, sired by Lane Ram (120); 2d dam, an Atwood ewe, bred by A. A. Farnsworth. Atwood and Robinson blood.

369 FORTUNE.

Bred by Geo. Hammond, 1874. Sire, Lane Ram (127); dam, an Atwood ewe, bred by E. Hammond & Son. Sold S. G Holyoke, Vt. Atwood blood.

370 NOONDAY.

Bred by E. Hammond & Son, 1868. Sire, Green Mountain (70); dam, bred by E. H. & S. Sired by Gold Drop (64); 2d dam, bred by E. H. Sire, Sweepstakes (32). Sold to W. L. Archer, W. Pa. Atwood blood.

371 BRICK.

Bred by W. L. Archer, W. Pa., 1870. Sire, Noonday (370); dam, by Archer's Fortune; 2d dam, by Archer's Henry. Sold J. M. Miller, Washington Co., Pa., and by him to D. M. Bailey & Bros.. Burgettstown, W. Pa., and John Moore, Short Creek, Har. Co , Ohio. Atwood blood.

372 SYMMETRY.

Bred by E. S. Dana, Cornwall, Vt., 1876. Sire, Allright (169); dam, a Robinson ewe, bred by T. S. Gibbs, Cornwall, Vt. Atwood and Robinson blood.

373 ROBBINS' RAM.

Bred by B. S. Fields, Cornwall, Vt. Sire, Golden Fleece (91); dam, by Charles I. Benedict, ram bred by W. S & E. Hammond; 2d dam, by a ram purchased of E. Hammond by John Sanford; 3d dam, by Wooster (16); 4th dam, an Atwood ewe brought from Conn. Sold Henry Robbins and S. E. Parkill, Cornwall. Atwood blood.

374 Q. C. RICH.

Bred by Q. C. Rich, Shoreham, Vt. Sire, Major (181); dam, a Rich or Robinson ewe. Sold J. A. Wright, Middlebury. Atwood and Rich blood.

375 THORP'S BARTON RAM.

Bred by F. D. Barton. Sire, King (129) ; dam, an Atwood ewe, bred by F. D. B. Sold H. Thorp, Charlotte, Vt. Atwood blood.

376 NORTH HERO.

Bred by H. Thorp. Sire, Thorp's Barton Ram (375) ; dam, an Atwood ewe, bred by H. T. Sold to C. B. Russell, North Hero, Vt. Atwood blood.

377 FLINT.

Bred by H. Thorp. Sire, a ram bred by H. W. Hammond, sired by Green Mountain (70); dam, by Black Top (90); G. D., bred by V. Wright. Sold to Flint Bros., Cal. Atwood blood.

378 H. THORP'S 1.

Bred by H. T 1875. Sire, Barton Ram (375) ; dam, twin sister to 377. Atwood blood.

379 H. THORP'S 22.

Bred by H. T. 1866. Sire, Barton ram (375) ; dam, an Atwood ewe, bred by H. T. Sired by Flint (377) ; 2d dam, sired by a ram bred by V. Wright, sired by Don Pedro (52). Atwood blood.

380 H. THORP'S 23.

Bred by H. T., 1876. Sire, Barton Ram (375) ; dam, an Atwood ewe, bred by H. T. Atwood blood.

381 PLATO.

Bred by Rockwell & Sanford. West Cornwall, Vt. Sire, Bonaparte (176) ; dam bred by S. S. Rockwell, by Eureka (58). Atwood and Robinson blood.

382 H. E. SANFORD'S 52.

Bred by Rockwell & Sanford, 1875. Sire, Plato (381) ; dam bred by S. S. Rockwell. Sold to Z. McFadden, Atlanta, Ill. Atwood and Robinson blood.

383 BONAPARTE.

Bred by J. T. & V. Rich, 1850. Sire and dam of Rich blood. Sold to C. & J. W. Rich. Lapeer Co., Mich. Rich blood.

384 CAPTAIN.

Bred by T. Stickney, 1852. Sire, Robinson (38) ; dam bred by T. Stickney. Sold to a Mr. Slayton, Lapeer Co., Mich. Stickney blood.

385 STICKNEY.

Bred by T. Stickney & Son. Sire, Fremont (126) ; dam, bred by T. S. & Son. Sold to J. T. Rich, Lapeer Co., Mich. Stickney blood.

386 CUTTING.

Bred by David Cutting, Shoreham, Vt. Sire and dam bred by D. Cutting. Sold S. W. Rich, Lapeer Co., Mich. Cutting blood.

387 STICKNEY No. 2.

Bred by T. Stickney & Son. Sire, Fremont (126); dam, bred by T. Stickney & Son. Sold to J. T. Rich, Lapeer Co., Mich. Stickney blood.

388 CAPT. JO.

Bred by E. Hammond. Sire and dam, bred by E. Hammond. Sold Joseph Harwood, West Rupert, Vt., and by him to Wm. M. Holmes, Greenwich, N. Y. Atwood blood.

389 GRUFFY.

Bred by J Harwood. Sire, an Atwood Ram, bred by E. Hammond; dam, Light-col. ewe, 3d; the dam of Sweepstakes (32). Atwood blood.

390 STEEL RAM.

Bred by Thos. S. Steel, Shushan, N. Y. Sire, Eureka (58); dam, bred by T. S. S. Sire, Golden Fleece (91); 2d dam, bred by P. Elitharp; sire America (35); 3d dam, bred by P. Elitharp. Atwood blood.

391 GREEN MOUNTAIN.

Bred by N. T. Sprague, Brandon, Vt. Sire, Peck Ram (144); dam, an Atwood ewe, bred by N. T. S. Sold to Elijah Smith, West Rutland, Vt., who sold him to J. H. Mead, same place, who sold him to Joel Baker, Lebanon, N. H. Robinson and Atwood blood.

392 BULL DOG.

Bred by F. H. Dean, Cornwall, Vt., 1868. Sire, Little Wrinkly (48); dam, an Atwood ewe, bred by F. H. Dean. Sold to J. H. Mead, West Rutland, who sold him to Isaac Wheaton, and his son, George Wheaton. Atwood blood.

393 YOUNG PRINCE.

Bred by J. H. Mead, 1872. Sire, Bull Dog (392); dam, Mead's old Queen ewe, Rich stock. Sold to A. J. Chillie, Meriden, N. H. Atwood and Rich blood.

394 PONY.

Bred by J. H. Mead, 1872. Sire, Bull-dog (392); dam, Mead's Pony ewe, bred by himself of Rich Stock; 2d dam bred by J. T. & V. Rich. Atwood and Rich blood.

395 BIG WRINKLY.

Bred by J. H. Mead, 1875. Sire Pony (394); dam bred by J. H. M.; sired by Green Mountain (391); 2d dam bred by J. T. & V. Rich. Sold to C. Horace Hubbard, Springfield, Vt. Atwood and Rich blood.

396 CHUNK.

Bred by J. H. Mead, 1878. Sire, Pony (294); dam, an ewe bred by J. H. M., of Rich stock. Atwood and Rich blood.

397 HULL'S ATWOOD.

Bred by Stephen Atwood, Conn., and sold to Alfred Hull, Wallingford, Vt., who used him as a stock ram for a number of years.

398 GEN. GRANT.

Bred by N. T. Sprague, Brandon, Vt. Sire, Peck Ram (144); dam, an Atwood ewe bred by N. T. S. Robinson and Atwood blood.

399 TOMMY SAYRES.

Bred by N. T. Sprague, 1863. Sire, Peck Ram (144); dam, an Atwood ewe bred by N. T. S. Sold to Merrill Bingham, Cornwall, Vt., and by him to A. Stocking. Robinson and Atwood blood.

400 HEENAN.

Bred by N. T. Sprague, 1863. Sire, Peck Ram (144); dam, an Atwood ewe bred by N. T. S. Sold to Merrill Bingham. Robinson and Atwood blood.

401 SILVER HORNS, Jr.

Bred by H. C. Burwell, 1874. Sire, Silver Horns (177); dam, Princess, by Elitharp & Burwell (79); 2d dam, an Atwood ewe bred by C. N. Hayward; sired by America (35). Sold to N. T. Sprague. Atwood and Robinson blood.

402 SPRAGUE'S BURWELL.

Bred by H. C. Burwell, 1876. Sire, Bismarck (221); dam, by Silver Ring (219); 2d dam bred by R. P. Hall. Sold to N. T. Sprague, Brandon, Vt. Atwood and Robinson blood.

403 PERKINS' SWEEPSTAKES.

Bred by A. E. Perkins, Pomfret, Vt. Sire, Stowell Ram (256); dam, an Atwood ewe bred by A. E. P. Sold to J. M. Campbell, Texas. Atwood blood.

404 CONSTITUTION.

Bred by A. E. Perkins, 1875. Sire, Sweepstakes (403); dam, an Atwood ewe bred by A. E. P. Atwood blood.

405 OLD DEACON.

Bred by A. E. Perkins, 1871. Sire, Champion (170); dam, an Atwood ewe bred by A. E. P. Sold to J. D. Patterson, California. Atwood blood.

406 A. E. PERKINS' No. 4.

Bred by A. E. Perkins, 1876. Sire, Stowell Ram (256); dam, an Atwood ewe bred by A. E. P. Atwood blood.

407 GREEN MOUNTAIN.

Bred by Geo. D. Bryant, Shoreham. Sire, Tottingham (40); dam, a Robinson ewe purchased of the Robinson estate. Robinson blood.

408 DEXTER.

Bred by O. & E S. Hall, East Randolph, Vt. Sire, Bull Dog (104); dam, an Atwood ewe, bred by E. Hammond, sired by Gold Drop (64). Sold J. H. & A. W. Peters, Bradford, Vt. Atwood blood.

409 YOUNG AMERICA.

Bred by Peters J. H. & A. W., 1876. Sire, Dexter (408); dam sired by Bull Dog (104); 2d dam bred by E. Hammond. Atwood blood.

410 BLACK BUCK.

Bred by O. & E. S. Hall 1872. Sire Bull Dog (104); dam an Atwood ewe, bred by E. Hammond. Sold J. H. & A. W. Peters. Atwood blood.

411 PEET'S FARNSWORTH RAM.

Bred by A. A. Farnsworth, Brooksville, Vt.. 1873. Sire, David (118); dam, an Atwood ewe, bred by A. A. F. Sold L. S. Peet, Cornwall, Vt. Atwood blood.

412 PECK & SONS' STICKNEY RAM.

Bred by T. Stickney & Son, 1873. Sire, Fremont (126); dam bred by T. S. & Son. Sold E. Peck & Sons, Geneva, Ill., 1876. Stickney blood.

413 LITTLE WRINKLY.

Bred by E. N. Bissell, Shoreham, Vt., 1874. Sire, Bissell's Stickney Ram (282); dam, an ewe bred by E. N. B. Sold to E. Peck & Sons, 1875. Stickney and Atwood blood.

414 HALL'S GOLD DROP.

Bred by O. & E. S. Hall, 1872. Sire, Bull Dog (104); dam bred by E. Hammond. Sired by Gold Drop (54). Atwood blood.

415 QUEEN RAM.

Breed by O. & E. S. Hall, 1873. Sire, Bull Dog (104); dam, a Queen ewe bred by E. Hammond, sired by Gold Drop (64.) Atwood blood.

416 NORTH STAR.

Bred by O. & E. S. Hall, 1875. Sire, Gold Drop (414); dam bred by O. & E. S. H., sired by Bull Dog (104); 2d dam bred by E. Hammond. Sold E. S. Fulsom, Tunbridge, Vt. Atwood blood.

417 DEFIANCE.

Bred by O. & E. S. Hall, 1873. Sire, Bull Dog (104); dam bred by E. Hammond. Sold W. B. Porter, North Tunbridge, Vt. Atwood blood.

418 STUB.

Bred by W. B. Porter, 1875. Sire, Gold Drop (414); dam bred by E. Hammond, sired by Bull Dog (104.) Atwood blood.

419 HAYES.

Bred by W. P. Porter, 1876. Sire, Gold Drop (414); dam bred by O. & E. S. Hall, from Geo. Hammond stock. Atwood blood.

420 GILT EDGE.

Bred by Peter Martin, Rush, N. Y., 1876. Sire, Triumph (106); dam bred by Peter Martin, sired by Green Mountain, Jr., (96); 2d dam bred by F. H. Dean, sired by Little Wrinkly (48). Sold Lyman Cate, Highland, Mich. Atwood blood.

421 BONAPART Jr.

Bred by Rockwell & Sanford, West Cornwall, Vt., 1872. Sire, Bonaparte (176); dam bred by S. S. Rockwell. Sold J. A. Childs, Weybridge. Atwood and Robinson blood.

422 J. A. CHILD'S No. 10.

Bred by J. A C., 1875. Sire, Bonaparte, Jr., (421); dam by Gold Dust (334); 2d dam by Stowell's Sweepstakes (47); 3d dam an Atwood ewe bred by E. S Stowell. Atwood and Robinson blood.

423 SMUGGLER.

Bred by A. E. Fuller, 1873. Sire, Stowell Ram (256); dam by Champion (170); 2d dam by Comet, Jr., (154); 3d dam by Wrinkley (150); 4th dam bred by W. R. Sanford. Sold to E. Townsend, Pavilion Centre, N. Y. Atwood blood.

424 WOOL MINE.

Bred by A. E. Fuller, 1875. Sire, Stowell Ram (240); dam sired by a ram bred by W. R. Sanford, sired by Comet (57). Atwood blood.

425 BLACK TOP.

Bred by A. E. Fuller, 1875. Sire, Perkins's Sweepstakes (403); dam by Champion, Jr., (174); 2d dam by Comet, Jr., (154). Atwood blood.

426 FORTUNE.

Bred by A. E. Fuller, 1876. Sire, Wool Mine (424); dam by Champion, Jr., (174); 2d dam by Comet, Jr. (154); 3d dam by Wrinkly (150); 4th dam, an Atwood ewe bred by A. E. F. Atwood blood.

427 YOUNG FREMONT.

Bred by Wm. McCauley, Cornwall, Vt. Sire, Fremont, Jr. (215); dam by Dean's Little Wrinkley (48); 2d dam bred by Wm. McCauley; 3d dam bred by S. S. Rockwell. Atwood, Stickney and Robinson blood.

428 BONNY.

Bred by O. P. Lee, Middlebury, Vt., 1875. Sire, Bonaparte (176); dam, an Atwood ewe bred by P. Elitharp, died 1877. Atwood and Robinson blood.

429 MASON'S SWEEPSTAKES.

Bred by E. Hammond. Sire, Sweepstakes (32); dam, an Atwood ewe bred by W. S. & E. Hammond. Sold to C. W. Mason, New Haven, and by him to Seth Langdon, same town. Atwood blood.

430 LANGDON'S CHERBINO RAM.

Bred by J. B. Cherbino, Weybridge, Vt. Sire, Drake Ram (235); dam, ewe bred by C. D. Lane. Atwood and Robinson blood.

431 H. LANE'S JONES RAM.

Bred by R. J. Jones, 1874. Sire, Allright (169); dam, an Atwood ewe bred by R. J. Jones. Sold H. Lane, Cornwall, Vt., and by him to Dr. Mills, of Ohio. Atwood blood.

432 GOLDEN FLEECE.

Bred by S. Jewett, Independence, Mo., 1872. Sire, Doty Ram (134); dam by one of Victor Wright's stock rams; 2d dam bred by B. W. Crane, of Hammond stock. Sold to J. R. Kinney, Mich. Atwood blood.

433 BUCK MOUNTAIN.

Bred by S. Jewett. Sire, Dea. James's Ram (52); dam, an Atwood ewe bred by B. W. Crane from Hammond stock. Atwood and Robinson blood.

434 CONSTITUTION.

Bred by J. B. Cherbino, Weybridge. Sire, Bonaparte (176); dam, a Robinson ewe bred by B. Myrick, Bridport, Vt. Sold to S. Jewett, Independence, Mo. Robinson and Atwood blood.

435 McCONNELL'S STICKNEY RAM.

Bred by T. Stickney & Son, 1874. Sire, Fremont, Jr. (215); dam bred by T. S. & Son. Sold to T. F. & C. D. McConnell, Ripon, Wis. Stickney blood.

436 McCONNELL'S HALL RAM.

Bred by Geo. H. Hall, Shoreham, Vt., 1875. Sire, Major (181); dam by Birchard's Wrinkly; 2d dam, an Atwood ewe by Seville (75); 3d dam, an Atwood ewe bred by R. J. Jones. Sold E. N. Bissell and by him to T. F. & C. D. McConnell. Atwood and Robinson blood.

437 E. D. BUSH RAM.

Bred by E. D. Bush, Shoreham, Vt. Sire, Fremont, Jr. (215); dam, an Atwood and Rich ewe. Sold T. F. & C. D. McConnell. Stickney and Atwood blood.

438 YOUNG BONNY.

Bred by S. S. Rockwell and H. E. Sanford, 1874. Sire, Bonaparte (176) ; dam bred by S. S. Rockwell. Sold W. H. & T. P. D. Matthews, Cornwall, Vt. Atwood and Robinson blood.

439 WANDERER.

Bred by H. W. Hammond, 1873. Sire, one of H. W. Hammond's stock rams; dam, an Atwood ewe, bred by H. W. H. Now owned by E. D. Mussey, Middlebury, Vt. Atwood blood.

440 MINGLE.

Bred by Wm. McCauley, Cornwall, Vt., 1876. Sire, Allright (169) ; dam bred by Wm. McC. of Atwood and Cock blood from S. S. Rockwell's flock. Atwood and Robinson blood.

441 J. T. STICKNEY'S 146.

Bred by J. T. Stickney, Shoreham, Vt., 1874. Sire, Fremont, Jr. (215) ; dam, bred by J. T. S., by Fremont (126). Stickney blood.

442 CENTENNIAL.

Bred by J. T. Stickney, 1875. Sired by Fremont, Jr. (115) ; dam, bred by J. T. Stickney, sired by Vermont (123). Stickney blood.

443 J. T. STICKNEY'S 22.

Bred by J. T. Stickney, 1875. Sire, Fremont, Jr. (215) ; dam, bred by J. T. S. Sold to Cherbino & Williamson. Stickney blood.

444 G. H. SMITH'S 2.

Bred by Geo. H. Smith, Bridport, Vt., 1876. Sire, Stub (222); dam, by Silver Horns (177) ; 2d dam, by Sea Lion (83) ; 3d dam, an Atwood ewe, bred by S. Benton, Cornwall, Vt. Atwood and Robinson blood.

445 OLD BLACK.

Bred by Victor Wright. Sire, one of V. Wright's stock rams ; dam, an Atwood ewe, bred by V. Wright Sold to Rollin J. Smith, Sudbury, Vt. Atwood blood.

446 SANFORD'S FORD RAM.

Bred by Charles R. Ford, Cornwall, Vt. Sire, Golden Fleece (91) ; dam, bred by E. S. Stowell. Sold Edgar Sanford. Atwood blood.

447 SANFORD'S BONAPARTE RAM.

Bred by H. E. Sanford. Sire, Bonaparte (176) ; dam, an ewe bred by S. S. Rockwell. Atwood and Robinson blood.

448 J. T. & V. R.'s No. 333.

Bred by J. T. & V. Rich, 1876. Sire, Hibbard (214) ; dam, bred by J. T. & V. R., from their Rich flock. Sold B. B. Tottingham, Shoreham, Vt. Atwood and Rich blood.

449 BALDWIN.

Bred by A. M. Baldwin, Whiting, Vt., 1872. Sire, Fremont, Jr. (215); dam, a Robinson ewe. Sold A. H. Hubbard, Whiting, Vt. Stickney blood.

450 FIELDS' EUREKA.

Bred by B. S. Fields, Cornwall, Vt. Sire, Eureka (58); dam, a Robinson ewe, bred by Elisha Rich, Whiting. Atwood and Robinson blood.

451 Dr. WRIGHT'S RAM.

Bred by Wm. P. Wright, Whiting, Vt., 1864. Sire, Fields' Eureka (450); dam, an Atwood and Remele ewe, bred by W. P. Wright. Atwood and Robinson blood.

452 HUGO.

Bred by Rollin J. Smith, 1874. Sire, Old Black (445); dam, bred by R. J. Smith, sired by Young Black Top (288); Atwood blood.

453 COMPANY RAM.

Bred by L. C. Remele, 1870. Sire, Loyal (310); dam, bred by L. C Remele, sired by Little Wrinkly (48); 2d dam, by Mountaineer (51). Sold to M. R. Atwood, Shoreham, by him to C. K. Williams, Whiting, Wm. Walker, Benson, G. Cutting, Shoreham, and finally to J. H. Paul, Wisconsin. Atwood and Robinson blood.

454 C. H. KETCHUM'S 51.

Bred by C. H. Ketchum, Whiting, Vt., 1875. Sire, Company Ram (453); dam, bred by W. P. Wright. Atwood and Robinson blood.

455 H. W. WALKER'S STOWELL RAM.

Bred by E. S Stowell, 1866. Sire, Golden Fleece (91); dam, an Atwood ewe bred by E. S. Stowell. Sold H. W. Walker, Shoreham, Vt. Atwood blood.

456 BACON'S STICKNEY RAM.

Bred by T. Stickney & Son. Sire, Fremont (126); dam bred by T. Stickney & Son. Sold to Bacon & Sprague, Waltham, Vt. Stickney blood.

457 H. W. WALKER'S STOCK RAM.

Bred by H. W. Walker, 1873. Sire, Bacon's Stickney (456); dam bred by H. W. Walker; died 1876. Stickney blood.

458 HOLMES'S RAM.

Bred by J. Holmes, Charlotte, Vt. Sire, Schofield ram bred by C. B. Cook, Charlotte, sired by one of H. Thorp's stock rams; dam, an Atwood ewe bred by C. B. Cook. Sold S. H. Weston, Winooski, Vt. Atwood blood.

459 WEBSTER'S BALDWIN RAM.

Bred by M. Baldwin, Whiting, Vt., 1875. Sire, Fremont. Jr. (215); dam, a Robinson ewe. Sold L. Webster, Shoreham, Vt. Stickney and Robinson blood.

460 WATTS'S ELLSWORTH RAM.

Bred by A. F. Ellsworth, Whiting, Vt., 1866. Sire, Ellsworth's Eureka (327); dam, an Atwood ewe, bred by A. F. E. Sold Emerson Watts, Whiting, Vt. Atwood blood.

461 RECTOR.

Bred by Rector Gage, Addison, Vt., 1875. Sire, Pat Henry (183); dam, an Atwood ewe, bred by R. G., sired by Little Wrinkly (48). Sold L. S. Wright, Weybridge, Vt. Atwood blood.

462 RICH'S STOCK RAM.

Bred by J. T. & V. Rich. Sire, Fremont (126); dam, one of the old Rich flock. Rich and Stickney blood.

463 BLACK TOP.

Bred by L. P. Clark, Addison, Vt. Sired by Chunkhead (182); dam, by Wrinkly (173); 2d dam, bred by N. A. Saxton. Atwood blood.

464 FINE-WOOL.

Bred by L. P. Clark. Sire, Kilpatrick (71); dam by Chunkhead (182); 2d dam, bred by Victor Wright. Atwood blood.

465 HINDS'S RAM.

Bred by Col. E. S. Stowell, and sold in the ewe to Albert W. Hind, Addison, Vt., 1865. Sire, Golden Fleece (91); dam, an Atwood ewe, bred by E. S. S. Sold Alonzo Baldwin, Whiting, Vt. Atwood blood.

466 ALONZO.

Bred by T. Stickney & Son. Sire, Fremont (126); dam, bred by T. Stickney & Son. Sold to Alonzo Baldwin. Stickney blood.

467 LITTLE HALL RAM.

Bred by G. H. Hall, 1868. Sire, Major (43); dam, an Atwood ewe, bred by G. H. H. Sold German Cutting, Shoreham, Vt., when a lamb; died 1878. Atwood blood.

468 WM. JARVIS.

Bred by F. H. Dean, 1876. Sire, Cassius (257); dam, an Atwood ewe, bred by F. H. Dean. Sold R. Lane, Cornwall, Vt., and by him to J. A. Wright, Middlebury, Vt. Atwood and Robinson blood.

469 JUNTA.

Bred by Moses Sheldon, Salisbury, Vt., 1877. Sire, Brooksville (354); dam, an ewe bred by Mrs. V. Wright. Sold R. Lane, Cornwall, Vt., and by him to J. A. Wright, Middlebury, Vt. Atwood blood.

470 J. T. & V. R.'s 301.

Bred by J. T. & V. Rich, 1873. Sire, Rich's Stock Ram (462); dam, bred by Messrs. Rich, from the old stock. Sold to J. A. Wright, Middlebury, and by him to Brant, Thompson, Kelley & Co., Ohio. Rich blood.

471 BANKER.

Bred by J. T. & V. Rich, 1875. Sire, Hibbard (214); dam, bred by J. T. & V. R., from their old stock. Rich blood.

472 J. T. & V. R.'s 331.

Bred by J. T. & V. Rich, 1876. Sire, Hibbard (214); dam, bred from their old stock. Rich blood.

473 J. T. & V. R.'s 350.

Bred by J. T. & V. Rich, 1876. Sire, Hibbard (214); dam, bred by J. T. & V. R., from their old stock; half interest sold to E. N. Bissell, Shoreham, Vt. Rich blood.

474 GEN. GRANT.

Bred by H. M. Perry, Shoreham, Vt., 1876. Sire, Little Hall Ram (467); dam, one of Cutting's best breeding ewes; died 1878. Cutting blood.

475 FORTUNE.

Bred by A. J. Towner, Shoreham, Vt., 1875. Sire, Snowflake (253); dam, a Robinson ewe. Sold F. & L. E. Moore, Shoreham, Vt. Robinson blood.

476 VICTOR.

Bred by Mrs. V. Wright, 1876. Sire, Pat Henry (183); dam, bred by V. Wright. Sold to W. C. Sturtevant, Weybridge, Vt. Atwood blood.

477 J. TOWLE'S 69.

Bred by J. Towle, Cornwall, Vt., 1873. Sire, Towle Ram (164); dam, an Atwood ewe, bred by R. P. Hall, sired by Don Pedro (119); Atwood blood.

478 J. TOWLE'S 70.

Bred by J. Towle, 1875. Sire, J. Towle's 69 (477); dam, an Atwood ewe, bred by R. P. Hall. Atwood blood.

479 J. TOWLE'S 71.

Bred by J. Towle, 1875. Sire, J. Towle's 69 (477); dam, an Atwood ewe, bred by R. P. Hall. Atwood blood.

480 L. P. CLARK'S 103.

Bred by L. P. Clark, Addison, Vt., 1874. Sire, Kilpatrick (71); dam, bred by L. P. C., sired by Sweepstakes (32); 2d dam, bred by N. A. Saxton. Sold E. C. Eells, and by him to T. J. Close, St. Clairsville, Ohio. Died 1877. Atwood blood.

481 L. P. CLARK'S 111.

Bred by L. P. Clark, 1876. Sire, General (210); dam, an Atwood ewe bred by L. P. Clark, sired by Kilpatrick (71); 2d dam by Chunk Head (182); 3d dam, Old Favorite by Green Mountain (70). Sold E. C. Eels and by him to John Allen, Bridgeport, Ohio. Atwood blood.

482 L. P. CLARK'S 124.

Bred by L. P. Clark, 1876. Sire, General (210); dam, by Kilpatrick (71); 2d dam by Little Wrinkly (48). Sold J. H. Close, St. Clairsville, Ohio. Atwood blood.

483 L. P. CLARK'S 108.

Bred by L. P. Clark, 1876. Sire, General (210); dam by Chunk Head (182); 2d dam, Old Favorite, by Green Mountain (70). Sold E. C. Eells and by him to J. M. Holmes & Son, Masterville, Ohio. Atwood blood.

484 L. P. CLARK'S 114.

Bred by L. P. Clark, 1876. Sire, General (210). Sold E. C. Eells, and by him to W. G. Markham, Avon, N. Y., and E. Townsend, Pavilion Centre, N. Y. Atwood blood.

485 L. P. CLARK'S 123.

Bred by L. P. Clark, 1875. Sire, Pat Henry (183); dam by Vigor (209); 2d dam, by Little Wrinkly (48). Sold E. C. Eells for J. H. Close, St. Clairsville, Ohio, and by him half interest to Thos. Healea & Son, Urichville, Ohio. Atwood blood.

486 L. P. CLARK'S 140.

Bred by L. P. Clark, 1876. Sire, Pat Henry (183); dam by Sweepstakes (32); 2d dam bred by Victor Wright. Sold E. C. Eells, and by him to J. H. Close, St. Clairsville, Ohio. Atwood blood.

487 L. P. CLARK'S 134.

Bred by L. P. Clark, 1876. Sire, General (210); dam, by Chunk Head (102); 2d dam bred by E. Hammond. Sold to J. H. Close and by him half interest to Thos. McLary, West Alexander, Pa. Atwood blood.

488 L. P. CLARK'S 138.

Bred by L. P. Clark, 1875. Sire, Pat Henry (183); dam, Atwood ewe bred by L. P. C. Sold E. C. Eells. Atwood blood.

489 L. P. CLARK'S 152.

Bred by L. P. Clark, 1875. Sire, Pat Henry (183); dam, Woolly Head, by Vigor (209); 2d dam, Old Favorite, by Green Mountain (70). Sold E. C. Eells, and by him to Thos. Healea, Urichville, Ohio, and J. M. Holmes, Harrison Co., Ohio, and the latter sold his interest to J. H. Close. Atwood blood.

490 L. P. CLARK'S 153.

Bred by L. P. Clark, 1875. Sire, Pat Henry (183); dam, by Chunk Head (182); 2d dam, Old Favorite, by Green Mountain (79). Sold E. C. Eells. Atwood blood.

491 GRANGER.

Bred by E. S. Stowell, 1877. Sire, Goliath (197); dam, Leonora, 3d, by Red Leg (109). Atwood blood.

492 L. P. CLARK'S 137.

Bred by L. P. Clark, 1875. Sire, Pat Henry (183); dam, an Atwood ewe bred by L. P. C. Sold E. C. Eells, and by him to John Moore, Short Creek, Ohio.

493 L. P. CLARK'S 112.

Bred by L. P. Clark, 1876. Sire, Pat Henry (183); dam, Old Favorite by Green Mountain (70); 2d dam by Sweepstakes (32); 3d dam by Black Top (90); 4th dam bred by N. A. Saxton. Atwood blood.

494 L. P. CLARK'S 107.

Bred by L. P. Clark, 1876. Sire, General (210); dam, Woolly Head by Vigor (209); 2d dam Old Favorite (see 493). Atwood blood.

495 L. P. CLARK'S 151.

Bred by L. P. Clark, 1878. Sire, General (210); dam, by Kilpatrick (71); 2d dam, by Little Wrinkly (48). This teg is called Moses and is full brother to (482.) Atwood blood.

496 GUNBOAT.

Bred by E. S. Stowell, 1876. Sire, Iron Clad (194); dam, Sukey 22d, by Red Leg (109). Atwood blood.

497 CORNWALL PRINCE.

Bred by E. S. Stowell, 1877. Sire, Goliath (197); dam, Princess 14th, by David (118). Atwood blood.

498 C. E. CRANE'S 101.

Bred by C. E. Crane, Bridport, Vt., 1876. Sire, Eureka, 3d (223); dam, by Doty Ram (134); 2d dam, bred by C. N. Hayward. Atwood blood.

499 REMELE'S 15.

Bred by W. R. Remele, 1876. Sire, Sam (368); dam, an Atwood ewe, bred by W. R. R. Sold to Messrs. Linville, Cloverdale, Cal. Died August, 1878. Atwood and Robinson blood.

500 E. S. STOWELL'S 274.

Bred by E. S. Stowell, 1875. Sire, Diamond Dust (122); dam, an Atwood ewe, bred by E. S. Stowell. Sold A. A. Farnsworth, Brooksville, Vt. Atwood blood.

501 C. D. LANE'S 274.

Bred by C. D. Lane, 1873. Sire, Buck Mountain (247); dam, bred by C. D. Lane. Sold J. A. Wright, Middlebury, Vt., who named him Ben Butler. Atwood and Robinson blood.

502 HAMBLIN'S LANE RAM.

Bred by Rollin Lane, Cornwall, Vt. Sire, Young Chunk (206); dam, an Atwood ewe, bred by Henry Lane. Sold to J. B. Hamblin. Atwood blood.

503 J. B. HAMBLIN'S 1.

Bred by J. B. Hamblin, Cornwall, Vt., 1873. Sire, Hamblin's Lane Ram (502); dam bred by J. B. Hamblin, sired by a ram bred by J. B. H, sired by Fields' Eureka (450); 2d dam, by Atwood ram, bred by Victor Wright. Sold to E. C. Eells, Middlebury, Vt. Atwood and Robinson blood.

504 M. J. ELLSWORTH'S 1.

Bred by M. J. Ellsworth, Cornwall, Vt., 1875. Sire, A. F. Ellsworth's Remele Ram (328); dam, bred by M. J. E. Mr. E. calls this ram Don Pedro. Atwood and Robinson blood.

505 VICTOR.

Bred by George Campbell, West Westminster, Vt., 1861. Sire, Old Grimes (49); dam, a Jarvis and Humphrey ewe bred by Geo. Campbell. This ram was taken to the international show at Hamburgh in 1863. He was awarded a first prize at that exposition in a competition of 291. Sold to Count Shur Thoss, of Silesia. Atwood and Jarvis blood.

506 VERMONT BOY.

Bred by George Campbell, 1861. Sire, Old Grimes (49); dam, a Jarvis and Atwood ewe bred by George Campbell. This ram was taken to Hamburgh with No. (505) and received a second-class award. Sold to Count Shur Thoss, of Silesia. Atwood and Jarvis blood.

507 LONG-WOOL.

Bred by S. E. Wheat, Putney, Vt. Sire, Golden Fleece (91); dam, ewe bred by S. E. Wheat. Atwood and Robinson blood.

508 FAVORITE.

Bred by S. E. Wheat, 1873. Sire, Longwool (507); dam, a Robinson ewe, bred by S. E. Wheat. Sold George Campbell, and by him to Richard Peters, Atlanta, Georgia. Atwood and Robinson blood.

509 LONGWOOL.

Bred by Rollin J. Smith. Sire, Old Black (445); dam by Cromwell (350); 2d dam by Sweepstakes (32.) Died 1878. Atwood blood.

510 A. A. FARNSWORTH'S 120.

Bred by A. A. Farnsworth, 1877. Sire, David (118); dam, an Atwood ewe, bred by A. A. F. Atwood blood.

511 ROBIN HOOD.

Bred by G. F. Martin, 1875. Sire, Smuggler (423); dam, bred by F. H. Dean, sired by Little Wrinkly (48). Now owned by breeder and D. P. Dewey, Grand Blanc, Mich. Atwood blood.

512 RODERICK DHU.

Bred by Rector Gage, 1875. Sire Fat Henry (183); dam bred by R. Gage. Sold to Peter Martin, Rush, N. Y. Atwood blood.

513 PETER MARTIN'S 201.

Bred by Peter Martin, 1877. Sire, Roderick Dhu (512); dam, bred by G. F. Martin, sired by Torrent (97) Atwood blood.

514 PETER MARTIN'S 202.

Bred by Peter Martin, 1877. Sire, Roderick Dhu (512); dam, P. Martin's 42, sired by Green Mountain, 2d (96); 2d dam bred by F. H. Dean, sire, Little Wrinkly (48). Atwood blood.

515 PETER MARTIN'S 203.

Bred by P. Martin, 1877. Sire, Little Phil (232); dam. P. M.'s 89, by Torrent (97); 2d dam, P. M.'s 4, bred by C. N. Hayward. Atwood blood.

516 PETER MARTIN'S 204.

Bred by P. Martin, 1876. Sire, Triumph (106); dam. P. M.'s 40, sired by Torrent (97). Atwood blood.

517 J. O. HAMILTON'S 51.

Bred by J. O. Hamilton, Bridport, Vt., 1872. Sire, Dea. James Ram (52); dam bred by J. O. Hamilton. Sold J. A. Wright, Middlebury, Vt., who named him Col. Humphreys. Robinson and Cutting blood.

518 J. O. HAMILTON'S 53.

Bred by J. O. Hamilton, 1876. Sire, J. O. Hamilton's 51 (517); dam, bred by J. O. H. Robinson and Cutting blood.   19

519 CLAY & WING'S 36.

Bred by Wales Bros. & Co., 1876. Sire, Pat Henry (188); dam bred by E. R. Clay. Sold Clay & Wing, Cornwall, Vt. Atwood blood.

520 CLAY & WING'S 37.

Bred by Wales Bros. & Co., 1876. Sire, General (210); dam bred by E. R. Clay. Sold Clay & Wing. Atwood blood.

521 CUSTER.

Bred by H. C. Burwell, 1876. Sire, Bismarck (221); dam, Lady Ironsides, by Elitharp & Burwell (79); 2d dam by Ironsides (80); 3d dam bred by C. N. Hayward, by America (35); 4th dam bred by P. Elitharp. Robinson and Atwood blood.

522 H. C. BURWELL'S 74.

Bred by H. C. Burwell, 1877. Sire, Bismarck (221); dam, Lady Hill, by Little Wrinkly (48); 2d dam bred by E. Hammond. Atwood and Robinson blood.

523 H. C. BURWELL'S 75.

Bred by H. C. Burwell, 1877. Sire, Bismarck (221); dam, an Atwood and Robinson ewe, bred by C. P. Morrison. Robinson and Atwood blood.

524 ACME.

Bred by H. C. Burwell, 1875. Sire, Black Hawk (220); dam, an Atwood ewe, bred by P. Elitharp. Sold to Cherbino & Williamson. Atwood blood.

525 L. S. BURWELL'S 22.

Bred by L. S. Burwell, Bridport, Vt., 1877. Sire, Acme (524); dam, an Atwood and Robinson ewe, bred by L. S. Burwell. Atwood and Robinson blood.

526 L. S. BURWELL'S 24.

Bred by L. S. Burwell, 1877. Sire, Constitution (366); dam bred by L. S. B. Sired by Elitharp & Burwell (79); 2d dam, a Robinson ewe, bred by G. W. Whitford. Atwood and Robinson blood.

527 L. S. BURWELL'S 27.

Bred by L. S. Burwell, 1877. Sire, Bismarck (221); dam bred by L. S. Burwell. Atwood and Robinson blood.

528 J. T. STICKNEY'S 142.

Bred by J. T. S., 1876. Sire, J. T. Stickney's 146 (441); dam by Vermont (123). Stickney blood.

529 J. J. CRANE'S 2.

Bred by J. J. Crane, Bridport, Vt , 1876. Sire, Eureka, 3d (223) ; dam an Atwood ewe, bred by J. J. Crane. Sold Forbes & Jones, Shoreham, Vt. Atwood and Robinson blood.

530 C. E. CRANE'S 103.

Bred by C. E. Crane, Bridport, Vt., 1876. Sire, Eureka, 3d (223) ; dam, an Atwood ewe, bred by C. E. C. Sold to H. C. Burwell, and by him to Forbes & Jones, Shoreham, Vt. Atwood and Robinson blood.

531 H. G. HIBBARD'S 7.

Bred by H. G. Hibbard, Orwell, Vt, 1877. Sire, Hibbard (214); dam, an Atwood ewe, bred by H. G. Hibbard. Sold to A. C. Martin, Benson, Vt., and a half interest by him to J. Forbes, Jr., Shoreham, Vt. Atwood blood.

532 H. W. WALKER'S 77.

Bred by H. W. Walker, Shoreham, Vt., 1876. Sire, H. W. W.'s stock ram (456) ; dam bred by H. W. W. Stickney blood.

533 DON JUAN.

Bred by L. C. Remelec, 1877. Sire, J. T. Stickney's 146 (441) ; dam bred by L. C R. Sired by one of T. Stickney & Son's stock rams. Sold to D. J. Wright, Shoreham. Stickney blood.

534 H. S. BROOKINS' 32.

Bred by Gustavus Cook, Shoreham, Vt., 1874. Sired by a ram bred by W. H. Cook, sired by one of F. & L. E. Moore's stock rams, from an ewe bred by J. T. & V. Rich. Dam of 534 by Fremont (126) ; 2d dam bred by J. T. & V. Rich. Sold to H. S. Brookins, Shoreham, and by him to J. A. Wright, Middlebury, Vt., who named him Rip Van Winkle. Robinson blood.

535 H. S. BROOKINS' 35.

Bred by H. S. Brookins, 1877. Sire H. S. Brookins' 32 (534) ; dam bred by Levi Wolcott, sired by Charlie (277) ; 2d dam bred by A. H. Hubbard. Atwood and Robinson blood.

536 STOGA (E. N. Bissell's 103).

Bred by M. R. Atwood, Shoreham, Vt , 1874. Sire, Hibbard (214) ; dam bred by L. C. Remele, sired by Fremont (126). Sold E. N. Bissell, Shoreham, Vt. Atwood and Robinson blood.

537 M. R. ATWOOD'S 7.

Bred by M. R. Atwood, 1876. Sire, J. T. Stickney's 146 (441) ; dam sired by sire of Co. Ram. Stickney and Robinson blood.

538 M. R. ATWOOD'S 50.

Bred by M. R. Atwood, 1876. Sire, J. T. Stickney's 146 (441); dam sired by one of T. Stickney & Son's stock rams. Stickney and Robinson blood.

539 J. T. & V. R.'S 357.

Bred by J. T. & V. Rich, 1877. Sire, Banker (471) ; dam bred by J. T. & V. Rich. Half interest sold to E. N. Bissell. Rich blood.

540 J. T. & V. R.'S 359.

Bred by J. T. & V. Rich, 1877. Sire, Centennial (442) ; dam bred by J. T. & V. Rich. Half interest sold to E. N. Bissell. Rich blood.

541 W. P. WRIGHT'S 38.

Bred by W. P. Wright, Whiting, Vt., 1876. Sired by C. H. Ketchum's 51 (454) ; dam bred by W. P. W., sired by Fremont, Jr. (215). Robinson blood.

542 M. R. ATWOOD'S 8.

Bred by M. R. Atwood, 1876. Sire, J. T. Stickney's 146 (441); dam bred by L. C. Remele. Sold F. G. Wright & C. K. Williams, Whiting, Vt. Stickney and Robinson blood.

543 LITTLE GREASY.

Bred by F. D. Barton, Waltham, Vt. Sire, Barton's Greasy (128) ; dam, an Atwood ewe, bred by W. R. Sanford. Sold O. C. Bacon, Waltham, Vt., and by him to Hiram Merrill, Addison, Vt. Atwood blood.

544 OSCAR.

Bred by R. J. Jones, 1876. Sire, Allright (169) ; dam bred by R. J. Jones. Sold to O. C. Bacon. Atwood blood.

545 JASON.

Bred by R. J. Jones, 1876. Sire, Allright (169) ; dam bred R. J. J. Sold O. C. Bacon. Atwood blood.

446 O. C. BACON'S 44.

Bred by O. C. Bacon, 1877. Sire, Allright (169) ; dam bred by O. C. B., sired by French & Mason (362) ; 2d dam bred by R. P. Hall. Atwood and Robinson blood.

547 CHUNK.

Bred by F. D. Barton, 1871. Sire, King (129) ; dam bred by W. R. Sanford, sired by Comet (57). Atwood blood.

548 PAYMASTER.

Bred by E. Hammond & Son. Sire, Gold Drop (64); dam, an Atwood ewe, bred by E. Hammond. Atwood blood.

549 NORTHERNER.

Bred by E. Hammond & Son. Sire, Gold Drop (64); dam, an Atwood ewe, bred by E. Hammond. Atwood blood.

550 ABE LINCOLN.

Bred by H. W. Hammond. Sire, Paymaster (548); dam, an Atwood ewe, bred by H. W. Hammond. Atwood blood.

551 ONWARD.

Bred by E. Hammond & Son. Sire, Bull Dog (104); dam, an Atwood ewe, bred by E. Hammond. Atwood blood.

552 LEVI.

Bred by Levi Wolcott, Shoreham, Vt. Sire, Charlie (253); dam a Robinson ewe, bred by L. Wolcott. Sold Merrill Bingham, Cornwall, Vt. Atwood and Robinson blood.

553 O. A. FIELD'S 1.

Bred by O. A. Fields, Cornwall, Vt , 1874. Sire, Levi (552); dam bred by O. A. F. Atwood and Robinson blood.

554 O. A. FIELDS'S 76.

Bred by O. A. Fields, 1877. Sire, O. A. Fields' (553); dam bred by O. A. F. Atwood and Robinson blood.

555 L. P. CLARK'S 110.

Bred by L. P. Clark, 1876. Sire General (210); dam, bred by L. P. C., by Vigor (209); 2d dam, Old Favorite by Green Mountain (70). Atwood blood.

556 L. P. CLARK'S 125.

Bred by L. P. Clark 1876. Sire, General (210); dam bred by L. P. C., by Little Wrinkly (48); 2d dam bred by L. P. C., by Little Wrinkly (48); 3d dam, Atwood ewe, bred by V. Wright. Atwood blood.

557 J. B. HAMBLIN'S 80.

Bred by J. B. Hamblin, 1877. Sire, M. J. Ellsworth's 1 (504); dam by Hamblin's Lane Ram (502); 2d dam bred by J. B. H. Atwood and Robinson blood.

558 M. J. ELLSWORTH'S 2.

Bred by M. J. E., 1876. Sire, Allright (169); dam, a Robinson ewe, bred by L. H. Payne. Atwood and Robinson blood.

559 M. J. ELLSWORTH'S 70.

Bred by M. J. E., 1877. Sire (558); dam bred by M. J. E. Atwood and Robinson blood.

560 R. GAGE'S 69.

Bred by R. Gage, Addison, Vt., 1877. Sire, General (210); dam, an Atwood ewe, bred by Victor Wright. Atwood blood.

561 G. H. SMITH'S 1.

Bred by G. H. Smith, 1877. Sire, Eureka, 3d (223); dam by Silver Horne (177); 2d dam by Sea Lion (83); 3d dam, an Atwood ewe, bred by S. Benton, Cornwall, Vt. A half interest sold to J. J. Crane, Bridport, Vt. Atwood and Robinson blood.

562 PRINCE.

Bred by E.S. Stowell, Cornwall, Vt., 1876. Sire, Ironclad (194); dam, Princess 14th, by David (118). Atwood blood.

563 LITTLE WRINKLY.

Bred by H. B. Wright, Shoreham, Vt. Sire, one of E. Hammond & Son's stock rams; dam, a Robinson ewe. A half interest was sold F. & L. E. Moore, finally sold to E. J. & E. W. Hardy, Osceola, Mich. Atwood and Robinson blood.

564 SAMMY.

Bred by F. & L. E. Moore. 1874. Sire. Duke (275); dam, a Robinson ewe. Sold to Wm. Ball, Mich., and by him to E. J. & E. W. Hardy, Osceola, Mich. Robinson blood.

565 NO NAME.

Breed by F. & L. E. Moore. Sire, Duke (275); dam, a Robinson ewe. Sold to E. J. & E. W. Hardy, Osceola, Mich. Robinson blood.

566 F. & L. E. MOORE'S 125.

Bred by F. & L. E. Moore, 1878. Sire, Banker (471); dam, a Robinson, Rich and Stickney ewe, bred by Gustavus Cook, Shoreham, Vt. Stickney, Robinson and Rich blood.

567 HASSAN.

Bred by C. R. Jones, Hubbardton, Vt., 1873. Sire, a pure Hammond ram, sired by Bull Dog (104), and out of a Hammond ewe. Sold Rollin J. Smith. Atwood blood.

568 DON JUAN.

Bred by Rollin J. Smith, 1877. Sire, Old Black (445); dam by Cromwell (350); 2d dam by V. Wright's Old Greasy (74); 3d dam bred by Victor Wright. Atwood blood.

569 RARUS.

Bred by George H. Hall, Shoreham, Vt., 1873. Sire, Major (181) ; dam, an Atwood ewe bred by G. H. H. Sold to L. J. Orcutt, Cummington, Mass., and by him to George Hammond, Middlebury, Vt. Atwood blood.

570 G. D. BUSH'S, 50.

Bred by G. D. Bush, North Orwell, Vt. Sire, J. Towle's 69 (477); dam, an Atwood ewe bred by G. D. Bush. Atwood blood.

571 SOUTH HERO.

Bred by H. Thorp, 1875. Sire, Thorp's Barton Ram (375) ; dam, an Atwood ewe bred by H. Thorp. Sold to H. Harrington, Keeler's Bay, Vt. Atwood blood.

572 H. G. HIBBARD'S 3.

Bred by H. G. Hibbard, 1875. Sire, Hibbard (214) ; dam, an Atwood ewe, bred by H. G. Hibbard. Sold to O. C. Martin, Benson, Vt., and by him to L. J. Wright, Weybridge, Vt. Atwood blood.

HISTORIES OF MERINO FLOCKS THAT HAVE BEEN SCATTERED.

No flock of Sheep whose pedigree and history we publish in this Register can trace their blood direct to importation, without passing through those that have ceased to exist as distinctive flocks; the founders and breeders of most of these flocks are now dead. Some of the flocks recorded have derived their blood from those of later date, that have become scattered. Therefore, to trace the pedigrees of flocks herein recorded intelligently, and direct to importation, it is necessary to give brief histories of many that have ceased to exist and have become scattered. The facts that can be ascertained in regard to many of these flocks at this day are very meager, but the researches of the Committee on Pedigrees and of Publication have been sufficient to convince them that all we shall mention are entitled to be considered and relied upon as having been founded and bred as pure-blooded Merino Sheep, direct and unmixed from the importations from Spain made in the years 1802, 1810 and 1811. There is a possibility that some importations were made in 1809, but we have found no reliable record of the same.

The first of these flocks through which the blood of our present Merino sheep can be traced was that of

COL. DAVID HUMPHREYS,

of Humphreyville, New Haven Co., Conn. The facts in regard to the importation of this flock of sheep and its foundation are given at length at the 23d and subsequent pages of this work, and it is not necessary to repeat them here. There is no doubt that Col. Humphreys kept the blood of these sheep pure and unmixed, and that the sheep he sold, or were sold by his agents, to Messrs. Atwood and others were pure when represented to be so ; and it is not necessary to consume time or space in discussing what cabana in Spain these sheep were selected from. The only *fact* we have found not published by Randall, Morrell, and others, is that contained in the contract of Col. Humphreys with Elihu Ives, which we publish at pages 23 and 24 in this work, "that they were extracted from Spanish Estramadura," "and ascertained by their pedigree to be of the purest and best race in Spain," which, according to Jarvis, "is the most material fact worth knowing." Col. Humphreys sold these sheep from year to year through his agents, and we find traces of them in different parts of the country; but it is through the sale of one ewe to Stephen Atwood that the most of the Humphreys blood comes to our flocks to-day. Col. Humphreys died in 1818, and his flock was soon after sold and scattered.

Following down with this line of blood we will next notice the flock of

STEPHEN ATWOOD, OF WOODBURY, LITCHFIELD COUNTY, CONN.

As before stated, Mr. Atwood purchased one ewe of Col. Humphreys. This was in 1813. This ewe with her progeny was bred to rams derived from Col. Humphreys' flock by Mr. Atwood's neighbors, a ram bought of Col. Humphreys by Younglove Cutler being the first one used. This way of breeding was practised until about 1830, after which Mr Atwood used rams of his own breeding except in one year, when he used a Saxon ram to a part of his ewes, but the cross not proving as profitable as his old flock, he soon disposed of and sold off all of this blood. From this flock were derived the flocks of Mr. Atwood's three sons, Chauncey, George and Eben, also those of Jerry Smith, O. P. Northrop, and perhaps others, all residing in the same neighborhood. All of these bred pure Atwood sheep and sold to the several parties that brought these sheep to Vermont. When Messrs Hammond and Hall made purchases of their Atwood sheep of these other parties, Mr. Stephen Atwood gave them certificates that they were bred pure from his flock.

Thirty years after Mr. Atwood commenced his flock he wrote: "Since I began my full-blood flock, I had three important properties in view to combine, viz.: constitution, quantity and quality; my success has at least been satisfactory to myself."

The cases we cite to show improvement in our article upon the improvement of Merinos, prove that Mr. Atwood was one of the earliest and most successful improvers of Merino sheep, and there is no doubt that sheep from his flock were in greater demand and brought higher prices from 1844 to 1850 than those from the flocks of other breeders, though from that time Mr. Atwood failed to keep pace in his improvements with Messrs. Hammond and other breeders in Vermont, who had derived their blood from his flock, for in later years he sent ewes to Vermont to be served by rams bred by Hammond, Stowell and perhaps others. As will by seen by reference to our list of stock rams and histories of flocks that are recorded in this Register, Mr. Atwood bred and furnished the foundation for many of the most celebrated flocks of Merino sheep in Vermont, and we believe there is scarcely a stock ram of any note in Vermont, that does not in whole or in part trace its pedigree to Mr. Atwood's flock.

Following this same line of blood down another step will bring us to the flock of the

MESSRS. HAMMOND, MIDDLEBURY, VT.

Previous to January, 1844, Messrs. W. S. & E. Hammond had been extensive wool-growers and sheep-breeders, having had a large

20

flock of sheep, mainly of Saxon blood ; but for a few years previous to the period of which we speak, heavier-wooled Merino rams had been used, the best that could be procured from the flocks of Vermont.

In January, 1844, Mr. E. Hammond, in company with Mr. R. P. Hall, of Cornwall, went to Connecticut and purchased twenty-nine ewes and three rams of Mr. Stephen Atwood and his neighbors. The rams and a part of the ewes were from Mr. Atwood's flock, and were of his own breeding. These ewes were divided after they were brought to Vermont, but the rams were owned in company.

In 1845, twenty-seven (one equal third) of Mr. Atwood's ewes, and one ram were purchased. Two more purchases were made in company with Mr. Hall, and one subsequently in which Mr. Hall was not interested. One of the purchases was the entire crop of ewe lambs raised by Mr. Atwood in one year. These purchases were all made of Mr. Atwood within a period of three years from the time of the first purchase in January, 1844, and formed the foundation of the flock of Atwood sheep that afterwards became so famous in the hands of Messrs. Hammond.

When the first purchase was made Mr. Atwood gave the following certificate :

"WOODBURY, LITCHFIELD Co , CONN., ⎞
        Jan. 27, 1844.   ⎠

This may certify that E. Hammond and R. P. Hall, of Addison Co., Vt., have this day purchased of me three full-blood Merino bucks, and of me and others, descendants of my flock, twenty-seven full-blood Merino ewes of the Paular breed, which originated from the celebrated flock imported from Spain by Col. Humphreys, of Derby, New Haven Co., Conn.

S. ATWOOD."

And Mr. Northrop the following :

"WOODBURY, Jan. 26, 1844.

This may certify that I have sold this day, to Messrs Hall & Hammond, of Addison Co., State of Vermont, fifteen pure-blooded Merino ewe sheep, which sprung from the stock of Stephen Atwood, of Woodbury, Conn.

OBADIAH P. NORTHROP."

W. S. &. E. Hammond were well situated to succeed in breeding a flock of improved sheep. The elder brother, William S., was an excellent manager and shepherd, but the younger partner and brother of the firm, Edwin Hammond, made the selections and directed the breeding of the flock. On one of the trips to Connecticut to purchase sheep, Mr. E. Hammond visited Newport, Rhode Island, and purchased five ewes of Joseph I. Bailey, and brought them home, and for a while they were kept with the flock. In 1854, in company with Mr. W. R. Sanford, some thirty or more ewes were purchased of Messrs. D. &. G. Cutting, being a selection or choice of their flock. These ewes were divided, Messrs. Hammond receiving one-half. A few years after a few Atwood ewes were purchased of L. D. Gregory.

In 1851, Mr. Wm. R. Sanford made a visit to France, Spain and Germany and examined the flocks of those countries, with a view of selecting and purchasing sheep, if any superior could be found. Messrs. Hammond, R. P. Hall, and W. R. Remele were engaged with Mr. Sanford in this enterprise, shared the expense and received each a portion of the sheep (equal to the amount each invested) that were selected and purchased by Mr. Sanford.   These sheep were two rams and twenty-three ewes, from the French flocks of Messrs. Cughnot and Gilbert, and six rams and twenty-five ewes from the Silesian flock of Louis Fischer, of Germany.   The portion of these sheep received by Messrs. Hammond were bred with the flock but a short time, a very few crosses being made upon the Silesian ewes with the Atwood stock rams of the flock, but the whole of these purchases, as well as all of the blood of these and of the Bailey and Cutting purchases, were sold out of the flock, leaving nothing but the Atwood blood.

A third-interest was purchased in 1849 in Old Black (No. 9), and he was used somewhat as a stock ram in the flock.   Atwood Ram (No. 12) was also used to a few ewes, and the stock rams of Mr. Sanford to a limited extent.

In 1858 the senior member of the firm died, and his portion of the flock was inherited by his son, Henry W. Hammond, a division of the flock being made in the year following, but the stock rams of both flocks were always used in each. In 1864 George, son of Edwin Hammond, became a partner with his father in the flock.   Edwin Hammond died Dec. 31st, 1870.   The flock descended to his son George, and was bred by him until 1874, when it was sold to L. J. Orcutt, Cummington, Mass., and taken to Massachusetts.   The portion the flock owned by H. W. Hammond, like the other, was also sold and taken to Ohio.

We publish nothing new when we say that Edwin Hammond was the leading breeder of his time, and made much greater and more rapid strides in the improvement of Merino sheep than any breeder that had preceded him.   For proof of this we refer to our article on the improvement of Merinos. That his improvements were appreciated by other flock masters, we only have to refer to the lists and pedigrees of stock rams and flocks which we publish in this register.   We find that list of stock rams is good evidence of improvement.   We find Old Black (9) sheared 14 lbs., his son, Wooster (16), bred by Mr. Hammond, sheared $19\frac{1}{4}$ lbs., and, following down through Old Greasy (18), Old Wrinkly (20) and Little Wrinkly (22) to Sweepstakes (32), we find the weight of fleece and per centum of wool nearly doubled in a period of about twelve years.

FLOCK OF WM. R. SANFORD, ORWELL, VT.

In 1830, Mr. Sanford purchased of Messrs. Grant & Jennison, of Walpole, N. H., twenty ewes that were bred by Consul Jarvis of his

pure Spanish Merino importation or blood. The ewes were bred to
a ram from the Cock flock, and after to two rams bred by Consul
Jarvis.   In 1845, Mr. Sanford bought a ram lamb of Messrs. Hammond, having a selection of the ram lambs of that year, and always
after that year used pure Atwood rams.   In 1846 or 7, Mr. Sanford
bought of Stephen Atwood three ram lambs and a few ewes from
J R. Nettleton's flock in Connecticut bred from the flockof Jacob N.
Blakeslee.   In 1849, thirteen ewes and the ram Old Black (9) were
bought of Stephen Atwood and his son George Atwood, thus laying
the foundation for the flock that subsequently became pure Atwood.
A few more ewes—eight or ten in number—were bought of Messrs.
Atwood at other times.   In 1851, Mr. Sanford made the journey to
Europe, before narrated in the history of Mr. Hammond's flock, and
received a part of the sheep he purchased in France and Germany,
and for a while he kept them on his place.   In 1854 he purchased of
W. R. Remele. of Middlebury, Vt , thirty-six ewes, being all of his
ewe lambs of 1853 and 4.   These were pure Atwood, as will be seen
by reference to the flock history of W. R. Remele.   The same year
Mr. Sanford made the purchase of Messrs. D. & G. Cutting, mentioned
in the history of Messrs. Hammonds' flock.   A portion of these ewes
were Atwood, but the larger portion were a part Atwood, and a part
Rhode Island blood.

The same year Mr. Sanford purchased seven yearling ewes and
ewe lambs that were bred by Mr. Abel P. Wooster, of West Cornwall, Vt.   They were sired by Wooster (16) and out of ewes purchased of W. S. & E. Hammond.

After the introduction of the Atwood sheep into the flock, the
ewes from this blood were retained in the flock, and those having
the Jarvis blood were sold off, as were the French and Silesian sheep
that were purchased in 1851.   The Cutting blood, or that part of the
Cutting purchase that were not pure Atwoods, were also sold off,
and the flock became pure Atwood and were so bred until 1874.
In addition to the rams mentioned heretofore in this narrative, many
of the stock rams of W S. & E. Hammond were used.   At one time
Mr. S. purchased a half interest in two of them (nos. 17 and 18),
as will be seen in our list of stock rams.   V. Wright's California
(43) was also used in the flock, Mr. Sanford purchasing a half interest in him.   In the few years preceding 1874, the flock was owned
jointly by William R. Sanford and his son, Charles Sanford.   In
1874, the flock was sold to L. J. Orcutt, Cummington, Mass.; Messrs.
Buell, of Sudbury, Vt., acting as Mr. Orcutt's agent, and receiving
a portion of the flock.

Many rams of rare excellence were bred in, and sold from this
flock, and great improvements were made from the original sheep
received from Mr. Atwood and others.   We believe Kilpatrick (71),
was the first Atwood ram that sheared a fleece of one year's growth,
that weighed over 30 pounds.

FLOCK OF R. P. HALL, CORNWALL, VT.

This flock was founded in 1844, by the purchase of rams and ewes of Stephen Atwood, and others, in company with Messrs. Hammond as narrated in the history of the Hammond flock. In 1845, W. R. Remele became a partner with Mr Hall in the ownership of the flock, and the partnership continued until 1849, when the flock was divided. In 1851 Mr. Hall was a partner with Messrs Hammond, Sanford & Remele, in the purchase of the French and Silesian sheep, but soon sold them from the flock.

In 1862, John Towle became a partner with Mr. Hall, in the ownership of the flock; four years after, Mr. Towle sold his interest in the flock to E. S. Stowell and Henry Manchester. In 1869, the flock was sold to John Towle, who continued to own the flock and their descendants. Mr. Hall sold the ewes to found many flocks of full-blood Atwood sheep, and bred and sold a large number of excellent stock rams. Besides the rams he purchased of Mr. Atwood, and owned in company with Messrs. Hammond, he used the stock rams of these gentlemen : Mr. Sanford, Victor Wright and R. Gleason to some extent.

FLOCK OF PROSPER ELITHARP, BRIDPORT, VT.

Mr. Elitharp commenced breeding pure blood Merino Sheep in 1835, and laid the foundation of his flock by the purchase of ewes of James Baker and D. Smith, of Bridport, Vt. A portion of them were bred by Leonard Beedle, and a part bred from his flock by Messrs. Baker & Smith. A few of them had crosses of Jarvis' blood, but most of them were pure from the Cock stock. A few ewes were also purchased from the flock of L C. Remele, and one from the Rich flock. These ewes were a cross of a Jarvis ram on ewes of the Cocks blood. In 1844, the foundation of the Atwood portion of the flock was laid by the purchase of the ram, Atwood (12), as narrated in the history of that ram, and two ewes of C. B. Cook of Charlotte, that were descended from ewes that Mr. Cook had purchased of Mr. Atwood, they were in lamb by Atwood (12). Soon after, more ewes were purchased of Mr. Cook. In 1846, three ewes were purchased of W S. & E. Hammond, that were purchased of Stephen Atwood. In 1847, Mr. Cook and Mr. Elitharp bought a few ewes of Mr. Stephen Atwood, and eleven ewe tegs of Chancy Atwood, and they were divided, Mr. Elitharp receiving one-half. On these ewes, Atwood (12) was used until he died in 1850. Rams of his get were used for a few years after; then Don Pedro (44); after the ram America (35), Comet (57) was also used to some extent. In 1864, Eureka (58) was used to 31 ewes, and in 1865 he was used to 10 more. After 1844, the Atwood rams were used principally to the old portion of the flock, until 1863. All that had this blood were sold to Mr. Elitharp's son, H. P. Elitharp. In 1873, Mr. Elitharp finally disposed of all his flock, selling the last ten that he had reserved to Otis P. Lee, of Middlebury, Vt. Mr. Elitharp was one of the very best judges of sheep of his time, a breeder of excellent judgment, and succeeded in making great improvements in his flock of sheep. Ewes of his breeding were in good demand and seldom disappointed the expectations of their purchasers.

FLOCK OF HON. WILLIAM JARVIS, WEATHERSFIELD, VT.

In our chapters upon the importation of Merinos from Spain, their introduction into Vermont, and improvemements of Merinos, we have so spoken of this flock that there seems to be little more to be given of it ; indeed, so much has been written of it by Mr. Jarvis and others that there is little that we can give that would be of interest, but it would be well to make a brief abstract or compilation from the various publications and republish them, to put in the history of this flock in a somewhat compact form, and incidentally introduce a few facts that may not be so generally known.

The flock was founded in 1811 by a selection from the importations of the Consul, made in 1810, and consisted of one-half Paulars, one-fourth Aguierries, one-eighth Escurials, and the other eighth Negrettis and Montarcos. They were gathered at Claremont, N. H., during the fall of 1810 and the early part of the winter following, and were brought across the Connecticut river and put upon the farm at Weathersfield. Vt., in April, 1811.

The several kinds or selections from the different Spanish Cabanas were bred separately until 1816, after which time the different bloods were not kept distinct, but bred together. May 4th, 1826, George and Thomas Searle, of Boston, Mass.. sold by auction at Brighton, 321 Saxon sheep and 58 lambs. Of these Mr. Jarvis purchased something more than 50, at prices from $32 50 to $137.50, the latter price being for a yearling ram. After this purchase Mr. Jarvis crossed the Saxon rams with the larger portion of the flock. But a hundred of the best of the Merino ewes were selected and bred to Merino rams only, thus preserving the best portion of the flock pure and unmixed with the Saxon. Later, Mr. Jarvis, finding the Saxon much less profitable and hardy than the Merino, selected out the Merino ewes that were left of those that had been crossed with the Saxon, and again bred them to Merino rams. Mr. Jarvis deeply regretted making this cross, and, so far as possible after, bred to Merino rams and their crosses. Mr. Jarvis made improvements in the quality of the wool of his Merinos.

The portrait of the ram Consul (1), which we give from a painting taken when he was owned by Mr. Sanford, and said to have been an excellent likeness, will show about the type of Merino bred by Mr. Jarvis twenty-five to thirty years after he imported them from Spain, and had bred the descendants of the different Cabanas together. It will also show wherein and how much we have improved upon the type of Merinos bred forty years ago.

It would be well, perhaps, to give the characteristics of the different Cabanas the descendants of which were bred together by Mr. Jarvis and constituted the blood of his flock of Merinos after 1817. Perhaps as good a description as any is that given by Mr. Jarvis himself in his letter to Mr. L. D. Gregory, of Weybridge, and before alluded to and quoted, although Dr. Randall thinks his description not as good as some others. It is taken from Mr. Morrill's American Shepherd, pages 73 and 74, and is as follows :

"I shall, in compliance with your wishes, give you a description of the different flocks sent to this country. The Paulars were undoubtedly one of the handsomest flocks in Spain. They were of middling height, round-bodied, well-spread, straight on the back, the neck of the bucks rising in a moderate curve from the withers to the setting on of the head ; their head handsome, with aquiline curve of the nose, with short, fine, glossy hair on the face, and generally hair on the legs, the skin pretty smooth, that is, not rolling up or doubling about the neck and body, as in some other flocks ; the crimp in the wool was not so short as in many other flocks, the wool was somewhat longer but it was close and compact, and was soft and silky to the touch, and the surface was not so much covered with gum. This flock was originally owned by the Carthusian friars of Paular, who were the best agriculturalists in Spain, and was sold by that order to the Prince of Peace when he came into power.

The Negretti flock were the tallest Merinos in Spain, but were not handsomely formed, being rather flat-sided, roach back, and the neck inclining to sink down from the withers ; the wool was somewhat shorter than the Paular and more crimped, the skin was more loose and inclined to double, and many of them were wooled on their faces and legs down to their hoofs. All the loose-skinned sheep had large dew-laps.*

The Aguierres were short-legged, round, broad-bodied, with loose skins, and were more wooled about their faces and legs than any other flock I ever saw. The wool was more crimped than the Paular, and less than the Negretti, but was thick and soft. This flock formerly belonged to the Moors of Spain, and at their expulsion was bought by the family of Aguierres. The wool in England was known as the Muros flock, and was highly esteemed. All the bucks of these three flocks had large horns.

The Escurials were about as tall as the Paulars, but not quite so round and broad, being in general rather more slight in their make. Their wool was crimped, but not quite so thick as the Paular or Negretti, nor were their skins so loose as the Negretti and Aguierres, nor had they so much wool on the face and legs.

The Moutarco have a considerable resemblance to the Escurials. The Escurial flock had formerly belonged to the crown, but when Philip II. built the Escurial palace, he gave them to the friars, whom he placed in a convent that was attached to the palace, as a source of revenue. Those four flocks were moderately gummed.

The Guadaloupe flock was rather larger in the bone than the two preceding, about the same height, but not quite so handsomely formed ; their wool was thick and crimped, their skins loose and doubling, their faces and legs not materially different from the two

*Mr. Jarvis in conversation with Prosper Elitharp stated that about one in ten of this flock were black.

latter flocks, but in general they were more gummed than either of the other flocks. In point of fineness there was very little difference between these six flocks, and as I have been told by well-informed persons, there is very little difference in this respect among the Leonesa Transhumantes in general. The Escurials, the Montarcos and the Guadaloupes were not in general so heavy horned as the other three flocks, and about one in six of the bucks were without horns—what is commonly called a polled buck."

As stated before, we believe Mr. Jarvis improved the quality of the fleeces from the original Spanish type, but we think the evidence is wanting to prove that he made any material improvement in other good qualities. As stated by Dr. Randall, we believe he was too willing to please the manufacturers, and bred out to too great extent the folds and oil from his flocks. It is possible that he may have increased the size of his sheep, and thus preserved on his pure Merinos the average weights of their fleeces, but if he increased their size he could not have preserved the relative per centum of wool to their live weight, for he only claimed in 1835 that the full-blood Merino part of the flock did not materially vary from the original weight. If he improved the form he hardly kept pace with Atwood, Cock and Rich, as the selections from his flock from 1835 to 1844 were not equal to those from the other flocks.

The practice of putting a ram "with twenty five to thirty-five ewes," instead of coupling each with a view to individual improvement or to remedy individual defects, as has been practised by the best breeders in later years, would probably account for the failure to reach the maximum of possible improvement. But if Consul Jarvis fell short of the highest success as a practical breeder and improver of Merino sheep attained by some other breeders of his and later times, none excelled him as noble public-spirited man, one entirely above the petty clap-trap that belittle the character of any breeder that practises it. His noble treatment of other importers in his later writings, some of whom had endeavored to defame the stock he imported, shows him to have been too noble a nature to have remembered aught of hatred and malice, and, although the greatest public service he performed for the United States was in importing such vast numbers of sheep from the best Cabanas in Spain, it was not the only evidence he gave of a noble public spirit. The greatest legacy he left Vermont was the memory of a noble, upright, intelligent, honest man.

It only remains for us to say that the flock of sheep he bred was soon scattered after his death, and that the farm at Weathersfield where he for so many years bred his Merinos with credit and profit, is not at this time pasturing or breeding a single pure-blooded Merino sheep.

FLOCK OF ANDREW COCK, FLUSHING, LONG ISLAND,

We have already published the account of the foundation of this flock in the letter of Judge Lawrence, which we publish on page 56, in the chapter on the "Introduction of Merino Sheep into Vermont."

From that letter we learn the first purchase was two Escurial ewes, imported by Richard Crowningshield, the next thirty were Paular sheep, and others were selected from different importations and added to the flock until they numbered about eighty.

The testimony of Judge Lawrence is, that they were bred pure and unmixed with any other than Spanish Merino blood.

In 1816 a few of these sheep were sold to Zebulon Frost and Hollett Thorn, and in 1823 what remained of this flock were sold to Leonard Beedle on accout of Jehial Beedle, Elijah Wright and the Hon. Charles Rich, of Shoreham. Vt., and the further history of the descendants of this flock will be found in the history of the Rich, Stickney, Robinson and other flocks.

THE FLOCK OF JACOB N. BLAKESLEE, WATERTOWN, CONNECTICUT.

On pages 46 and 47 we give the history of the foundation of Mr Blakeslee's flock. They probably sprung from the Infantado, Guadaloupe and Negretti Cabanas. Mr. Blakeslee, like Mr. Atwood, Consul Jarvis and others, introduced Saxon blood into his flock, and like Mr. Atwood, not liking the cross, he soon after sold off all of this blood.

Mr. Blakeslee died in 1877, and we are not informed what became of the part of the flock left at his death. The most marked excellence of this flock was in the fineness and style of wool and fluidity and fineness of the oil, rather than in their constitutions or heavy fleeces. In the last two qualities they were below the average.

FLOCK OF DAVID BUFFUM AND WILLIAM I. BAILEY, NEWPORT, RHODE ISLAND.

This flock was founded in 1810, from an importation by Capt. Paul Cuffe and advertised in the Providence Gazette of Sept. 8th, 1810 :

"To be sold at public auction on the 21st of 9th month inst, at 10 o'clock, A. M., at the house of David Buffum, in Newport, seventy-four Merino rams and ewes, warranted of the pure Merino breed, shipped by William Jarvis, Esq., American Consul at Lisbon.

PAUL CUFFE.
ISAAC CORY.

Newport, 9th Mo. 7, 1810."

There seems to be good reason to believe that David Buffum purchased the foundation of this flock from this importation, and that the flock of George Irish and William Bailey were descended from them.

There are many circumstances and much proof to substantiate this, probably quite as good as any that can be found at this day in regard to the purity of any flock or the line of blood of any family of Merino sheep.

We have statements in regard to the matter signed by David D. Buffum and Thos. B. Buffum, grandsons of David Buffum, Sen., and sons of David Buffum, Jr. The following certificate was given us at Newport, R. I., April 13, 1877 :

"It is my impression that the Merino sheep sold by Joseph I. Bailey or other farmers in this vicinity, were descended from sheep imported from Spain by Paul Cuffe, and sold to my grandfather, David Buffum, previous to 1812. My father was married in 1812, and I have heard him say he had some clothes made from the wool to be married in.

(Signed),   THOS. B. BUFFUM."

Below this certificate on the same paper, we also have the statement of David D. Buffum (an elder brother of Thos. B. Buffum), that his recollections were more vivid than those of his brother, that the facts were as stated in his brother's certificate. We also have a letter from Mr. Thomas R. Hazard, a Quaker gentleman of 80 years of age or over, that was written to us last January in reply to one from us requesting information in regard to the early flocks of Merino sheep in Rhode Island. In that letter Mr. Hazard says :

"I know that David Buffum, Esq., father of David Buffum Jr. (deceased), and grandfather of the present Thomas Buffum, was one of the first (if not the very first) farmer in Rhode Island who owned Merino sheep."

Mr. Hazard in the same letter names other parties that had full-blood Merino sheep in Newport and South Kingston as early as during the war of 1812, and also mentions some 7-8th-blooded Merino sheep owned by Joseph Congdon that were brought from Shelter Island thus proving that the flocks of some were pure-bred and those of others were high grades.

Another portion of Mr. Hazard's letter, though not relating to these sheep, is interesting as confirming the description of some of the flocks of Spain by Consul Jarvis. It is as follows :

"George P. Hazard, of Boston Neck, also had Merino sheep at about the same time, which he used to show me and explain their various qualities, under the heads of Polar (or Paular), Nigretti and Escurial breeds, the latter, as he said, being from the royal flock in Spain. The Polar breed, I remember, was stout in form, with a good length of wool, but not as fine as the Escurial."

We have also the statement of Mrs. Mary Irish, now nearly eighty years old, a daughter of Mr. William Bailey, who says "she remembers well the excitement over the importation of the Merino breed of sheep and the purchase of them by her father and others on the Island," but she does not remember from whom they were bought. "Her father's, Mr. Buffum's and Mr. George Irish's farms joined each other."

Mr. Thos. Buffum stated to us that the sheep of Mr. William Bailey were descended from the flock of his grandfather, David Buffum. As to the flock of Mr. Bailey, we have the sworn statement before William P. Sheffield, notary public of Newport, R. I., as follows:

"I, Joseph I. Bailey, of the city and county of Newport, in the State of Rhode Island, on oath depose and say that I am seventy years of age and upwards; that the name of my father was William Bailey, who resided in Middletown, adjoining to Newport, and that his occupation was that of a farmer, and upon his decease I succeeded him in the business and in the title of a part of his homestead farm; that when in life my father owned a flock of sheep that was imported into Newport from Spain; that I owned the descendants of these sheep; that many years since I sold some of these sheep to go to the State of Vermont. I do not remember the exact time when these sheep were sold by me, or the names of the persons to whom I sold them. In all I sold several hundred to go to Vermont. I now remember that I sold sheep to A. L. Bingham, as well as to other persons.

(Signed).   J. I. BAILEY.
Newport, Nov. 18, 1878.

Then personally appeared before me the before-named Joseph I. Bailey and made oath to the truth of the before-written statement by him subscribed. I further certify that the said Joseph I. Bailey has been personally known to me for many years, and that I know that he is the person described in and who subscribed the said affidavit.

In witness whereof I have hereto set my hand and official seal, the day and year above written.

{ SEAL. }

WILLIAM P. SHEFFIELD,
Notary Public."

In a letter written after this depositon was made, Mr. Bailey says, in thinking of it more he remembered the "name of Hammond, and a man that belonged to the Quaker or Friend persuasion." In the same letter he says:

"When I gave up farming, I think I sold Mr. A. L. Bingham my entire flock, something over one hundred and fifty ewes; in the month of March they were heavy with lamb."

In our investigations we have found that Mr. Hammond, Mr. A. L. Bingham and two Quaker gentlemen, did buy sheep of Mr. Bailey, thus confirming his statement. Mr. Bundy, who bought with Messrs. Murray & Munger in 1835, is believed to have been a Quaker, and Mr. R. T. Robinson, whose narrative we give at page 53, also was a member of that persuasion. We think the evidence we have adduced is sufficient to prove that these sheep stand on a par with the best, as to the purity of the blood from which they were derived, and that no better evidence exists to sustain the purity of any other flock, or line of blood. As is the case in the Humphreys sheep, we can find no evidence to show from what Cabana in Spain these came   It only remains to complete the history of these flocks, to say that Mr. Thomas Buffum now and for a number of years has been a breeder of Southdown sheep, and that the flock of Mr. Bailey, as stated above, was closed out to Mr. A. L. Bingham. Mr. Bailey died soon after making the affidavit accompaning this history.

FLOCK OF N. A. SAXTON, WALTHAM, VT.

The foundation of this flock was commenced soon after W. S. & E. Hammond commenced breeding Atwood sheep by a purchase of a few pure Atwood ewes of them. These ewes were usually served by the stock rams of Messrs. Hammond, with an occasional purchase of a stock ram from the same flock. America (35) was one of the most celebrated of these. Rams bred within the flock were used to a great extent. Those of Mr. Saxton's neighbors bred from the flocks of Messrs. Hammond and Sanford were used in the later history of the flock, or in 1870–71 and '72, the last year using the stock rams of O. C. Bacon. After Mr. Saxton died the flock was sold out to different individuals, the last being sold, according to a memorandum kept at his place, Nov. 25th, 1874.

FLOCK OF E. R. ROBINSON, SHOREHAM, VT.

Mr. Robinson commenced his flock in 1836 by a purchase of pure Cock sheep to the number of about a hundred of Charles Rich of Shoreham, it being the larger portion of the flock he had left from his portion of the division of his father's (Hon. Charles Rich's) flock, which took place in 1828, or four years after the death of Hon. Charles Rich.

Mr. Robinson continued to breed these sheep in the same line or Cock family until 1845, when he introduced a strain of Atwood blood by using the ram Atwood (12) to twenty ewes and purchasing the ram Elitharp (13) of his breeder, Prosper Elitharp. This ram was used as the stock ram of the flock for a few years, or until the Old Robinson Ram (38) was bred within the flock, although Atwood (12) was used to a considerable extent in 1846. An addition of thirty ewes was made to the flock in 1848 by a purchase of Mr. Prosper Elitharp. These ewes were mainly if not all bred by Mr. E. and combined the blood of the Cock, Jarvis and Atwood flocks.

Soon after the introduction of the French Merinos into Vermont, Mr. Robinson procured the use of a ram of this blood, which he used to a third of his ewes one year. These ewes were selected so as to represent a fair average of the flock. The result of this cross was a large increase of the size of carcass, but a decided decrease in the average weight of fleece, thus very largely decreasing the average of wool to live weight, and very materially increasing the cost of the wool, while the quality was not so good as that grown on the other portion of the flock. This cross, proving so unsatisfactory, was soon weeded out and sold from the flock.

Erastus R. Robinson died in 1854, and in 1856 the flock was divided ; a portion falling to the share of his widow, Mrs. Sallie D.

Robinson, and a portion to his son, Darwin E. Robinson. As will appear in the histories and pedigrees of several of the flocks hereafter recorded, both these portions of the flock were leased upon shares to other parties, and made the foundations of several of those flocks. E. R. Robinson was an excellent judge of sheep, a very judicious breeder, and made great improvements in the flock which he established, and bred with so much credit and profit. Sheep of his breeding were in demand, and no better certificate could be given to prove a sheep meritorious and pure-blooded, than one that certified it was a pure-bred Robinson sheep.

There are a few flocks more that have ceased to exist as distinctive flocks, through which the pedigrees of some that are registered trace, but the facts in regard to them can be stated so briefly, that they will be given in the histories of those flocks.

FLOCK REGISTER OF IMPROVED SPANISH MERINO SHEEP.

The numbers in brackets refer to our list of stock rams.

FLOCK 1.

Owned by MELVIN R. ATWOOD, Richville, Addison Co., Vt.

18 RAMS.

Marked with metallic labels in ears, having thereon M. R. Atwood and numbers 1 to 9 and 38 to 46 and 50.

28 EWES.

Marked same as rams, with numbers 10 to 31 and 32 to 37.

PEDIGREE.

Descended from importations from Spain through the flocks of Andrew Cock, Long Island; David Humphreys, Stephen Atwood, Conn.; William Jarvis, Weathersfield; Hon. Charles Rich, his sons, Charles and J. Thurman Rich and grandsons, J.T.& V. Rich, Leonard Beedle, E. R. Robinson, T. Stickney & Son, L. C. Remele, J. T. Stickney, G. H. Hall, Shoreham; John Q. Stickney, Whiting; Lyman Webster, Sudbury; Seneca Root, Hubbardton; R.P.Hall, John Towle, R. J. Jones, Cornwall; Baker & Smith, Prosper Elitharp, Bridport; C. B. Cook, Charlotte; Messrs. Hammond, Middlebury; Victor Wright, Weybridge; W. R. Sanford, H. G. Hibbard, Orwell; Ward M. Lincoln, Brandon; E. Porter, Rutland.

EWES PURCHASED.

Flock commenced in 1864 by a purchase of six ewes of L. C. Remele. In 1870 eight ewes were purchased from the same flock. These ewes combined the blood of the Atwood, Jarvis and Rich flocks.

RAMS PURCHASED OR USED.

Those bred by L. C. Remele and within the flock most of the time. In 1873 J. Q. Stickney's Young Fremont (215) to one ewe. Same year, ram owned by German Cutting, bred by G. H. Hall, to one ewe. In 1874, Hibbard (214) to two ewes. In 1875, J. T. Stickney's 146 (441) to a few ewes. In 1876, six ram tegs were purchased of L. C. Remele. They were sired by J. T. Stickney's 146 (441). In 1876 and 1877 M. R. Atwood's 7 (537), and 50 (538) were used.

―――o―――

FLOCK 2.

Owned by R N. & O. F. ATWOOD, RICHVILLE, ADDISON Co., VT.

14 RAMS.

Marked with metallic labels in ears, having thereon R. N. & O. F. A., and numbers 46 to 53.

57 EWES.

Marked same as rams, with numbers from 1 to 39 and 54 to 71.

PEDIGREE.

Descended from importations from Spain, through the flocks of David Buffum, William Bailey, Joseph I. Bailey, Rhode Island; David Humphreys, Stephen Atwood, Conn.; Andrew Cock, Long Island; William Jarvis, Weathersfield; S. T. Baker, A. F. Ellsworth, Whiting; Messrs. Hammond, Middlebury; Messrs. Rich, Leonard Beedle, L. C. Remele, E. R. Robinson, T. Stickney & Sons, J. M. Ormsbee, D. & G. Cutting, Shoreham; Ward M. Lincoln, Brandon; Baker & Smith, Prosper Elitharp, Bridport; C. B. Cook, Charlotte.

HISTORY,

This flock was founded in 1847 or '48 by R. N. Atwood, and another branch in 1865, by O. F. Atwood. In the first case, R. N. Atwood purchased of D. & G. Cutting a few ewes of Rhode Island blood. The rams of D. & G. Cutting, J. T. & V. Rich and T. Stickney were used, and in 1861 a ram was purchased of C. K. Williams that was sired by D. Cutting's Wooster ram; dam, a Cutting ewe. This ram was used two years. In 1865 a few ewes were taken to Ellsworth's Eureka (327). After this the stock rams of L. C. Remele and M. R. Atwood were used to some extent. In 1872 this branch of the flock was consolidated with the other, which was commenced in February, 1865, by O. F. Atwood, who purchased of Ward M. Lincoln sixteen ewe tegs of Atwood blood. Six of these were reserved as breeding stock, the remainder were sold. On these ewes were used the stock rams of John Towle, of Atwood blood, and one bred by L. C. Remele, of Atwood blood. In the summer of 1869, O. F. Atwood sold his sheep to his brother, H. C. Atwood, who bred them three years, using a ram received with the ewes and Fremont (126). In the spring of 1872 R. N. Atwood purchased these sheep of H. C. Atwood, thus consolidating the two flocks. In 1873 a half interest in the flock was sold by R N. Atwood to O. F. Atwood, and they are now joint owners. Since 1873 rams from the flocks of L. C. Remele and M. R. Atwood have been used, and two bred from them, and also two of Joseph T. Stickney's have been used.

——o——

FLOCK 3.

Owned by P. C. ABBEY, Essex, Chittenden Co., Vt.

6 RAMS,

Marked by metallic labels in ears, having thereon P. C. Abbey and Numbers 41 to 46.

40 EWES.

Marked same as rams, with numbers 1 to 40.

PEDIGREE.

Descended from the Humphreys importation from Spain, through the flocks of Stephen Atwood, Conn ; Messrs Hammond, Middlebury ; Victor Wright, Weybridge ; Henry Thorp, Charlotte ; Chapman & Henry Giddings, Fairfax.

EWES PURCHASED.

Flock commenced Oct. 2d, 1865, by a purchase of five Atwood ewes of Chapman & Henry Giddings. In 1865 one Atwood ewe of Dan Giddings, that was bred by Victor Wright. Nov. 21, 1871, sixteen Atwood ewes were purchased of Henry Giddings.

RAMS PURCHASED OR USED.

H. W. Hammond's Kearsarge (55) and another of his stock rams, Long Wool (112), Dan Gidding's Young Kearsarge, bred by H. W. Hammond, Henry Thorp's Green Mountain Ram, Vanderbilt (261), purchased of Henry Thorp, Oct. 1870, and Thorp, bred by Henry Thorp and purchased of him Oct., 1874. Also California, bred within the flock, sired by Vanderbilt, and ram sired by Thorp. There are now in the flock three ewes of the purchase of Henry Giddings, Nov., 1871, one sired by Kearsarge (55), two ewes by Long Wool, two ewes by Young Kearsarge, two ewes by H. W. Hammond's stock ram, named above, nine ewes by Vanderbilt, one ewe by Henry Thorp's Green Mountain Ram, one ewe and one ram by California, eighteen ewes and five rams by Thorp. and one ewe by ram by the latter.

FLOCK 4.

Owned by CHARLES E. ABELL, ORWELL, ADDISON Co., VT.

50 EWES,

Marked by metallic labels in ears, having thereon C. E. Abell, and numbered 1 to 50.

PEDIGREE.

Descended from importations from Spain through the flocks of David Humphreys, Stephen Atwood, Connecticut; Andrew Cock, Long Island; William Jarvis, Weathersfield; Messrs. Hammond, Middlebury; Leonard Beedle, Messrs. Rich. L. C. Remele, E. R. Robinson, T. Stickney & Son, Shoreham; Baker & Smith, Prosper Elitharp, Bridport; H. S. Root, Benson; W. R. Sanford, J. H. Thomas, H. G. Hibbard, B. L. & J. W. Buell, R. D. Hall, Orwell; Ward M. Lincoln, Brandon; Ebenezer Porter, Rutland.

EWES PURCHASED.

Flock commenced Jan. 1, 1864, by a purchase of twenty-seven ewe tegs of R. D. Hall, of Atwood and Rich blood. They were bred from stock procured of J. H. Thomas, who bred them from stock procured from W. R. Sanford and Messrs. Rich. These ewe tegs were sired by the stock rams of W. R. Sanford.

RAMS PURCHASED OR USED.

Comet (57), Kilpatrick (71); a ram owned by O. S. Branch, bred by W. R. Sanford. A ram was purchased of D. B. Buell, and used, he was sired by Sanford (135); dam, bred by J. H. Thomas. The next was an Atwood ram purchased of H. G. Hibbard, sired by Gold Drop (64); next a ram purchased of Messrs. Catlin & Hack, of Orwell, that was bred by W. R. Sanford. A ram was next purchased of Henry Root, of Benson, that was sired by H. Root's stock ram by Fremont (126); dam, an Atwood and Robinson ewe. Since that time, rams bred by Messrs. Buell, from W. R. Sanford stock have been used.

———o———

FLOCK 5.

Owned by C. NEWELL ALWARD, PATASKALA, LICKING Co., OHIO.

1 RAM.

Marked by metallic labels in ears, having thereon C. N. Alward, and No. 213.

22

14 EWES.

Three marked by metallic labels in ears, having thereon C. Miner and Nos. 1, 2, 3. Also 11 marked on labels, J. C. Alward, and Nos. 2, 5, 9, 15, 39, 45, 46, 72, 79, 90, 96. Also 5 marked on labels C. N. Alward, and Nos. 203, 204, 206, 209, 210.

PEDIGREE.

Descended from importations from Spain through the flocks of David Humphreys, Stephen Atwood, Connecticut; David Buffum, Joseph l. Bailey, Rhode Island; Andrew Cock, Long Island;, William Jarvis, Weathersfield; Messrs. Rich, E. R. Robinson, Leonard Beedle, L. C. Remele, E. A. Birchard, J. M. Ormsbee, D. & G. Cutting, Shoreham ; George Campbell, Westminister West; W. R. Sanford, Orwell ;, Edgar Sanford, S. S. Rockwell, Cornwall ; Baker & Smith, P. Elitharp, Barney Myrick, F. D. Doty, C. N. Haywood, J. J. Crane, H. C. Burwell, C. P. Crane, C. C. Miner, Bridport; Messrs Hammond, Middlebury ; S. T. Baker, Whiting.

EWES PURCHASED.

Flock commenced Oct. 12, 1866, by a purchase of three ewes of Atwood and Jarvis blood of S. S. Mathews, that were bred by George Campbell, of Westminister West, Vermont. In 1877, three Atwood and Robinson ewes were purchased of J. A. Wright, that were bred by C. C. Miner, Bridport, Vt.

RAMS USED.

On the Campbell ewes a ram that was sired by E. Keller's Nub. His dam was bred by E. Hammond. In 1875, a ram called Addison, bred in Vermont, purchased of Weiant & Higbee, of Licking Co., Ohio. This ram was sired by Elitharp & Burwell (79) ; his dam was bred by P. Elitharp.

———o———

FLOCK 6.

Owned by E. N. BISSELL, East Shoreham, Addison Co., Vt.

127 RAMS.

One hundred and eight marked with metal labels in ears, having thereon E. N. Bissell, and Nos. 100 to 143, 180 to 217, and 238 to 258. Nineteen marked with metallic labels in ears having thereon J. T. & V. R., and Nos. 313, 314, 316, 318, 319, 324, 329, 330, 332, 334, 336, 340 to 342, 349, 351 to 354.

223 EWES.

One hundred and forty-six marked with labels same as rams, having thereon E. N. Bissell, and Nos. 1 to 99 (minus 16, 17, 18, 21), 149 to 179, and 218 to 237. Seventy-seven ewes are marked with labels having thereon J. T. & V. R., and Nos. 3, 11, 12, 18, 22, 30, 40, 45 to 47, 53, 56, 62, 66, 67, 72, 76, 80 to 83, 85 to 93, 95, 97, 98, 100, 101, 103, 106, 108, 109, 110, 111, 112, 113, 114, 116, 117, 141, 145 to 150, 152, 153, 155.

PEDIGREE.

Descended from importations from Spain through the flocks of David Humphreys, Stephen Atwood, Jacob N. Blakeslee, John Nettleton, Connecticut; Andrew Cock, Long Island; William Jarvis, Weathersfield; Hon. Charles Rich, his sons, Charles and J. Thurman Rich, and grandsons, J. T. & V. Rich, Jasper Barnum; T. Stickney & Son, L. C. Remele, E. R. Robinson, Lucius Robinson, George H. Hall, S. L. Bissell, Shoreham; W. R. Sanford. H. G. Hibbard, Orwell; Ward M. Lincoln, Jesse Hines, Brandon; E. Porter, Rutland; Messrs. Hammond, Middlebury; N. A Saxton. Waltham; R. P. Hall, E. S. Stowell, John Towle, R. J. Jones, Cornwall; Baker & Smith, Prosper Elitharp, Bridport; C. B. Cook, Charlotte; A H. Hubbard, Whiting.

EWES PURCHASED.

Flock commenced August 18, 1848. Six ewes (Atwood, Rich and Jarvis blood) of L. C. Remele. Nov. 9, 1848, twenty ewes were purchased of W. R. Sanford. Their blood was Blakeslee; they were bred by John Nettleton. Nov 25, 1848, twenty Robinson ewes were purchased of Jasper Barnum; in Oct., 1863, E. N. Bissell, five ewes of Stephen Atwood; three ewes of Chancy Atwood; nine ewes of George Atwood, and six ewes of Jerry Smith, all bred from the flock of Stephen Atwood. The same year three Atwood ewes were purchased of E. Hammond. In 1866, E. N. Bissell purchased of S. L. Bissell his flock, and the two were consolidated. In 1868, ten ewes were purchased of A. H. Hubbard (their blood was Atwood, Jarvis and Rich); April, 1875, thirty-five Robinson ewes were purchased of Lucius Robinson. Same year, ten breeding ewes and fifteen ewe tegs were purchased of J. T. & V. Rich; Nov. 22, 1876, nineteen yearling ewes were purchased of J. T. & V. Rich; December, 1876, fourteen ewe tegs were purchased of J. T. & V. Rich; Dec. 15, 1877, ten ewe tegs and thirteen breeding ewes were purchased of V. Rich.

RAMS PURCHASED OR USED.

In 1848, Sanford (62) was purchased by S. L. Bissell. This ram was used for five years, except in 1853 five ewes were served

by Old Robinson (38) ; in 1854–55 and 56, an Atwood ram was used
that was purchased in company with G. Cutting of Jesse Hinds ; in
1857–58 and 59, York State (145) was used ; in 1861, Monitor (142)
was purchased of E. Hammond by S. L. Bissell, and he was used to
a large portion of the flock until 1865 ; in 1863, a ram teg was purchased of S. Atwood and another of George Atwood by E. N. Bissell ; in 1863, Sweepstakes (32), Silver Mine (61), and Gold Drop
(64) were used ; in 1864, Sweepstakes (32), Kearsarge (55), Gold
Drop (64), Paymaster (547) and Abe Lincoln (550) were used ; also
Dew Drop (59), Golden Fleece (91) and Towle ram (164). In 1865
Golden Fleece (91), Dew Drop (59) and Dean's Little Wrinkly (48),
and Hubbard (165), and No. 1 (146.) In 1866, the last named ram
and 52 (147) ; in 1866, 67, 68, 69 and 70, 52 (147), and Addison
Chief (148) were used ; in 1871, Gen. Fremont (126) and Stickney
ram (282) were used. This last named ram was used in 1872, '73,
'74 and '75, except eight ewes to a ram bred in the flock, sired by
(282) and a part of the ewes in the last year to Hibbard (214) ; in
1876, this ram was purchased, also two Atwood rams of George H.
Hall. Most of the ewes were served in that year by Hibbard (214) ;
a few by Stoga (212) and a few by Banker (471) ; Aug. 30, 1877,
thirteen one and two year-old rams were purchased of V. Rich. In
1877, Banker (471), J. T. & V. R.'s 331 (472) and J. T. & V. R.'s 350,
(473) were used.

------o------

FLOCK 7.

Owned by H. C. BURWELL, Bridport, Vt.

27 RAMS.

Marked with metallic labels in ears, having thereon H. C.
Burwell, and numbers 64 to 81 and 83 to 91.

97 EWES.

Marked same as rams, and numbers 1 to 63 minus 51, and 101
to 128, minus 115, and 129 to 136.

PEDIGREE.

Descended from importations from Spain, through the flocks of
David Humphreys, Stephen Atwood, Conn.; Andrew Cock, Long
Island ; William Jarvis, Weathersfield ; Messrs. Hammond, Middle-
bury ; Messrs. Rich, L. C. Remele, Leonard Beedle, E. D Bush, A.
C. Harris, Alvin Clark, E. R. Robinson, E. A. Burchard, Shoreham ;
Victor Wright, Weybridge ; Baker & Smith, P. Elitharp, C. N.
Hayward, J. J. Crane, D. E. & James Hill, Barney Myrick, Ezra
Fitch, L. S Burwell, James Howe, Bridport ; C. B. Cook, Charlotte ;

Nelson Richards, Panton ; C. W. Hinds, G. W. Whitford, C. P. Morrison, Addison ; W. R. Sanford, Orwell ; E. S. Stowell, Cornwall

EWES PURCHASED.

Flock commenced Nov. 28, 1862, by a purchase of one Atwood ewe, first choice of James Howe's flock, sire, one of V. Wright's stock rams ; dam bred by E. S. Stowell. In March, 1863, two ewe tegs were purchased of C W. Hindes ; they were Atwood and Rich blood ; they were sired by America (85) ; their dams were bred by C. N. Hayward. August 20, 1864, two ewe lambs were purchased of James Hill ; they were Atwood and Rich blood ; they were sired by the Fitch Ram (92). March, 1869, nine one and two-year-old Atwood ewes were purchased of P. Elitharp ; they were the first choice from his flock of that age. Nov. 2d, 1872. forty breeding ewes, of Robinson and Atwood blood, were purchased of A. C. Harris.

Fall of 1878, twelve Atwood ewes were purchased of D. Edgar Hill ; they were the first choice of his flock, and were bred direct from the flock of E. Hammond & Son ; eight of them were sired by Little Wrinkly (48), four of them by a ram bred by Mr. Hill, sired by Little Wrinkly ; his dam, a Hammond ewe.

July 15, 1877, eight ewe lambs and seven yearling ewes were purchased of L. S. Burwell ; they were sired by Bismarck (221) and Stub (222) ; their dams were Robinson blood, bred by G. W. Whitford.

RAMS PURCHASED OR USED.

Silver Ring (219) was purchased and used three years ; in 1871 Elitharp & Burwell (79), and used two years. Black Hawk (220) was also purchased and used to a few ewes. Besides these the following stock rams have been used : Sea Lion (83), Ironsides (80), Dew Drop (59), Golden Fleece (91), Bonaparte (176). The first was used three years, Ironsides two years, Dew Drop and Golden Fleece one year and Bonaparte two years. Of rams raised within the flock, the following have been used : Silver Horns (177), two years ; Bismarck (221), Stub (222) and Eureka, 3d (223), for the four years past ; Custer (521) was used to a few ewes in 1877.

---o---

FLOCK 8.

Owned by OSCAR C. BACON, Vergennes, Addison Co., Vt.

1 RAM.

Marked in ear with metallic label having thereon O. C. Bacon, and No. 44.

50 EWES.

Marked same as ram and Nos. 3, 4, 6, 8, 10, 14, 17, 19 to 22, 24, 25, 27 to 29, 33, 34, 36, 38, 41, 42, 45 to 49, 52, 57, 59, 65, 66, 69, 71, 81, 83 to 92, 94, 95, 98, 102, 105, 106, 118.

PEDIGREE.

Descended from importations from Spain through the flocks of David Humphreys, Stephen Atwood, Connecticut; Andrew Cock, Long Island; William Jarvis, Weathersfield; R. P. Hall, R. J. Jones, Cornwall; Messrs. Hammond, Middlebury; W. R. Sanford, Orwell; F. D. Barton, N. A. Saxton, Waltham; Messrs. Rich, Leonard Beedle, Alvin Clark, E. R. Robinson, L. C. Remele, T. Stickney & Son, Shoreham; Baker & Smith, Prosper Elitharp, Bridport; W. R. Sanford, Orwell; C. B. Cook, Charlotte; J. Marsh, Hinesburgh; A. A. Farnsworth, Brookeville.

EWES PURCHASED.

Flock commenced in 1854 by a purchase of Atwood ewes of R. P. Hall; next a purchase of ten, and again of twenty-three Atwood ewes of N A. Saxton. In 1869 five ewes were purchased from the same flock. In 1876, eight Atwood ewes were purchased of F. D. Barton.

RAMS PURCHASED OR USED.

In 1853, a ram teg was purchased of N. A. Saxton, that was used on the ewes purchased of R. P. Hall in 1854. Old Greasy (18) and Longwool (21), were used to a few ewes; in 1863, thirteen ewes were served by Eureka (58); in 1864, eleven ewes were served by Prince (69); in 1866, two yearling rams were purchased of N. A. Saxton, and in 1868 two ram tegs were purchased of the same. All the blood introduced from the Saxton flock was Atwood. Sea Lion (83) was used two seasons to ten ewes each season. Bacon's Farnsworth (156) was purchased and used; Cross Tom (36) was used to some extent. In 1870, two Atwood rams were purchased of F. D. Barton and another in 1874; one of these was Little Greasy (543); in 1874, Bacon's Stickney ram (456) was purchased and used; in 1876, two Atwood ram tegs, Oscar (544) and Jason (545) were purchased, and have been used to a portion of the ewes since. In 1876, a part of the ewes were served by Allright (169). Of the Nos. recorded in this flock 3, 4, 6, 8, 10, 14, 17, 19, 24, 25, 29, 34, 36, 38, 41, 44, 45, 47, 49, 57, 59, 66, 69, 84, 86, 87, 94, are pure Atwoods, while the remaining Nos. 20, 21, 22, 27, 28, 33, 42, 46, 48, 52, 65, 71, 81, 83, 85, 90, 91, 92, 95, 98, 102, 105, 106, 118, have a cross of Robinson and Stickney blood.

FLOCK 9.

Owned by H. MERLE BOTTUM, SHAFTSBURY, BENNINGTON CO., VT.

33 RAMS.

Marked by metallic labels in ears, having thereon H. M. B., and Nos. 88 to 120.

93 EWES.

Marked same as rams and Nos. 1 to 4, 7 to 10, 12 to 14, 16, 19 to 22, 24 to 40, 42 to 56, 58, 59, 61 to 66, 73 to 87, 129 to 150.

PEDIGREE.

Descended from importations from Spain through the flocks of David Humphreys, Stephen Atwood, Connecticut: Andrew Cock, Long Island; William Jarvis, Weathersfield; Messrs. Hammond, Middlebury; Charles I. Benedict, Arlington; N. & N. Bottum, Shaftsbury; Leonard Beedle, Messrs. Rich, L. C. Remele, E. R. Robinson, Shoreham; Baker & Smith, P. Elithorp, Bridport; C. B. Cook, Charlotte; J. H. Mead, Rutland.

HISTORY.

The present owner came into the possession of the flock in 1873, by a division of a joint interest in a flock bred up by N. & N. Bottum, from stock of 22 year-old-ewes, and 18 ewe tegs purchased (by N. & N. B.) of Charles I. Benedict. The yearling ewes were sired by Messrs. Hammond's stock rams of 1852, and the ewe tegs were sired by a ram bred by E. Hammond. All were out of dams sold C. I B., in 1852 by Messrs. Hammond; in 1856, N. & N. B. purchased of W S & E. Hammond twenty yearling and one two-year-old ewes; in 1859, they purchased of E. Hammond five two-year-old old ewes in lamb; in 1863, they purchased of the same a yearling ewe in lamb by Gold Drop (64); from this ewe was raised their Prince of Gold Drops (153), for three years their stock ram. Bottums' Lawrence (138), Benedict & Bottum (151), Benedict (152), Bottums' Green Mountain (227), and other rams bred within the flock were used. Longwool (21,) Little Wrinkly (22), Lawrence ram (24), Kearsarge (55), Gold Drop (64) and Green Mountain (70) were used, and none other than of Hammond blood, until the division of the flock, in 1873, three years after the death of Norman Bottum, father of the present owner of the branch of the flock herein recorded. At the division of the flock in the Spring of 1873, H. Merle Bottum selected fifty, as his share of the flock of one hundred and thirty then on hand.

RAMS USED BY H. M. B.

In 1875, Pony (394); in 1876, a ram bred within the flock called Beau was used; in 1877, this ram and one sired by Pony (394).

FLOCK 10.

Owned by ALONZO BALDWIN & SON, Whiting, Addison Co., Vt.

16 RAMS.

Eleven marked with metallic labels in ears, having thereon A. Baldwin, and Nos. 27 to 37, and five marked on labels E. A. Baldwin, and Nos. 31 to 35.

76 EWES.

Thirty-one marked on labels A. Baldwin, and Nos. 1 to 26 and 38 to 41, also forty-five marked on labels E. A. Baldwin, and Nos. 1 to 30, and 36 to 50.

PEDIGREE.

Descended from importations from Spain through the flocks of A. Cock, Long Island; David Humphreys, Stephen Atwood, Connecticut; William Jarvis, Weathersfield; Leonard Beedle, Messrs. Rich, E. R. Robinson, L. C. Remele, Lucius Robinson, T. Stickney & Son, Shoreham; Baker & Smith, Prosper Elitharp, Bridport; O. B. Cook, Charlotte; E. S. Stowell, Cornwall; Lyman Webster, Sudbury; Seneca Root, Hubbarton; J. Q. Stickney, Whiting.

EWES PURCHASED.

Flock commenced in 1860 by a purchase of twenty-five ewes of Lucius Robinson that were bred pure from the flock of E. R. Robinson.

RAMS PURCHASED OR USED.

Eureka (58). In 1865, Hindes (465) was purchased, and used two years; in 1875, a ram was purchased of T. Stickney & Son. After that, one bred and owned by J. Q. Stickney, was used. Except these, rams bred within the flock have been used.

——o——

FLOCK 11.

Owned by DENNISON BLACKMER, Brandon, Rutland Co., Vt.

5 RAMS.

Marked with metallic labels in ears, having thereon D. Blackmer, and Nos. 45 to 49.

44 EWES.

Marked same as rams and Nos. 1 to 44.

PEDIGREE.

Descended from importations from Spain through the flocks of Andrew Cock, Long Island; David Humphreys, Stephen Atwood, J. N. Blakeslee, John Nettleton, Connecticut; William Jarvis, Weathersfield; W. R. Sanford, Orwell; Messrs. Hammond, Middlebury; Messrs. Rich, Leonard Beedle, L. C. Remele, E. R. Robinson, E. A. Birchard, T. Stickney & Son, Jasper Barnum, George D. Bryant, E. G. Farnum, D. E. Robinson, S. L. Bissell, E. N. Bissell, Geo. H. Hall, James Forbes, Jr., Shoreham; Baker & Smith, P. Elitharp, Bridport; C. B. Cook, Charlotte; R. P. Hall, E. S. Stowell, Edgar Sanford, R. J. Jones, Cornwall; Julius G. Barker, Leicester; Lucius Merriam, Fred. H. Farrington, Brandon.

EWES PURCHASED.

Flock commenced in 1871, by a purchase of eleven ewes of Lucias Merriam; they were of Robinson blood, bred by Mr. M. from ewes purchased of E. A. Birchard, and rams bred by E. A. Birchard and T. Stickney & Son. Nov. 20, 1876, twenty ewes were purchased of Fred. H. Farrington that were purchased of Lucius Merriam, the same day.

RAMS PURCHASED AND USED.

In 1871, a ram purchased of Julius G. Barker, bred by Mr. B. In 1873, a ram from the same flock was purchased; in 1875, a ram bred by Fred H. Farrington was purchased and used to a small extent; in 1876, Snow Flake (277) was purchased in company with F. H. Farrington, and used also in 1877. Ewes numbered 39, 40, and 44, and rams 46, 47 and 48, were sired by him. In 1876, Cassius (257) was used to two ewes; he sired ewes 37 and 41. Sanford's Bonaparte (447) was used to three ewes. He sired ewes 42 and 43, and ram 49. Hibbard (214) was used to three ewes he sired; ewes 35 and 38, and ram 45. Ram purchased of F. H. Farrington sired ewe 36.

---o---

FLOCK 12.

Owned by M. K. BARBOUR, Bridport, Vt.

15 RAMS.

Marked with metallic labels in ears, having thereon M. K. Barbour and numbers 44 to 55, and 65 to 67.

53 EWES.

Marked same manner as rams, with numbers 1 to 43, 56 to 64 and 68.

23

PEDIGREE.

Descended from importations from Spain through the flocks of David Humphreys, Stephen Atwood, Conn.; Andrew Cock, Long Island; William Jarvis, Weathersfield; Leonard Beedle, Messrs. Rich, E. R. Robinson, E. A. Birchard, L. C. Remele, Shoreham; Baker & Smith, P. Elitharp, C N. Hayward, Barney Myrick, J. J. Crane, H. C. Burwell, George N. Payne, D. F. Doty, Bridport; Messrs. Hammond, Middlebury; W. R. Sanford, Orwell; Nelson Richards, Panton.

*EWES PURCHASED.

Flock commenced Sept. 16, 1874, by a purchase of seven ewes of C. N. Hayward, of Atwood and Rich blood, the first largely predominating. Oct. 21, 1875, fifteen ewes of Atwood and Robinson blood were purchased of George N. Payne.

August 13, 1877, fourteen ewes, sameblood, were purchased of Mr. Payne.

RAMS USED.

Eureka, 3d (223), one bred by H. C. Burwell, belonging to the estate of C. N. Hayward, and one bred in the flock, sired by the last-named ram.

——o——

FLOCK 13.

Owned by the estate of A. M. BALDWIN, WHITING, ADDISON Co., VT.

10 RAMS.

Marked with metallic labels in ears, and having thereon A. M. Baldwin and numbers 41 to 50.

40 EWES.

Marked with labels same as rams and numbers 1 to 40.

PEDIGREE.

Descended from the importations from Spain through the flocks of Andrew Cock, Long Island; David Humphreys, Stephen Atwood, Conn.; William Jarvis, Weathersfield, Messrs. Rich, Leonard Beedle, E. R. Robinson, L. C. Remele, T. Stickney & Son, Shoreham; Messrs. Hammond, D. Hooker, S. W. Remele, Middlebury; R. P. Hall, John Towle, F. Hooker, Cornwall; N. A. Saxton, Waltham; A. A. Farnsworth, Brooksville; J. Q. Stickney, Whiting; Seneca Root, Hubbardton; Lyman Webster, Sudbury.

EWES PURCHASED.

The flock was commenced Oct. 11, 1856, by a purchase of thirty ewe tegs of Messrs. E.R. Robinson by William and Alonzo Baldwin, and the same day William purchased of Alonzo his share in this purchase. William Baldwin was father of A. M. Baldwin. These ewes were bred to the rams of Tyler Stickney and F. Hooker, and one sired by L. C. Remele's stock ram. John Towle's stock rams were used; Little Wrinkly (48). One bred within the flock, sired by one of J. Towle's rams, was used. At the death of William Baldwin the flock was divided between A. M. Baldwin and J. Q. Stickney. Since the division, the ram Fremont, Jr. (215), and rams sired by him have been used.

———o———

FLOCK 14.

Owned by EDSON A. BIRCHARD, SHOREHAM, ADDISON CO., VT.

5 RAMS.

Marked in ears with metallic labels, having thereon E. A. Birchard and numbers 1 to 5.

82 EWES.

Marked same as rams, and numbered 6 to 87.

PEDIGREE.

Descended from the importations from Spain through the flocks of Andrew Cock, Long Island; David Humphreys, Stephen Atwood, Conn.; William Jarvis, Weathersfield; Hon. Charles Rich and his son, Charles Rich, Leonard Beedle, L. C. Remele, E. R. Robinson, George H. Hall, T. Stickney & Son, Shoreham; Messrs. Hammond, O. Severance, Middlebury; W. R. Sanford, Orwell; Victor Wright, Weybridge; R. P. Hall, E. S. Stowell, R. J. Jones, Cornwall; John Q. Stickney, Whiting; Lyman Webster, Sudbury; Seneca Root, Hubbardton; Baker & Smith, Prosper Elitharp, Bridport; C. B. Cook, Charlotte.

ORIGIN.

Flock commenced in 1856 by a lease upon shares of sixty-two ewes of the widow of E. R Robinson. From these ewes and the Lute Robinson Ram (39) the foundation of the flock was obtained. No other ewes have been added to the flock.

RAMS USED.

Besides the Lute Robinson Ram (39) mentioned above, several bred within the flock, among which are Tottingham (40), Old Ethan (67) and Wrinkly(78); rams used from other flocks have been Major (181), Fremont, Jr. (215), and Charlie (253).

FLOCK 15.

Owned by G. D. BUSH, North Orwell, Addison Co., Vt.

13 RAMS.

Marked with metallic labels in ears, having thereon G. D. Bush and number 50 and 89 to 100.

49 EWES.

Marked same as rams and numbers 1 to 49.

PEDIGREE.

Descended from importations from Spain through the flocks of David Humphreys, Stephen Atwood, Conn.; W. R. Sanford, J. C. Thomas, O. S. Branch, E. D. Griswold, Orwell; Messrs. Hammond, Middlebury; R. P. Hall, John Towle, Cornwall.

EWES PURCHASED.

Flock commenced in the Fall of 1869, by a purchase of eight Atwood ewes of J. C. Thomas. They were bred by Mr. Thomas from an ewe purchased in 1862 of W. R. Sanford, and by using the stock rams of Mr. Sanford.

Subsequently six Atwood ewes were purchased of James C. Conkey, Orwell, that he purchased of the estate of O. S. Branch. Mr. Branch bred these ewes from stock that he purchased in 1860 of W. R. Sanford. Jan. 11, 1877, sixteen Atwood ewes were purchased of E. D. Griswold, that, like the others, were bred pure from the flock of W. R. Sanford.

RAMS USED.

The Thomas ewes were bred two years to Blucher (77), for two years after to ram sired by him from one of the Thomas ewes; in 1873 and '74 to J. Towle's 69 (477), in 1875 and '76 G. D. Bush's 50 (570) was used.

——o——

FLOCK 16.

Owned by LEWIS S. BURWELL, Bridport, Addison Co., Vt.

12 RAMS.

Marked in ears with metallic labels, having thereon L. S. Burwell and numbers 20 to 31.

19 EWES.

Marked same as above and numbers 1 to 19.

PEDIGREE.

Descended from importations from Spain, through the flocks of David Humphreys and Stephen Atwood, Conn; Andrew Cock,

Long Island ; William Jarvis, Weathersfield ; Messrs. Rich, Leonard Beedle, L. C. Remele, Erastus R. Robinson, Alvin Smith, Bela Howe, Shoreham ; C. B. Cook, Charlotte ; Messrs. Hammond, Middlebury ; Baker & Smith, P. Elitharp, C. N. Hayward, Barney Myrick, H. C. Burwell, Bridport ; N. A. Saxton, John H. Sprague, Waltham ; Nelson Richards, Panton ; C. G. Seeger, G. W. Whitford, Addison ; W. R. Sanford, Orwell ; E. S. Stowell, Cornwall.

EWES PURCHASED.

Flock commenced Spring of 1872, by a purchase of twenty-two ewes of Atwood and Robinson blood, of G. W. Whitford, Addison.

RAMS USED.

A ram bred by the late C. N. Hayward and used four years, or until 1875, in that year Bismarck (221) was used to a part of the ewes. In 1876 Bismarck (221), Stub (222), Black Top (225), Constitution (434) and Acme (524), were used.

———o———

FLOCK 17.

Owned by HENRY A. BASCOM, RICHVILLE, SHOREHAM, ADDISON CO., VT.

14 RAMS.

Marked by metallic labels in ears, having thereon H. A. Bascom, and numbers 1 to 14.

37 EWES.

Marked same as rams, and numbers from 15 to 51.

PEDIGREE.

Descended from importations from Spain through the flocks of David Humphreys, Stephen Atwood, J. N. Blakeslee, John Nettleton, Conn.; Thomas Buffum, William Bailey, Joseph I. Bailey, Rhode Island ; Andrew Cock, Long Island ; Messrs. Rich, Leonard Beedle, E. R. Robinson, T. Stickney & Son, D. & G. Cutting, S. L. Bissell, A. B. Treadway, Shoreham ; S. T. Baker, J. Q. Stickney, Whiting ; W. R. Sanford, Orwell ; Victor Wright, Weybridge ; R. P. Hall, E. S. Stowell, J. Towle, B. Casey, Cornwall ; Lyman Webster, Sudbury ; Seneca Root, Hubbardton : Messrs. Hammond, Middlebury.

ORIGIN.

Flock commenced in 1852 by A. B. Bascom, father of the present proprietor, who leased a number of ewes upon shares of S. L. Bissell. These ewes were of Atwood, Rich, Rhode Island and Blakeslee blood.

RAMS USED.

Nearly all of D. & G. Cutting's stock rams; in 1858 an Atwood ram, bred by V. Wright, and Sweepstakes (32); in 1859 to 1863, Golden Fleece (91), Gen. Grant (258), Vermont (130), Fremont, Jr. (215), Bascom's Sweepstakes; since that time, Old Peter (264), Stickney Ram (282); and in 1874 and 1875 a ram bred in the flock, sired by Stickney Ram (282).

——o——

FLOCK 18.

Owned by H. S. BROOKINS, RICHVILLE, SHOREHAM, ADDISON Co., VT.

27 RAMS.

Marked by metallic labels in their ears, having thereon H. S. B. and numbers 32, 33, 43 to 50, 61 to 63, 86 to 99.

72 EWES.

Marked same as rams, and numbers 1 to 31, 34 to 42, 51 to 60, 64 to 85.

PEDIGREE.

Descended from importations from Spain, through the flocks of Andrew Cock, Long Island; David Buffum, Wm. Bailey and J. I. Bailey, Rhode Island; David Humphreys, Stephen Atwood, Conn.; Messrs. Rich, L. C. Remele, Leonard Beedle, E. R. Robinson, M. W. C. Wright, Melvin R. Atwood, George Bryant, Levi Woolcott, Gustavus Cook, T. Stickney & Sons, F. & L. E. Moore, James M. Ormsbee, D. & G. Cutting, Shoreham; J. Sheldon, Fairhaven; Messrs. Hammond, W. R. Remele, D. Hooker, S. W. Remele, Middlebury; A. A. Farnsworth, Brooksville; E. S. Stowell, F. Hooker, N. B. Douglas, Cornwall; A. H. Hubbard, S. T. Baker, Whiting; Baker & Smith, P. Elitharp, Bridport.

EWES PURCHASED.

A few ewes were received from the flock of the late M. W. C. Wright. Aug. 10, 1874, twenty ewes were purchased of E. G. Farnham that he had purchased of N. B. Douglas. The blood of these ewes was Atwood, Cutting and Rich. April 21st, 1875, one ewe of Robinson blood was purchased of Schuyler Doane. Sept.

13th, 1876, four ewes were purchased of L. C. Remele. Sept. 25th, 1876, sixteen ewes, a part of the old Robinson flock, were purchased of Mrs. E. R. Robinson. Oct. 24, 1876, ten ewes were purchased of L. C. Remele.

RAMS PURCHASED AND USED.

April 10th, 1875, H. S. Brookin's 32 (534) was purchased and was used in 1875, '76 and '77.

——o——

FLOCK 19.

Owned by JULIUS G. BARKER, Leicester, Addison Co., Vt.

16 RAMS.

Marked with metallic labels in ears, having thereon J. G. Barker and numbers 35 to 50.

34 EWES.

Marked same as rams, and numbers from 1 to 34.

PEDIGREE.

Descended from importations from Spain through the flocks of Andrew Cock, Long Island; David Humphreys, Stephen Atwood, Conn.; William Jarvis, Weathersfield; Messrs. Rich, Leonard Beedle, L. C. Remele, E. R. Robinson, E. A. Birchard, G. D. Bryant, E. G. Farnham, Shoreham; Messrs. Hammond, S. W. Remele, David Hooker, Middlebury; A. A. Farnsworth, Brooksville; N. A. Saxton, Waltham; Baker & Smith, P. Elitharp, Bridport; C. B. Cook, Charlotte; R. P. Hall, Franklin Hooker, Cornwall; H. M. Graves, Salisbury.

EWES PURCHASED.

Flock commenced in 1868 by a purchase of ewes and a ram of H. M. Graves. They were descendents of fourteen ewes purchased in 1862 of L. C. Remele by Loyal C. Graves, father of H. M. Graves, who resold them to the latter before he bred them. H.M.Graves bred them to the Tottingham Ram (40) in 1862, and from the lambs resulting he selected a ram that he used exclusively until he sold the flock to Mr. Barker.

RAMS PURCHASED OR USED.

After Mr. B. purchased the flock he used the same ram until 1871, when he purchased a half interest in Young America, a ram bred by E. G. Farnham, of Robinson blood. This ram was used in 1871, '72, '73 and '74. In 1875 and '76 Wrinkly (292) was used.

FLOCK 20.

Owned by H. C. BROWN, Whiting, Addison Co., Vt.

27 RAMS.

Marked in ears by metallic labels having thereon H. C. BROWN and numbers 1 to 27.

53 EWES.

Marked same as rams, and numbers 28 to 80.

PEDIGREE.

Descended from importations from Spain through the flocks of A. Cock, Long Island ; David Humphreys, Stephen Atwood, J. N. Blakeslee, John Nettleton, Conn ; William Jarvis, Weathersfield ; W. R. Sanford, H. G. Hibbard, Orwell ; Messrs. Hammond, Middlebury ; Messrs. Rich, Leonard Beedle, E. A. Birchard, T. Stickney & Son, E. R. Robinson, W. H. & Reuben Cook, William Cook, E. R. Cudworth, Shoreham ; A. H. Hubbard, J. Q. Stickney, William Baldwin, Elisha Rich, Whiting ; Ebenezer Porter, Rutland ; N. A. Saxton, Waltham ; R. P. Hull, J. Towle, Henry Lane, S. S. Gibbs, Cornwall ; Baker & Smith, Prosper Elitharp, Bridport ; C. B. Cook, Charlotte ; Lyman Webster, Sudbury ; Seneca Root, Hubbardton.

EWES PURCHASED.

Flock commenced in 1860, by a purchase of twenty ewes of Edwin Cudworth. Ten were bred by J. T. & V. Rich, and ten by Cudworth from ewes that were purchased of W. R. Sanford, of Blakeslee blood.

RAMS USED.

In 1861 and '62, Elisha Rich's Sweepstakes (53) ; in 1863, ram bred by William Cook, of Rich blood ; in 1864, ram bred by W. H. & Reuben Cook, sire, Tottingham (40), dam, a Robinson ewe ; also a ram owned by A. H. Hubbard, bred by Henry Lane, of Atwood blood ; in 1865-6, Hubbard's Rich Ram (269); in 1867, Holmes & Jones Ram (265); in 1868-9, Towle Ram (164); in 1870-1, Hubbard (165) was used ; in 1872-3 an Atwood and Robinson Ram, bred by A. H. Hubbard, and Fremont, Jr.(215) were used ; in 1874, Hibbard (214); in 1875, a ram bred within the flock, sired by Fremont, Jr. (215); in 1876, the same and nineteen ewes to Hibbard (214).

———o———

FLOCK 21.

Owned by ELMER BARNUM, Shoreham, Addison Co., Vt.

2 RAMS.

Marked with metallic labels in ears, having thereon E. Barnum and numbers 1 and 2.

31 EWES.

Labeled same as rams and numbers 3 to 31.

PEDIGREE.

Descended from importations from Spain, through the flocks of Andrew Cock, Long Island; David Humphreys, Stephen Atwood, J. N. Blakeslee, John Nettleton, Conn.; William Jarvis, Weathersfield; Messrs. Hammond, Middlebury; Messrs. Rich, Leonard Beedle, E. R. Robinson, L. C. Remele, E. A. Birchard, T. Stickney, Denny & Harris, E. D. Bush, Gasca Rich, Shoreham; Baker & Smith, P. Elitharp, Bridport; Seneca Root, Hubbardton; Lyman Webster, Sudbury.

EWES PURCHASED.

September 13, 1864, six ewes were purchased of Denny & Harris. Sept. 19, 1864, ten ewes were purchased of Gasca Rich. The former were from Rich and Robinson dams, and sired by Gold Drop (64) and Tottingham (40). The last ten were Rich, Atwood and Blakeslee blood.

RAMS USED.

Some of Messrs. Hammond's stock rams at first; some of the E. A. Birchard stock rams, E. D. Bush's stock rams; otherwise rams bred within the flock, sired by these.

———o———

FLOCK 22.

Owned by JOHNSON S. BENEDICT, CASTLETON, RUTLAND Co., VT.

16 RAMS.

Marked by metallic labels in ears, having thereon J. S. B. and numbers 101 to 116.

71 EWES.

Labeled same as rams, with numbers 1 to 71.

PEDIGREE.

Descended from importations from Spain, through the flocks of Andrew Cock, Long Island; David Humphreys, Stephen Atwood, J. N. Blakeslee, John Nettleton, Conn.; Messrs. Rich, Leonard Beedle, E. R. Robinson, D. E. Robinson, S. L. Bissell, T. Stickney & Son, L. C. Remele, Jasper Barnum, B. B. Tottingham, E. A. Birchard, Shoreham; Messrs Hammond, Middlebury; V. Wright, Weybridge; W. R. Sanford, Orwell; N. A. Saxton, Waltham; O. S. Rumsey, Hubbardton; R. P. Hall, John Towle, Cornwall; Baker & Smith, Prosper Elitharp, Bridport; C. B. Cook, Charlotte. 24

EWES PURCHASED.

Flock commenced Sept. 23d, 1853, by a purchase of three ewes of E. D. Bush, of Shoreham, that were bred by E. R. Robinson. Dec. 25, same year, ten Atwood ewes were purchased of C. S. Rumsey. Oct. 23, 1862, twenty-three Atwood, Rich and Blakeslee ewes were purchased of S. L. Bissell. Jan. 31, 1865, three ewe lambs were purchased of B. B. Tottingham. They were of Rich blood.

RAMS PURCHASED OR USED.

Sept. 23, 1853, a yearling ram was purchased of J. T. & V. Rich. Jan. 15, 1864, a ram teg was purchased of D. E. Robinson. These rams were used for a number of years each, and, after these, rams bred within the flock, with the following exceptions: a ram belonging to Capt. H. Ainsworth, bred by T. Stickney, was used to a few ewes; in 1864, seven ewes were served by Don Pedro (119); in 1866–7, Little Wrinkly (48) was used to nine ewes; in 1872, seven ewes were served by an Atwood ram, bred by John Towle, owned by C. L. Barber, of Castleton.

———o———

FLOCK 23.

Owned by D. J. BROWN, Whiting, Addison, Co., Vt.

6 RAMS.

Marked with metallic labels in ears, having thereon D. J. Brown and numbers 11 to 16.

10 EWES.

Labeled same as rams, with numbers 1 to 10.

PEDIGREE.

Descended from importations from Spain through the flocks of David Buffum, William Bailey, Joseph I. Bailey, Rhode Island; Andrew Cock, Long Island; David Humphreys, Stephen Atwood, J. N. Blakeslee, John Nettleton, Connecticut; William Jarvis, Weathersfield; Messrs. Rich, D. & G. Cutting, J. M. Ormsbee, Leonard Beedle, E. R. Robinson, T. Stickney & Son, T. Brookins, M. W. C. Wright, E. A. Birchard, Shoreham; J. Q. Stickney, S. T. Baker, H. C. Brown, Whiting; Baker & Smith, Prosper Elitharp, Bridport; W. R. Sanford, H. G. Hibbard, Orwell; Messrs. Hammond, Middlebury; E. Porter, Rutland; R. P. Hall, Henry Lane, John Towle, S. S. Gibbs, Cornwall; C. B. Cook, Charlotte.

EWES PURCHASED.

Flock commenced in the Fall of 1871, by a purchase of five ewes of Thurman Brookins. Their blood was Rich, Atwood and Rhode Island.

RAMS USED.

The stock rams of H. C. Brown, all the years but one; that year Hibbard (214) was used.

———o———

FLOCK 24.

Owned by WILLIAM BALL, HAMBURG, LIVINGSTON Co., MICH.

5 RAMS.

Marked with metallic labels in ears, having thereon W. Ball and numbers 1 to 5.

100 EWES.

Labeled same as rams, and numbers 1 to 100.

PEDIGREE.

Descended from importations from Spain through the flocks of Andrew Cock, Long Island; David Humphreys, Stephen Atwood, Jacob N. Blakeslee, Conn.; Thomas Buffum, William Bailey, Joseph I. Bailey, Rhode Island; William Jarvis, Weathersfield; Leonard Beedle, Messrs. Rich. E. R. Robinson, L. C. Remele, E. A. Birchard, Denny & Harris, E. D. Bush, E. G. Farnham, B. B. Tottingham, James Forbes, Jr., F. & L. E. Moore, T. Stickney & Son, Darwin E. Robinson, F. B. Douglas (now of Whiting), A. B. Treadway, E. Barnum, H. W. Jones, D. &. G. Cutting, J. N. North, L. Catlin, Wm. Cook, J. M. Ormsbee, Shoreham; Messrs. Hammond, W. R. Remele, Oliver Severance, Middlebury; S. T. Baker, Whiting; Ward M Linclon, N. T. Sprague, Brandon; C. S. Rumsey, Hubbardton; W. R. Sanford, Linus Wilcox, O. S. Branch, J. T. Branch, D. B. & J. N. Buell, O. H. Bascom, H. T. Cutts, H. G. Hibbard, Orwell; R. P. Hall, John Towle, Cornwall; E. Porter, Rutland; Baker & Smith, Prosper Elitharp, C. N. Hayward, W. W. Winchester, Bridport; C. B. Cook, Charlotte; N. A. Saxton, Waltham; Nelson Richards, Panton; Victor Wright, Weybridge; E. Porter, Rutland; Alfred Hull, Wallingford.

EWES PURCHASED.

Flock commenced in 1864 or 1865 by a purchase of twenty-one ewes of F. & L. E. Moore. They were sired by Small Tom (88). In 1874, in company with E. W. Hardy, nineteen ewe tegs were purchased of E. D. Bush, twenty-five ewe tegs of B. B. Tottingham, and twenty-nine of James Forbes, Jr. Also in 1874–75, sixty ewes and ewe tegs were purchased of H. W. Jones. All these combined the blood of the Cock, Jarvis and Atwood flocks. In 1875 thirty-five ewe tegs were purchased of L. E. Moore, bred by Wm. Cook and James Forbes, Jr. These last combined the blood of the same flocks as the others. A part of all these purchases were sold, and a part of all were retained as breeding stock in the flock.

RAMS PURCHASED OR USED.

Little Corporal (289), Addison Chief (278), Don Pedro (276), Little Giant, McClellan and Rutherford B. (280), bred by F. & L. E. Moore, were purchased and have been used to breed up the present flock.

——o——

FLOCK 25.

Owned by W. H. BINGHAM, West Cornwall, Addison Co., Vt.

5 RAMS.

Marked with metallic labels in ears, having thereon W. H. Bingham, and Nos. 91 to 95.

90 EWES.

Labeled same as rams and numbers 1 to 90.

PEDIGREE.

Descended from importations from Spain through the flocks of Andrew Cock, Long Island ; David Humphreys, S. Atwood, J. N. Blakeslee, J. Nettleton, Conn.; David Buffum, William Bailey, Joseph I. Bailey, Rhode Island ; William Jarvis, Weathersfield, Messrs. Rich, Leonard Beedle, E. R. Robinson, Jasper Barnum, S. L. Bissell, D. & G. Cutting, H. A. Bascom, F. & L. E. Moore, J. M. Ormsbee, L C. Remele, Shoreham ; Messrs. Hammond, J. Piper, Middlebury ; R. P. Hall, J. Towle, S. S. Rockwell, E. Sanford, S. B. Ives, B. F. Haskell and Datus Haskell, O. A. Fields, Cornwall ; S. T. Baker, J. Q. Stickney, Whiting ; Seneca Root, Hubbardton ; Lyman Webster, Sudbury ; P. Elitharp, Baker & Smith, Bridport ; C. B. Cook, Charlotte.

EWES PURCHASED.

Flock commenced in 1869 by a purchase of ten ewes and two ewe tegs, of Atwood and Rich blood, of S. B. Ives, bred from ewes from the flock of S. S. Rockwell. Six of these ewes were sired by Eureka (58); four and the tegs were sired by the stock rams of R. J. Jones.

About the same time of the above purchase thirty-one ewes and four ewe tegs were purchased of B. F. Haskell. These were a flock that had been bred by Datus Haskell and the sons of A. L. Bingham in company. The original ewes were bred by H. A. Bascom, of Shoreham, and sold by him to E. C. Eells and by him to D. Haskell. The blood of these ewes was Atwood, Cutting, Robinson and Blakeslee. On these ewes were used rams of Atwood and Robinson blood, Prince (86), for two years, a ram bred by David Cutting, and one year a ram bred in the flock, sire by the last-named ram was used.

RAMS USED BY W. H. BINGHAM.

The ram named last in the history of the Haskell flock was purchased with them. In 1871-2 Bingham's Moore Ram (288) was used. In 1873 Keeler Ram (287) was used. In 1874 a ram bred by O. A. Fields, of Atwood & Stickney blood; in 1875 a ram bred within the flock and one bred by E. Sanford, were used both of Atwood and Robinson blood, were used. In 1876 a ram bred by O. A. Fields, of Atwood and Stickney blood, was used.

———o———

FLOCK 26.

Owned by GEORGE D. BRYANT, West Cornwall, Addison Co., Vt.

13 RAMS.

Marked with metallic labels in their ears, having thereon G. D. Bryant and numbers 35 to 47.

34 EWES.

Labeled same as rams and numbers 1 to 34.

PEDIGREE.

Descended from importations from Spain through the flocks of Andrew Cocks, Long Island; D. Humphreys. Stephen Atwood, Conn.; William Jarvis, Weathersfield; Messrs. Rich, Leonard Beedle, E. R. Robinson. L. C. Remele, E. A. Birchard, James Forbes, Jr. (for extended pedigree of this flock see J. Forbes, Jr.'s, pedigree), Shoreham; Baker & Smith, P. Elitharp, W. W. Winchester, Bridport; Messrs. Hammond, S. W. Remele, David Hooker, Middlebury; A. A. Farnsworth, Brooksville; R. P. Hall, F. Hooker, Cornwall.

EWES PURCHASED.

Flock commenced in November, 1867, by a purchase of twenty ewes of Mrs. Erastus R. Robinson, being descendants of the old Robinson flock.

RAMS USED.

Tottingham & Birchard (202) was used in 1868, then rams bred within the flock until 1874, when Snow Flake (277) was used. Then rams bred within the flock until 1876, when Wrinkly (292) was used.

———o———

FLOCK 27.

Owned by THURMAN BROOKINS, East Shoreham, Vt.

25 RAMS.

Marked with metallic labels in their ears, having thereon T. Brookins, and numbers 1 to 13, 114 to 125.

114 EWES.

Labeled same as rams, and numbers 14 to 113 and 126 to 139.

PEDIGREE.

Descended from importations from Spain through the flocks of Thomas Buffum, William Bailey, Joseph I. Bailey, Newport, Rhode Island; Andrew Cock, Long Island; David Humphreys, Stephen Atwood, Connecticut; William Jarvis, Weathersfield; S. T. Baker, Whiting; Hon. Charles Rich and his sons Charles and J. Thurman Rich, and grandsons, J. T. & V. Rich, D. & G. Cutting, T. Stickney & Son, Leonard Beedle, E. R. Robinson, J. M. Ormsbee, M. W. C. Wright, Shoreham; Baker & Smith, Prosper Elitharp, Bridport; E. Hammond, Middlebury.

EWES PURCHASED.

Flock commenced in 1847 or '48, by a purchase of thirteen ewes of Rhode Island blood, of D. & G. Cutting, and the year after another purchase of four ewes of David Cutting, of the same blood. In 1855 one ewe of Jarvis, Atwood and Rich blood was purchased of M. W. C. Wright.

RAMS.

To breed up the present flock the stock rams of D. & G. Cutting, Tyler Stickney & Sons, E. Hammond, E. R. Robinson and J. T. & V. Rich, have been used, besides those bred within the flock from the above ewes and rams.

FLOCK 28.

Owned by C. E. BUSH, Shoreham, Addison Co., Vt.

12 RAMS.

Marked with metallic labels in their ears, having thereon C. E. Bush, and numbers 1 to 12.

64 EWES.

Fifty-four labeled same as rams and numbers 17 to 70 ; five labeled same as rams, with J. T. & V. R. and numbers 19, 32, 34, 35, 63 on their labels ; five that are marked on their labels G. H. Hall and numbers 5, 7, 14, 31, 38 ; also twelve that are marked J. Hinds and numbers 64 to 76.

PEDIGREE.

Descended from importations from Spain through the flocks of A. Cock, Long Island; D. Humphreys, Stephen Atwood, J. N. Blakeslee, J. Nettleton, Conn.; Messrs. Rich, Leonard Beedle, L. C. Remele, T. Stickney & Son, Rich & Cudworth, A. C. Harris, J. N. North, E. R. Robinson, G. H. Hall, Shoreham ; John Q. Stickney, Whiting ; Messrs. Hammond, Middlebury ; Baker & Smith, P. Elitharp, Bridport ; Lyman Webster, Sudbury ; Seneca Root, Hubbardton ; Charles Merrill, Addison ; C. B. Cook, Charlotte ; R. P. Hall, R. J. Jones, Cornwall ; W. R. Sanford, J. H Thomas, Orwell ; J. Hinds, Brandon.

EWES PURCHASED.

In March, 1865, ten pure Rich ewes were purchased of J. H. Thomas. The same year eleven pure Atwood ewes were purchased of E. C Everest, of Vergennes, Vt., that were bred by Charles Merrill. Twenty-one were also purchased of A.C. Harris, bred from the stock rams of Messrs. Hammond, on Rich ewes sold Mr. Harris by E. D. Bush. Oct. 31, 1876, five breeding ewes were purchased of V. Rich. Nov. 6, 1876, ten ewes were purchased of Geo. H. Hall, six of which were Atwood blood, and four Atwood and Robinson. During the same month seven ewe tegs were purchased of the estate of E. R. Cudworth, of Blakeslee and Rich blood. April, 1878, twelve Atwood ewes were purchased of E. D. Hinds, Brandon.

RAMS USED.

At first the stock rams of Messrs. Hammond, and for some years rams sired by them. In 1873 and '74, Fremont, Jr. (215) was used, and since that time rams that were sired by Fremont Jr. (215).

FLOCK 29.

Owned by L. P. BARNES, Whiting, Addison Co., Vt.

7 RAMS.

Marked with metallic labels in ears, having thereon L. P. Barnes and numbers 44 to 50.

26 EWES.

Labeled same as rams and numbers 1 to 26.

PEDIGREE.

Descended from importations from Spain through the flocks of David Humphreys, Stephen Atwood, Conn.; A. Cock, Long Island; William Jarvis, Weathersfield; Messrs. Rich, Leonard Beedle, E. R. Robinson, L. C. Remele, Lucius Robinson, Shoreham; Messrs. Hammond, D. Hooker, S. W. Remele, Middlebury; N. A. Saxton, Waltham; A. A. Farnsworth, Brooksville; Victor Wright, Weybridge; R. P. H. H. Hall, E. S. Stowell, F. Hooker, John Towle, Cornwall; A. Baldwin, A. H. Hubbard, Whiting; Baker & Smith, Prosper Elitharp, Bridport; C. B. Cook, Charlotte.

EWES PURCHASED.

Flock commenced about the year 1858 by a purchase of five ewes of Lucius Robinson.

In 1864 two ewes were purchased of John Towle, of Atwood blood.

RAMS USED.

Sweepstakes (32) was used to a limited extent. John Towle's stock rams and Little Wrinkly (48) were used. Ram owned by Merrill Bingham, bred by E. S. Stowell, Wrinkly(292), Towle (164), Dew Drop (59), Horace Greeley (291), Eureka (58), rams bred within the flock, sired by Dew Drop (59), and Don Pedro (331), were all used to breed up the present flock.

——o——

FLOCK 30.

Owned by D. B. BUELL, Sudbury, Addison Co., Vt.

8 RAMS.

Marked with metallic labels in their ears, having thereon D. B. Buell and numbers 1 to 8,

72 EWES.

Labeled same as the rams and numbers 1 to 72.

PEDIGREE.

Descended from Col. David Humphreys' importation from Spain through the flocks of Stephen Atwood, Conn.; W. R. Sanford, Orwell; R. P. Hall, John Towle, Cornwall; Messrs. Hammond, Middlebury; Victor Wright, Weybridge.

EWES PURCHASED.

In 1862 four Atwood ewes were purchased of W. R. Sanford; in 1871 twelve Atwood ewes were purchased of the same; in 1874 thirteen ewes were purchased from the same flock at the time it was sold out to L. J. Orcutt; in 1873 sixteen Atwood ewes were purchased of John Towle.

RAMS USED.

In 1862 the stock rams of W. R. Sanford; in 1863 an interest in Sanford (135) was purchased, and he was used for two seasons. Green Mountain (46) was then purchased and used two years. Buell's Sanford (293) was used until 1870. An Atwood ram was purchased of W. R. Sanford and used for two years. Since 1872 Buell's Towle Ram (290) has been used as the stock ram of the flock.

———o———

FLOCK 31.

Owned by B. L. BUELL, Sudbury, Rutland Co , Vt.

15 RAMS.

Marked with metallic labels in their ears, having thereon J. W. Buell and numbers 1 to 15.

101 EWES.

Labeled same as rams, and numbers from 1 to 101.

PEDIGREE.

Descended from Col. David Humphreys' importation from Spain through the flocks of Stephen Atwood, Conn.; W. R. Sanford, Orwell; Messrs. Hammond, Middlebury; Victor Wright, Weybridge; R. P. Hall, John Towle, Cornwall.

EWES PURCHASED.

Five Atwood ewes were purchased of W. R. Sanford in 1862; ten more were purchased in 1864; in 1871 thirteen, and in 1874 thirteen more, all of the same blood and from the same flock. These last were a part of the flock when sold out to Mr. Orcutt, the Messrs. Buell acting as Mr. Orcutt's agents. Thirteen ewes were also purchased of Mr. J. Towle in 1873.   25

RAMS USED.

In 1862 the stock rams of W. R. Sanford ; in 1873–4, Sanford (135) was used, an interest in him having been purchased of Mr. Sanford ; Green Mountain (46) was then purchased in company with D. B. Buell, and he was used two seasons ; Buell's Sanford (293) was used until 1871, when a ram was purchased of W. R. Sanford and used until 1872 ; since that time Buell's Towle ram (290) has been used as the stock ram of the flock.

——o——

FLOCK 32.

Owned by O. H. & W. O. BASCOM, ORWELL, ADDISON Co., VT.

12 RAMS.

Marked with metallic labels in their ears, having thereon O. H. & W. O. B., and numbers 67 and 101 to 111.

61 EWES.

Labeled same as rams, and numbered 1 to 17, 22, 24 to 29, 32, 34, 36, 37, 39, 42, 45, 48, 51, 55 to 57, 59, 62, 65, 70 to 84, 112 to 117.

PEDIGREE.

Descended from Col.David Humphreys' importations from Spain, through the flocks of Stephen Atwood, Conn.; W. R. Sanford, Orwell ; R. P. Hall, John Towle, Cornwall ; Messrs. Hammond, Middlebury ; Victor Wright, Weybridge ; A. H. Hubbard, Whiting.

EWES PURCHASED.

The foundation of this flock was in ewes purchased of W. R. Sanford at several times, but the exact number and dates of the purchases cannot be given. Some of the ewes purchased of Mr. Sanford had Jarvis blood in them, but in the year 1860 the flock was reduced very low and to pure Atwood sheep only, those having any of the Jarvis blood being at that time sold from the flock.

RAMS USED.

None but pure Atwood rams have ever been used in the flock. Comet (57), Kilpatrick (71) and Blucher (77) were used to a considerable extent ; after that, rams that were sired by these rams ; after these, Hubbard ram (165), and he has been the stock ram since.

FLOCK 33.

Owned by MERRILL BINGHAM, West Cornwall, Addison Co., Vt.

10 RAMS.

Marked with metallic labels in their ears, having thereon M. Bingham and numbers 1, 2, 10, 13, 25, 27, 28, 29, 30, 31.

127 EWES.

Labeled same as rams, with numbers 101 to 221, 223 to 228.

PEDIGREE.

Descended from importations from Spain through the flocks of David Humphreys, Stephen Atwood, Conn.; Andrew Cock, Long Island; William Jarvis, Weathersfield; R. P. Hall, E. S. Stowell, S. S. Rockwell, Wm. McCauley, Cornwall; Messrs. Rich, Leonard Beedle, T. Stickney & Sons, E. R. Robinson, L. C. Remele, Levi Wolcott, Shoreham; P. Elitharp, C. N. Hayward, Bridport; N. A. Saxton, Waltham; A. Richards, Panton; Messrs. Hammond, Middlebury; Victor Wright, Weybridge; Lyman Webster, Sudbury; Seneca Root, Hubbardton; J. Q. Stickney, Whiting.

EWES PURCHASED.

In 1860 or '61 sixty Atwood ewes were purchased of Henry Cross, then living in Waltham. These were ewes bred from ewes purchased of R. P. Hall by Mr. Cross a number of years previous, Mr. Cross buying the entire crop of ewe lambs raised by Mr. Hall in one year, besides a few old breeding ewes. These ewes were bred to pure Atwood rams bred by Victor Wright, Mr. Hall, Messrs. Hammond, and, perhaps, one or two others. At the time of the sale to Mr. Bingham, no others but pure Atwood rams had been used. This is the foundation of the flock entered for record so far as the ewes purchased.

RAMS USED.

An Atwood ram was purchased of N. A. Saxton and used for a number of years; also rams of the same blood of Victor Wright and E. S. Stowell; Little Wrinkly (48) was used to a number of ewes; a ram bred by C. N. Hayward was purchased and used; Eureka (58) was used to a few ewes; Levi (552) was purchased and used a few years; a ram purchased of William McCauley, sired by Fremont, Jr. (215), dam by Little Wrinkly (48) has been the stock ram for the last two years.

FLOCK 34.

Owned by CHARLES I. BENEDICT, Arlington, Bennington Co., Vt.

1 RAM.

Marked with metallic label in his ear, having thereon C. I. B. and No. 1.

50 EWES.

Labeled same as rams and Nos. 1 to 50.

PEDIGREE.

Descended from the importation from Spain by Col. David Humphreys, through the flocks of Stephen Atwood, Conn.; Messrs. Hammond, Middlebury.

EWES PURCHASED.

In 1862 twenty-five Atwood ewes were purchased of W S. & E. Hammond. Feb. 9th, 1865, Queen ewe, No 4, was purchased of E. Hammond & Son, and the same year four more ewes were purchased of E. Hammond & Son.

RAMS USED.

First, a ram purchased of Messrs. Hammond, since that rams bred within the flock. Benedict & Bottum (151), Benedict (152) and Prince of Gold Drops (153) were the most noted. Other rams bred within the flock have been used, but we have not received their pedigrees.

————o————

FLOCK 35.

Owned by GEORGE CAMPBELL, Westminster West, Windham Co., Vt.

13 RAMS.

Marked with metallic labels in their ears, having thereon G. Campbell and numbers 20, 21, 29, 30, 33, 65, 143, 150, 151, 152, 153, 181, 184.

100 EWES.

Labeled same as rams, and numbers 2, 2, (6), 7, 10, 13, 19, 22 24, 27, 29, 31, 34, 37, 38, 43, 46, 54, 55, 57, 67, 68, 69, 72, 73, 74, 78 81, 82, 86, 86 (3), 86 (5), 88 to 92, 96 to 99, 101, 105, 107, 118, 125, 134, 165, 175, 177, 179, 182, 182 (3), 182 (4), 182 (5), 183, 183 (3) 186, 186 (3), 186 (4), 188, 191, 192, 204, 205, 205 (4), 206, 209, 214, 222,

248, 251, 252, 253, 254, 254 (0), 258, 260, 263, 287, 288, 291, 296, 302, 304, 304 (4), 304 (5), 305, 308, 313, 321 to 328, 330, 334, 335.*

PEDIGREE.

Descended from importations from Spain through the flocks of David Humphreys, Stephen Atwood, Conn.; Andrew Cock, Long Island; William Jarvis, Weathersfield; Thomas Buffum, William Bailey, Joseph I. Bailey, Rhode Island; Messrs. Hammond, Middlebury; W. R. Sanford, Orwell; Messrs. Rich, L. C. Remele, E. R. Robinson, J.M. Ormsbee, D.& G. Cutting, E.G. Farnham, Shoreham; S. T. Baker, Whiting; E. Sanford, E. S. Stowell, Henry Lane, Cornwall; Nathan Cushing, Woodstock; Charles Blood, Mark Crawford, Daniel Mason, S. E. Wheat, Putney; Ebenezer Bridge, A. E. Fuller, Pomfret; M. C. Roundy, Rockingham; P. Elitharp, Bridport; M. C. Roundy, Rockingham.

EWES PURCHASED.

Flock commenced in 1839 by a purchase of twenty ewes of Humphreys blood, and twenty of Mark Crawford, of Jarvis and Humphreys blood.

1847, five ewes of Atwood and Jarvis blood were purchased of Nathan Cushing.

In 1862 five choice Atwood ewes were purchased of E. Hammond, and one of the same blood of H. W. Hammond. The same year three Atwood ewes were purchased of W. R. Sanford, and fifteen of Atwood, Rich and Jarvis blood of Edgar Sanford.

In 1864 or 65 a half interest in ten Atwood ewes, bred by Col. E. S. Stowell, was purchased of M. C. Roundy. These ewes were a selection of two years of Col. Stowell's breeding. A year or two later the remaining interest with their increase was purchased of Mr. Roundy.

RAMS USED.

The rams that have been used to breed up the flock have been : In 1839, a pure Jarvis ram was purchased of Daniel Mason. In 1842 another was purchased of William Jarvis. These rams and some bred from them were used until 1847. In 1846 a ram lamb was purchased of Ebenezer Budge. This ram was called Old Pomp (30). This ram lived until he was fifteen years old, and, with rams sired by him, was used until Young Wooster (28) was purchased. In 1848 Old Grimes (49) was purchased, and was used for a number of years. One or two years later another Atwood ram lamb was purchased of W. S. & E. Hammond. In 1865 eleven ewes were served by Golden Fleece (91), and a ram of Atwood blood was used belonging to A. E. Fuller.

---

*Where there are duplicate numbers, the numbers in the brackets following are on the labels.

A ram bred within the flock, sired by Golden Fleece, was used three years. In 1871, or 2 a ram was purchased of Mr. Dimmick, of Lyme, N. H. This ram was bred by Col. Stowell, was sired by Golden Fleece (91), and was in the ewe when she was purchased of Col. Stowell by M. D. This ram was named Boxer. In 1874 Favorite (508) was purchased, and has been used as the stock ram of the flock since.

——o——

FLOCK 36.

Owned by J. E. CASWELL, Shoreham, Addison Co., Vt.

15 RAMS.

Marked with metallic labels their in ears, having thereon J. E. Caswell, and numbers 22 to 26, 39 to 48.

37 EWES.

Labeled same as rams, and numbers 1 to 21, 27 to 38, 49 to 52.

PEDIGREE.

Descended from importations from Spain through the flocks of A. Cock, Long Island; David Humphreys, Stephen Atwood, J. N Blakeslee, John Nettleton, Conn.; William Jarvis, Weathersfield; Messrs. Rich, G. Rich & E. R. Cudworth, T. Stickney & Son, J. T. Stickney, Shoreham; Messrs. Hammond, Middlebury; N. A. Saxton, Waltham; W. R. Sanford, Orwell; L. P. Clark, S W. Smith, Addison; Victor Wright, Weybridge; N. A. Saxton, Waltham.

EWES PURCHASED.

In 1869 nine ewes were purchased of C. H. Bowker, of Orwell. They were bred from Blakeslee, Jarvis and Rich blood, by Gasca Rich and E. R. Cudworth. Sept 16, 1876 twelve Atwood ewes were purchased of S. W. Smith; five of the twelve were purchased of N. A. Saxton, and seven were bred from Saxton ewes and sired by Victor (209).

RAMS USED.

Upon the ewes first purchased a ram bred and owned by J. T. Stickney was used. The stock rams of Geo. H. Hall, Don (338) and Ben (339), also Pat Henry (183) and General (210) have been used in later years.

FLOCK 37.

Owned by LYMAN CATE, Highland, Oakland Co., Mich.

4 RAMS.

Marked with metallic labels in their ears, having thereon L. Cate and numbers 1, 2, 3, 4.

18 EWES.

Six of them labeled same as the rams and numbers 5 to 10, twelve of them having on their labels H. M. Perry and numbers 60, 62 to 72.

PEDIGREE.

Descended from importations from Spain, through the flocks of David Humphreys and Stephen Atwood, Conn.; David Buffum, Wm. Bailey J. I. Bailey, Rhode Island ; Messrs. Hammond, Middlebury ; S. T. Baker, Whiting ; J. M. Ormsbee, Messrs. Cutting, H. M. Perry, George H. Hall, Shoreham ; R. P. Hall, R. J. Jones, Cornwall ; W. R. Sanford, Orwell ; Victor Wright, Weybridge ; N. A. Saxton, Waltham.

EWES PURCHASED.

Flock commenced 1878, by a purchase of twelve ewes of Halladay & Farnham that were bred from the old Cutting flock and registered by H. M. Perry, of East Shoreham, Vt. The ewes were in lamb to Gen. Grant (474) and the ewes marked L. Cate and numbered 5 to 10 are the produce of these ewes and Gen. Grant (474).

RAMS PURCHASED.

Gilt Edge (420) was purchased of Peter Martin, and is No. 1 of the flock.

—o—

FLOCK 38.

Owned by J. Q. CASWELL & SONS, Shoreham, Addison Co., Vt.

8 RAMS.

Marked with metallic labels in their ears, five of them having thereon J. Q. C. & Sons and numbers 42 to 46, and three of them J. E. C. and numbers 25, 45, 47.

46 EWES.

Labeled same as rams, forty-five of them having J. Q. C. & Sons and numbers 1 to 41, 47 to 51 ; also one marked J. E. C. and number 49.

PEDIGREE.

Descended from importations from Spain through the flocks of David Humphreys, Stephen Atwood, J. N. Blakeslee, J. R. Nettleton, Conn.; Andrew Cock, Long Island; William Jarvis, Weathersfield; Messrs. Rich, L. C. Remele, Gasca Rich, E. R. Cudworth, T. Stickney & Son, J. T. Stickney, M. W. C. Wright, J. E. Caswell, G. H. Hall, Shoreham; W. R. Sanford, Orwell; L. P. Clark, S. W. Smith, Addison; N. A. Saxton, Waltham; Victor Wright, Weybridge; W. P. Wright, Whiting.

EWES PURCHASED.

One was purchased of J. V. Sanford, that was bred by L. C. Remele. One was purchased of E. N. Townsend that was bred by W. P. Wright. One was purchased of J. E. Caswell and thirty-nine were purchased of Stephen Barnum, that were bred by Mr. Barnum from stock he purchased of the late M. W. C. Wright.

These ewes were all purchased in the Fall of 1877 or Spring of 1878.

RAMS USED.

G. H. Hall's Don (338) and Ben (339).

———o———

FLOCK 39.

Owned by CHERBINO & WILLIAMSON, Middlebury, Addison Co., Vt.

68 RAMS.

Marked with metallic labels in their ears, having thereon C. & W. and numbers 24, 25, 276, 277, 279, 281, 285, 286, 288, 290, 291, 292, 294, 295, 302, 303, 306, 307, 313, 315 to 317, 320, 322, 324 to 326, 329, 332, 333, 341, 342, 353, 355, 360, 362, 366 to 369, 523 to 548.

355 EWES.

Labeled same as rams and numbers 1 to 23, 67 to 80, 136 to 265, 271, 272, 275, 278, 280, 282, 283, 284, 287, 289, 296, 298, to 300, 304, 305, 308, 310 to 312, 314, 318, 319, 321, 323, 327, 328, 330, 331, 334, 335, 337 to 340, 343 to 353, 354, 356 to 359, 361, 385 to 522.

PEDIGREE.

Descended from importations from Spain, through the flocks of David Humphreys, Stephen Atwood, Conn.; Andrew Cock, Long Island; Messrs. Hammond, W R. Remele, J. Davenport, H. B. Wright, U. D. Twitchell, Middlebury; Victor Wright, A. W. Dana,

Weybridge ; Messrs. Rich, L.C. Remele, E. R. Robinson, E.A. Birchard, T. Stickney & Son, J. T. Stickney, A. C. Harris, E. D. Bush, Shoreham ; R. P. Hall ; E. S.Stowell, John Towle, C.D. Lane, Henry Lane, C. J. Benedict, M. J. Ellsworth, Cornwall ; Baker & Smith, P. Elitharp, J. J. Crane, C. N. Hayward, H._ C. Burwell, B. Myrick, E. Wilcox, F. D. Doty, D. E. Hill, Bridport ; W. R. Sanford, Orwell ; N. A. Saxton, O. C. Bacon, Waltham ; A. A. Farnsworth, New Haven ; A. Lawrence, Hinesburgh.

EWES PURCHASED

The foundation of this flock was in 37 ewes purchased of J. T. Stickney, a number of C. D. Lane (a few only of which could be identified), a few of J. Davenport, that were pure Atwoods, and thirty Robinson ewes that were purchased of B. Myrick and E. Wilcox ; also a few Atwood ewes were purchased of U. D. Twitchell ; in March, 1876, twenty-three Atwood ewes were purchased of W. R. Remele ; in the Fall of 1876 twelve Atwood and Robinson ewes were purchased of II. B. Wright that were bred from the flock of A. C. Harris ; the same Fall nineteen Atwood ewes were purchased of John Towle ; July 21, 1877, one hundred and twenty ewes were purchased of Charles I. Benedict. These last were bred by Mr. Benedict from ewes purchased of P. Elitharp and A. W. Dana, and by using rams raised by W. R. Remele, Messrs. Hammond, Messrs. Rich, E. S. Stowell aud L. C. Remele.

RAMS USED.

Those used by Messrs. Cherbino & Williamson have been : The Lane Ram (120), Jim (234), Drake Ram (235), Buck Mountain (247), Bonaparte (176), Silver Horns (177), Young Sam (342) and Acme (524).

———o———

FLOCK 40.

Owned by ALBERT CHAPMAN, Middlebury, Addison Co., Vt.

8 RAMS.

Marked with metallic labels in their ears, having thereon A. Chapman, and numbers 3, 13, 18, 20, 22, 26, 28, 29.

18 EWES.

Eleven labeled same as rams and numbered 1, 2, 4, 5, 6, 11, 14 to 17, 19, 21, 23, 27, 30, 31.

One marked on label O. & E. S. H. and number 28.

PEDIGREE.

Descended from importations from Spain through the flocks of David Humphreys, Stephen Atwood, Conn.; Andrew Cock, Long Island; William Jarvis, Weathersfield; Messrs. Hammond, W. R. Remele, S. W. Remele, Oliver Severance, Middlebury; Victor Wright, Weybridge; W. R. Sanford, Orwell; R. P. Hall, R. J. Jones, E. S. Stowell, Cornwall; W. B. Cook, Henry Thorp, Charlotte; Ezra Meech, E S. Rowley, Shelburn; A. A. Farnsworth, C. D. Mason, New Haven; Messrs. Rich, E.R. Robinson, Shoreham; Prosper Elitharp, H. C. Burwell, D. F. Doty, J. J. Crane, C. N. Hayward, Bridport; N. A. Saxton, Waltham; L.P. Clark, Addison; O. & E. S. Hall, Randolph.

EWES PURCHASED.

From a flock of Atwood ewes purchased of John H. Sherman, Charlotte, in August, 1871, two ewes were selected. These sheep were bred by E. S. Rowley from pure Atwood sheep he purchased of V. Wright, Henry Thorp and Ezra Meech, all descended unmixed from the flock of Messrs. Hammond. Nov. 24, 1874, one Atwood ewe was purchased of Mrs. Victor Wright that was bred by V. Wright. This ewe was in lamb by Phil. Sheridan (172); the produce was an ewe lamb, now in flock. In the same year an Atwood ewe was purchased of George Hammond that was bred by C. D. Mason; in 1875 an Atwood ewe that was bred by Prosper Elitharp was purchased of James Howe, Bridport; Nov. 15, 1876, four Atwood ewes were purchased of S. W. Remele; Feb., 1878, an Atwood ewe descended from Hammonds' Old Queen ewe was purchased of O. & E. S. Hall. She was in lamb by Hall's Gold Drop (414), and the produce was an ewe lamb, now in the flock.

RAMS USED.

Bonaparte (176), Silver Horn,, (177) and Allright (169) (to one ewe), Bismarck (221), Stub (222), General (210).

———o———

FLOCK 41.

Owned by D. W. CHILDS, GREEN HORN, COLORADO.

35 RAMS.

Marked with metallic labels in their ears, having thereon D. W. Childs, and numbers 262 to 296.

141 EWES.

Labeled same as rams and numbers 1 to 141.

PEDIGREE.

Descended from importations from Spain by Col. Humphreys through the flocks of Stephen Atwood, Conn.; Messrs. Hammond, M. T. Shackett, Middlebury; R. P. Hall, F. H. Dean, E. S. Stowell, J. Towle, Cornwall; W. M. Gage, Ferrisburgh; Victor Wright, A. J. Stowe, Weybridge; L. P. Clark, Addison; N. A. Saxton, Waltham; P. Elitharp, D. F. Doty, C. N. Hayward, J. J. Crane, Bridport.

EWES PURCHASED.

Nov. 4th, 1868, thirty-six Atwood ewes were purchased of M. T. Shackett, that were bred pure and direct from the flocks of E. Hammond and R. P. Hall. Among these ewes was one of Mr. Hammond's Queen family, for which Mr. Shackett paid Mr. Hammond $625; Feb. 27, 1873, twenty-six Atwood ewe tegs were purchased of F. H. Dean. The produce from these two lots of sheep have been kept marked, and are now labeled distinct from each other. These purchases were made in company with Col. A. J. Childs, now of St. Louis, Mo., who was for a few years a partner with D. W. Childs in the ownership of the flock.

RAMS USED.

At first, Gold Dust (334) and Doty Ram (134); since these were used, King Charles (335), Alexander (336) and Wonder (337). All were bred within the flock. In 1877, to a part of the flock two yearling rams, bred within the flock, and the remainder the rams abovenamed.

——o——

FLOCK 42.

Owned by C. F. CHURCH, WHITING, ADDISON CO., VT.

13 RAMS.

Marked with metallic labels in their ears, and having thereon C. F. Church and numbers 101 to 113.

58 EWES.

Labeled same as rams, and numbers 1 to 58.

PEDIGREE.

Descended from importations from Spain through the flocks of David Humphreys, Stephen Atwood, Jacob N. Blakeslee, J. R. Nettleton, Connecticut; Andrew Cock, Long Island; David Buffum, William Bailey, Joseph I. Bailey, Rhode Island; William Jarvis, Weathersfield; Messrs. Rich, Leonard Beedle, E. R Robinson, T.

Stickney & Son, D. & G. Cutting, J. M. Ormsbee, L. C. Remele, E. A. Birchard, Cudworth & Rich, Baker & Smith, Prosper Elitharp, B. Myrick, J. J. Crane, C. N. Hayward, H. C. Burwell, D. F. Doty, Bridport; R. P. Hall, J. Towle, F. H. Dean, F. Hooker, S. S. Rockwell, Cornwall; W. R. Sanford, G. D. Bush, Orwell; Fayette Holmes, Lyman Webster, Sudbury; Seneca Root, Hubbardton; R. Gleason, Benson; E. Cook, S. T. Baker, J. Q. Stickney, A. H Hubbard, Whiting; Messrs. Hammond, D. Hooker, S. W. Remele, Middlebury; N. A. Saxton, Waltham; J. Hinds, Brandon; A. A. Farnsworth, Brooksville.

EWES PURCHASED.

Sixteen ewes of Atwood, Blakeslee and Rich blood were purchased of Davis Rich, bred by Cudworth & Rich; six ewes of H. Graves, purchased of and bred by F. Hooker; seven ewes from G. D. Bush, of Atwood & Robinson blood; nine ewes of E. Cook, Robinson and Cutting blood; two ewe tegs purchased of F. Holmes, of Atwood and Rich blood.

RAMS USED.

Dan Holmes' Atwood ram, sired by Silver Mine, 2d(84); a ram bred by A. H. Hubbard, sired by Young Fremont (215); some of J. T. Stickney & Son's stock rams. In 1877 two yearling Atwood and Robinson rams, bred by F. H. Dean, were purchased of E. D. Searl and used; two ram tegs were purchased of E. Cook in 1877.

---o---

FLOCK 43.

Owned by JOHN A. CHILDS, WEYBRIDGE, ADDISON CO., VT.

12 RAMS.

Marked with metallic labels in their ears, having thereon J. A. Childs, and numbers 1 to 12.

48 EWES.

Labeled same as rams and numbers 13 to 62.

PEDIGREE.

Descended from importations from Spain through the flocks of David Humphreys, Stephen Atwood, Conn.; Andrew Cock, Long Island; Wm. Jarvis, Weathersfield; Messrs. Hammond, Middlebury; R. P. Hall, John Towle, S. S. Rockwell, E. S. Stowell, Cornwall; Messrs. Rich, Leonard Beedle, E. R. Robinson, L. C. Remele, Shoreham; Baker & Smith, P. Elitharp, C. N. Hayward, J. J. Crane,

B. Myrick, D. F. Doty, Bridport; W. R. Sanford, Orwell; V. Wright, Weybridge.

EWES PURCHASED.

In 1862 fifteen Atwood ewes were purchased of E. S. Stowell, that were bred by Messrs. Hammond, or bred from ewes purchased of them.

RAMS USED.

For a few years Stowell's Sweepstakes (47). He was followed by Gold Dust (334); a ram bred by V. Wright was used one year, but the produce was all sold. These three rams were all bred from Atwood blood. A ram was next used that was bred by Edgar Sanford, his blood was Atwood and Robinson. He was followed by Bonaparte, Jr. (421). J. A. Childs' No. 10 (422) is now the stock ram of the flock and has been used two years. Other rams have been used, bred within the flock.

———o———

FLOCK 44.

Owned by LUMAN P. CLARK, ADDISON, VT.

52 RAMS.

Marked with metallic labels in ears, having thereon L. P. Clark, and Nos. 101 to 140, 151 to 162.

82 EWES.

Labeled same as rams and numbers 1 to 76, 80 to 85.

PEDIGREE.

Descended from the Humphreys importation from Spain through the flocks of Stephen Atwood, Conn.; Messrs. Hammond, Middlebury; N. A. Saxton, Waltham; Victor Wright, Weybridge; W. R. Sanford, Orwell.

EWES PURCHASED.

Flock commenced by a purchase of three Atwood ewes of N. A. Saxton, one of them bred by E. Hammond; also two Atwood ewes of the late Victor Wright, and in 1874 four Atwood ewes of Mrs. Victor Wright that were bred by Mr. Wright before his death.

RAMS USED.

Victor Wright's Black Top (90), Sweepstakes (32), Gold Drop (64), and Green Mountain (70). N. A. Saxton's Thousand Dollar Ram and Dean's Little Wrinkly (48) were each used to one or more ewes. Kilpatrick (71), bred by W. R. Sanford, was purchased and used for a number of years until his death.   Rams bred within the flock have been used as follows : Vigor (209), Chunk Head (182), Patrick Henry (183), and General (210) also two or three other rams bred within the flock have been used to a small extent.

——o——

FLOCK 45.

Owned by CHARLES M. CLARK, WHITEWATER, WISCONSIN.

239 EWES.

Twenty marked with metallic labels in their ears having thereon A 65 and numbers 1 to 10, 12 to 21. Ninety-four with labels marked C. M. Clark and numbers 2, 9, 12, 13, 14, 17 to 21, 23, 27. 31, 35, 38, 52, 53, 54, 58, 61, 62, 63, 69, 70, 73, 78, 87, 89, 92, 93, 98, 100, 102, 103, 104, 106, 108, 109, 111, 114, 115, 117, 119, 121, 122, 123, 125, 133, 135, 139, 142 to 145, 147, 156, 158, 160, 163, 167, 168, 171, 173, 175, 178, 179, 181, 184, 185, 188 to 191, 194, 199, 206, 208, 210, 212 to 216, 218, 220, 221, 223 to 226, 229, 234, 250, 252. Eighty with labels marked C M C & B, and numbers 1 to 12, 14 to 26, 28 to 32, 34 to 65, 67, 73, 75 to 90.   Forty-seven with labels marked C M C & C and numbers 1, 2, 6 to 10, 12, 13, 15, 18, 19, 22, 24 to 28, 31 to 33, 37, 38, 40, 42 to 44, 47 to 51, 53, 55 to 58, 64, 67, 68, 70 to 73, 75, 76, 211.

PEDIGREE.

Descended from the importations from Spain through the flocks of David Humphreys, Stephen Atwood, Conn.; Andrew Cock, Long Island;   William Jarvis, Weathersfield ; Messrs. Hammond, S. W. Remele, Middlebury ; Victor Wright, Weybridge ;   Ward M. Lincoln, Brandon ; E. Porter, formerly of Rutland, later of Wisconsin; Messrs. Rich, L. C. Remele, E. R. Robinson, T. Stickney & Son, F. D. Douglass, E. D. Bush, E. Barnum, Alvin Clark, E. G. Farnham, Shoreham; J. Sheldon, Fair Haven ; Baker & Smith, P. Elitharp, C. N. Hayward, Bridport ; N. Richards, Panton ; R. P. Hall, A. J. Wooster, E. S. Stowell, E. Sanford, H. Lane, Cornwall ; John H. Paul, Waukesha, Wis.; W. R. Sanford, S. L. Stevens, Orwell ; W. M. Holmes, Greenwich, N. Y.; C. B. Cook, Charlotte ; S. B. Lusk, S. S. Lusk, E. B. Pottle, E. Townsend, Western N. Y.

EWES PURCHASED.

Flock was commenced in the fall of 1857 by a purchase of a flock of ewes and lambs that had been bred by Ebenezer Porter from

Atwood stock purchased of Ward M. Lincoln. The ewes were taken from Rutland, Vt., to Wisconsin by Mr. Porter in 1856. The lambs were sired by Young Matchless, a ram bred by A. J. Wooster, West Cornwall, Vt., from an ewe purchased of Messrs. Hammond and sired by one of their stock rams. March 3, 1876, twenty Rich or Robinson ewes were purchased of E. D. Bush. They were bred by E. Barnum, S. L. Stevens and E. D. Bush. These ewes are the ones marked A 65 and numbered 1 to 21 minus 11.

RAMS USED.

In 1857 and 1858, Young Matchless, the same used by Mr. Porter. In 1859 and two years following, Robinson Rich (323). In 1862 and 1863 Washoe (136). In 1864, '65, '66 and '67, Young Ironsides (82). In 1862 Green Mountain (304) was used to a part of the ewes; the same year Columbus, a ram bred by J. H. Sprague, sired by Sea Lion (83), was used to part of the ewes. In 1868 and '69, a ram called Grimes, bred by D. D Williams, of Vermont, and sold by him to E. S. Lake and by him to J. H Paul; he was sired by Old Grimes (49). In 1870 a ram sired by the latter dam by Washoe (136) was used. In 1871 one ewe was served by Robinson, a ram bred by J. H. Paul, sire and dam bred by F. D. Douglass, and the remainder by Golden Fleece, Jr, sired by Golden Fleece (91), dam bred by V. Wright, and the same ram was used in 1872. Since 1872 a ram bred by L. C. Remele, of Atwood, Jarvis and Rich blood, was used for two years; a ram bred by E. D. Bush, sired by Fremont Jr. (215), was used to a few ewes; Prince (326) and Royal Prince (325) have been used two seasons. In 1873 Clark's Cook Ram (324) was used; except the above, rams bred within the flock have been used.

———o———

FLOCK 46.

Owned by CLAY & WING, CORNWALL, ADDISON CO., VT.

2 RAMS.

Marked with metallic labels in their ears, having thereon Clay & Wing, and numbers 36 and 37.

35 EWES.

Labeled same as rams with numbers 1 to 35.

PEDIGREE.

Descended from the importation from Spain by Col. David Humphreys, through the flocks of Stephen Atwood, Conn.; Messrs. Hammond, E. D. Munger, E. R. Clay, Middlebury; C. B. Cook, Charlotte; N. A. Saxton, Waltham; Henry Lane, Cornwall, Victor Wright, Weybridge; A. J. Stowe, Weybridge; J. Hinds, E. D. Hinds, Brandon; W. R. Sanford, Orwell; L. P. Clark, Addison.

HISTORY.

This flock was commenced in 1862, by E. R. Clay, who purchased ten Atwood ewe tegs of Henry Lane; four years after, five Atwood ewes were purchased of E. Hammond & Son, and about the same time one Atwood ewe was purchased of H. W. Hammond; in 1871 one Atwood ewe was purchased of Geo. Hammond; in 1873 eighteen yearling ewes and ewe tegs were purchased from the Atwood flock of E. D. Hinds; in 1876 a few Atwood ewes were purchased of A. J. Stowe that were bred from the flock of E. D. Munger, who bred his flock from Messrs. Hammond and Hinds.

RAMS USED.

The first were the stock rams of Messrs. Hammond, Sweepstakes (32), Gold Drop (64), Green Mountain (70) and Silver Mine (61); rams bred within the flock: Jim (234), Clay's Green Mountain (344). A ram bred by L. P. Clark was used, the last rams used being Clay & Wing's 36 (519) and Clay & Wing's 37 (520). E. R. Clay bred the flock up to 1871, when he leased it to breed upon shares to other parties. In 1878 a selection from the flock of 35 ewes and the two rams last-named was made by Mr. Clay, and Messrs. Wing & Son, of Cornwall, became partners and owners in the flock.

———o———

FLOCK 47.

Owned by LYMAN CLARK, JR. Addison, Addison Co., Vt.

9 RAMS.

Marked with metallic labels in their ears, having thereon L. Clark and numbers 1 to 9.

27 EWES.

Labeled same as rams, with numbers 10 to 36.

PEDIGREE.

Descended from importation from Spain, by Col. David Humphreys, through the flocks of Stephen Atwood, Conn.; Messrs. Hammond, Middlebury; Victor Wright, Weybridge; W. R. Sanford, Orwell; N. A. Saxton, Waltham; E. S. Stowell, Cornwall; Luman P. Clark, Addison.

EWES PURCHASED.

Nov. 11, 1865, five Atwood ewe tegs were purchased of N. A. Saxton.

RAMS USED.

In 1866, Golden Fleece (91) and Little Wrinkly (48); a ram sired by the latter was more or less used until 1872; in 1871–2

Chunk Head (182) was used ; in 1873 Kilpatrick (71) and one other of L. P. Clark's stock rams ; in 1874, Patrick Henry (183); in 1875, General (210); in 1876 and '77 Victor Emanuel (242).

———o———

FLOCK 48.

Owned by ELISHA COOK, LEICESTER JUNCTION, ADDISON Co., VT.

16 RAMS.

Marked with metallic labels in their ears, having thereon E. Cook and numbers 35 to 50.

34 EWES.

Labeled same as rams and numbers 1 to 34.

PEDIGREE.

Descended from importations from Spain through the flocks of Andrew Cock, Long Island; D. Humphreys, Stephen Atwood, Conn.; David Buffum, Wm. Bailey, Joseph I. Bailey, Rhode Island; Wm. Jarvis, Weathersfield ; Messrs. Hammond, Middlebury ; V. Wright, Weybridge ; Messrs. Rich, L. C. Remele, E. R. Robinson. L. Robinson, Leonard Beedle, T. Stickney & Son, J. M. Ormsbee, D. & G. Cutting, Shoreham ; Baker & Smith, P. Elitharp, Bridport ; C. B. Cook. Charlotte ; W. R. Sanford, Orwell ; E. S. Stowell, Cornwall ; S. T. Baker, Joel Barlow, Whiting.

EWES PURCHASED

May, 1871, a flock of ewes were purchased of Joel Barlow that were bred pure from the flock of Lucius Robinson. One ewe bred from the flock of W. R. Sanford was purchased of C. Jones.

RAMS USED.

Gen. Fremont (126) to a few ewes ; a ram was hired of the late German Cutting ; one was hired of Col. E. S. Stowell ; otherwise those bred within the flock.

———o———

FLOCK 49.

Owned by W. H. COOK & RALPH BROWN, RICHVILLE, ADDISON Co., VT.

6 RAMS.

Marked with metallic labels in their ears, having thereon W. H. Cook & R. B., numbers 1 to 6.

44 EWES. ·

Labeled same as rams and numbers 7 to 50.

PEDIGREE

Descended from importations from Spain through the flocks of Andrew Cock, Long Island ; David Humphreys, Stephen Atwood, Conn.; William Jarvis, Weathersfield ; L. C. Remele, E. R. Robinson, T. Stickney & Son, A. B. Treadway, F. & L. E. Moore, Leonard Beedle, Shoreham ; Baker & Smith, Prosper Elitharp, Bridport; C. B. Cook, Charlotte.

EWES PURCHASED.

In 1862 one ewe, of Rich blood, was purchased of A. B. Treadway ; during the Fall of 1876, fifteen ewes were purchased of T. Stickney & Son.

RAMS USED.

On the first-named ewe and her female descendants the stock rams of J. T.& V. Rich, T. Stickney & Son, F.& L.E. Moore, and those bred from them ; in 1876 Fremont, Jr. (215) was used, and the same year a yearling ram was purchased, bred by L. C. Remele.

———o———

FLOCK 50.

Owned by E. S. CORLISS, NEPEUSKUN, WINNEBAGO Co., WIS.

16 RAMS.

Marked with metallic labels in their ears, having thereon E. S. Corliss and numbers 1 to 16.

52 EWES.

Labeled same as rams, and numbers 17 to 68.

PEDIGREE.

Descended from importatations from Spain through the flocks of Andrew Cock, Long Island ; David Humphreys, Stephen Atwood, J. N. Blakeslee, J. R. Nettleton, Conn ; Wm. Jarvis, Weathersfield; Messrs. Rich, E. R. Robinson, Leonard Beedle, T. Stickney & Son, L. C. Remele, S. L. Bissell, E. N. Bissell, Shoreham ; C. B. Cook, Charlotte ; Baker & Smith, P. Elitharp, Bridport ; W. R. Sanford, H. G. Hibbard, Orwell ; Messrs. Hammond, Middlebury ; F. D. Barton, Waltham ; Ward M. Lincoln, J. Hinds, Brandon ; E. Porter, Rutland ; R. P. Hall, E. S. Stowell, Cornwall.

EWES PURCHASED.

In 1874 twenty-three ewes and three ewe tegs were purchased of E. N. Bissell ; their blood was Atwood, Rich and Blakeslee ; the same day thirty-four Atwood ewes were purchased of F. D. Barton, in company with C. D. McConnell, E. S. Corliss receiving one-half.

RAMS USED.

Three ram tegs were purchased of J. T. & V. Rich in 1874, they were sired by Hibbard (214), dams, Rich ewes; an Atwood ram was purchased of F. D. Barton at the same time that the purchase of ewes was made of him; Corliss Stickney Ram (298) was purchased and has been used as the principal stock ram of the flock, though the Barton ram has been used to some extent. Corliss' Stickney Ram (298) was lambed in 1874, not 1876, as is printed erroneously in our list of stock rams.

———o———

FLOCK 51.

Owned by HENRY CORLISS, RIPON, WISCONSIN.

10 RAMS.

Marked with metallic labels in their ears, nine having thereon H. Corliss and numbers 1, 2, 4, 7, 8, 10, 18, 80, 94, and one having on his label J. T. S. and No. 103.

99 EWES.

Labeled same as rams, seventy-two having on their labels H. Corliss, and numbers 3, 5, 6, 9, 11, 12, 13, 15, 16, 17, 19 to 79, 95, and twenty-seven having on their labels J. T. S. and numbers 2, 47, 50, 51, 56, 57, 58, 60, 63, 66, 67, 69 to 84.

PEDIGREE.

Descended from importations from Spain, through the flocks of David Humphreys, Stephen Atwood, J. N. Blakeslee, J. Nettleton, Conn.; Andrew Cock, Long Island; William Jarvis, Weathersfield; Messrs. Rich, Joseph Barnum, T. Stickney & Son, L. C. Remele, E. R. Robinson, Lucius Robinson, Geo. H. Hall, S. L. Bissell, E. N. Bissell, J. T. Stickney, James Forbes, Jr., Elmer Barnum, C. E. Bush, A. C. Harris, E. Fanning, Miner Jones, E. R. Cudworth, Shoreham; W. R. Sanford, H. G. Hibbard, Orwell; Ward M. Lincoln, Jesse Hinds, Brandon; E. Porter, Rutland; Messrs. Hammond, Middlebury; N. A. Saxton, Waltham; R. P. Hall, E. S. Stowell, John Towle, R. J. Jones, Cornwall; Baker & Smith, P. Elitharp, Bridport; C. B. Cook, Charlotte; A. H. Hubbard, J. Q. Stickney, Whiting; Seneca Root, Hubbardton; L. Webster, Sudbury.

EWES PURCHASED.

February, 1877, eighty-eight ewes were purchased of E. N. Bissell, of East Shoreham, Vt. Thirty-seven of them were bred by himself, and combined the Cock, Atwood, Blakeslee and Jarvis strains of blood. Thirteen were bred by E. Barnum, eight were bred by James Forbes, Jr., three by E. Fanning, three by C. E. Bush, six by J. T. Stickney, seven by M. Jones, and eleven by E. R. Cudworth. All these, bred by the last seven breeders combined the blood of Cock, Atwood and Jarvis. A part of these ewes were in lamb by J. T. Stickney's 146 (441) and Centennial (442), and the lambs dropped in the Spring of 1877 were sired by them.

RAMS PURCHASED OR USED.

In 1874 a ram of Stickney blood was purchased of T. Stickney & Son; Feb., 1877, one ram of Stickney blood was purchased of E. N. Bissell, that was bred by J. T. Stickney; and two of Rich blood that were bred by J. T. & V. Rich.

----o----

FLOCK 52.

Owned by O. COOK, Whitewater, Wis.

30 RAMS.

Marked with metallic labels in their ears, having thereon O. Cook, the letter A. and numbers from 86 to 100, 501, 505, 506, 508, 510, 512, 521, 525, 534, 536, 552, 553, 556, 565, 568.

113 EWES.

Marked same as rams and numbered from 1 to 85, 500, 503, 504, 511, 516, 518, 524, 527, 528, 530, 531, 532, 535, 537, 538, 539, 547, 548, 550, 551, 554, 555, 561, 562, 563, 565, 566, 569.

PEDIGREE.

Descended from importations from Spain through the flocks of David Humphreys, Stephen Atwood, J. N. Blakeslee, J. R. Nettleton, Conn.; Andrew Cock, Long Island; David Buffum, William Bailey, Joseph I. Bailey, Rhode Island; William Jarvis, Weathersfield; Messrs. Hammond, Middlebury; Messrs. Rich, L. C. Remele, Leonard Beedle, E. R. Robinson, T. Stickney & Son, D. & G. Cutting, J. M. Ormsbee, Henry W. Walker, H. A. Bascom, S. L. Bissell, Shoreham; Baker & Smith, Prosper Elitharp, Bridport; S. T. Baker, Whiting; Victor Wright, Weybridge; Rollin Gleason, Benson; O. C. Bacon, Waltham; Jesse Hinds, Dennison Blackmer, Brandon; C. S. Rumsey, Albert Bresee, M. Barber, C. Roach, Hubbardton; W. R. Sanford, V. V. Blackmer, Orwell; E. S. Stowell, H. F. Dean, S. S. Rockwell, Cornwall; L. C. Smith, Whitewater, Wis.; Joseph Marsh, A. Lawrence, Hinesburgh.

EWES PURCHASED.

Flock commenced in the Spring of 1864, by a purchase of five ewes of A. F. Knox, of Whitewater, Wis. They were bred by T. Stickney & Son, and were in lamb by his stock rams. They were Atwood, Jarvis, and Cock blood. In the Fall ot 1867, three ewes were purchased of A. F. Knox, bred by L. C. Remele, Atwood, Cock and Jarvis blood. In the Fall of 1871, nine ewes were purchased of L. C. Smith, one was bred by Rollin Gleason, three by O. C. Bacon, three by Chester Roach, two by Dennison Blackmer, all of Atwood blood ; also, same day, eight ewes bred from the above by rams of Atwood blood ; Eldorado and Cosset (45), bred by W. R. Sanford, and Rattler (303) bred by V. V. Blackmer.

RAMS PURCHASED OR USED.

In 1864, Cosset (45) (This ram was called Gold Mine after he went to Wisconsin); in 1865, same ram and Rattler (303); in 1866, Green Mountain (304) and Rattler (303); in 1867–68–69, Green Mountain Boy (305); in 1870–71, Elitharp. No.5 (307); in 1872–73–74, G. Cutting, bred by G. Cutting; in 1874, a part of the ewes were served by Wrinkly (308); in 1875–76 Wrinkly (308) and Walworth (302).

———o———

FLOCK 53.

Owned by REUBEN COOK, RICHVILLE (in Shoreham), ADDISON Co., VT.

21 RAMS.

Marked with metallic labels in their ears, having thereon R. Cook, and numbers 1 to 19, 30, 31.

110 EWES.

Labeled same as rams, and numbers 20 to 29, 32 to 131.

PEDIGREE.

Descended from importations from Spain, through the flocks of A. Cock, Long Island; David Humphreys, Stephen Atwood, Conn.; Wm. Jarvis, Weathersfield ; Messrs Rich, T. Stickney & Son, E. R. Robinson, Leonard Beedle, L. C. Remele, Shoreham ; C. B. Cook, Charlotte ; Baker & Smith, P. Elitharp, Bridport.

EWES PURCHASED.

In Oct., 1863, four breeding ewes were purchased of J. T. & V. Rich, and about the same time, six ewes were purchased of A. B. Treadway, that were bred by J. T. & V. Rich.

RAMS USED.

Rams from the flocks of J. T. & V. Rich and T. Stickney & Son ; otherwise, rams bred within the flock.

FLOCK 54.

Owned by C. B. COOK, Charlotte, Chittenden Co., Vt.

14 RAMS.

Marked with metallic labels in their ears, having thereon C. B. Cock and numbers 37 to 50.

36 EWES.

Labeled same as rams and numbers 1 to 36.

PEDIGREE.

Descended from the importation from Spain by Col. David Humphreys, through the flocks of Stephen Atwood, Conn.; Messrs Hammond, Middlebury; Victor Wright, Weybridge; W. R. Sanford, Orwell; Henry Thorp, Charlotte.

RAMS PURCHASED.

Flock commenced in Oct., 1841, by a purchase of twenty-three ewes of Stephen Atwood; in Jan., 1845, six ewes were purchased of the same, and five Atwood ewe lambs of his son, Chancy Atwood.

RAMS USED.

At the time of the first purchase of ewes, Cook Ram (14) was purchased of S. Atwood; January, 1845, three ram tegs were purchased of S. Atwood, one of these, Black Buck (15), was used for some years; in 1847-48-49, Atwood ram (12) was used; after this the stock rams of Messrs. Hammond were used with those bred within the flock until 1859, when a half interest in America (35) was purchased; for seven years past Atwood rams, bred by Henry Thorp, have been used.

————o————

FLOCK 55.

Owned by B. W. COPE, Smithfield, Jefferson Co., Ohio.

11 RAMS.

Marked with metallic labels in their ears, having thereon B. W. Cope, and numbers 1, 85 to 89, 95 to 98, 101.

39 EWES.

Labeled same as rams, and numbers 1 to 34, 90 to 94.

PEDIGREE.

Descended from importations from Spain through the flocks of David Humphrey's, Stephen Atwood, Conn.; A. Cock, Long Island; Wm. Jarvis Weathersfield; Messrs. Hammond, Middlebury; Victor

Wright, Weybridge ; C. B. Cook, Charlotte ; W. R. Sanford, Or-
well ; Messrs. Rich, Leonard Beedle, L. C. Remele, E. R. Robinson,
E. A. Birchard, Shoreham ; Baker & Smith, P. Elitharp, B. Myrick,
H. C. Burwell, C. N. Hayward, J. J. Crane, D. F. Doty, J. Hill, E.
Fitch, Bridport ; G. J. Hollenbeck, Hoosick, N. Y.; N. A. Saxton,
Waltham ; C. D. Lane, Henry Lane, Cornwall ; R. Perriue, Patter-
son's Mills, Pa.; L. P. Clark, Addison.

EWES PURCHASED.

Nov., 1875, one Atwood ewe was purchased of R. Lane, one
Atwood and Robinson ewe of Cherbino & Williams that was bred
by C. D. Lane, and one Atwood ewe of L. P. Clark ; Jan., 1876, eight
Atwood and Robinson ewes were purchased of H. C. Burwell, and
eighteen of Cherbino & Williamson ; the last were Atwood and
Robinson blood.

RAMS USED.

Advance (316), B. W. Cope's 1 (317), Young Hammond (315),
B. W. Cope's 83 (318), have all been used, the last to but one ewe.

---o---

FLOCK 56.

Owned by S. B. M. COWLES, NEW HAVEN, ADDISON Co., VT.

1 RAM.

Marked with metallic label in his ear, having thereon S. B.
Cowles and No. 50.

49 EWES.

Labeled same as ram and No. 1 to 49.

PEDIGREE.

Descended from Col. David Humphreys' importation from Spain,
through the flocks of Stephen Atwood, Conn.; Messrs. Hammond,
Middlebury ; W. R. Sanford, Orwell ; Victor Wright, Weybridge ;
N. A. Saxton, F. D. Barton, Waltham ; Martin Cowles, New Haven.

EWES PURCHASED.

Ten Old Atwood ewes were purchased by Martin Cowles, father
of S. B. M. Cowles, of W. S. & E. Hammond, about 1848-49 ; Sept.
25, 1862, three Atwood ewes were purchased of N. A. Saxton.

RAMS USED.

Rams of Atwood blood bred by Messrs. Hammond, N. A. Saxton and F. D. Barton ; also some bred within the flock, the only
exception being five ewes one season to a Robinson ram, the produce from which were all sold.

FLOCK 57.

Owned by BYRON W. CRANE, Bridport, Addison Co., Vt.

16 RAMS.

Marked with metallic labels in their ears, having thereon B. Crane, and numbers 62, 64, 70, 72 to 76, 82, 85, 90, 93, 94, 98, 99, 100.

92 EWES.

Sixty-seven labeled same as rams and numbers 1 to 44, 57, 60, 61, 63, 65 to 69, 71, 77 to 81, 83, 84, 86 to 89, 91, 92.

Twenty-five marked with labels having thereon C. D. Lane and numbers 12, 15, 21, 23, 26, 28, 31, 34, 37, 43, 47, 49, 50, 53, 64, 55, 68, 104, 111, 168, 174, 176, 258, 261, 264.

PEDIGREE.

Descended from importations from Spain through the flocks of David Humphreys, Stephen Atwood, Conn; A. Cock, Long Island; Wm. Jarvis, Weathersfield; Messrs Hammond, Middlebury; Victor Wright, Weybridge; R. P. Hall, R. J. Jones, Cornwall; Messrs. Rich, D. E. Robinson, G. H. Hall. L. C. Remele, Shoreham; W. R. Sanford, Orwell; Baker & Smith, P. Elitharp, C. N. Hayward, Bridport; Nelson Richards, Panton; F. D. Parton, Waltham.

EWES PURCHASED.

In 1863, two Atwood ewes were purchased of the estate of L. A. Johnson, Weybridge, that were bred by E. Hammond; in 1874, fifteen ewes were purchased of E. B. Jewett, that were bred by Geo. H. Hall; they were sired by rams bred by F. D. Barton.

RAMS USED.

The two ewes first purchased were bred to the stock rams of Victor Wright for three years, then rams bred within the flock were used until 1874, with the exception of one year, when a Robinson ram was used, bred by D. E Robinson; since 1874 a ram bred and owned by the estate of C. N. Hayward has been used.

———o———

FLOCK 58.

Owned by CASIUS P. CRANE, Bridport, Addison Co., Vt.

62 RAMS.

Fifty-three marked with metallic labels in their ears, having thereon C. P. Crane, and numbers 1 to 31 and 37 to 58; nine marked with labels having thereon H. C. Burwell and numbers 64 to 72.

101 EWES.

Labeled same as rams, and numbered 1 to 60 and 88 to 91, 93 to 130.

PEDIGREE.

Descended from importations from Spain through the flocks of David Humphreys, Stephen Atwood, Conn.; Andrew Cock, Long Island; William Jarvis, Weathersfield; Messrs. Hammond, Cherbino & Williamson, E. R. Clay, U. D. Twitchell, O. P. Lee, Middlebury; W. R. Sanford, Orwell; C. B. Cook, Charlotte; Baker & Smith, P. Elitharp, C. N. Hayward, D. F. Doty, B. Myrick, H. C. Burwell, J. J. Crane, Bridport; Nelson Richards, Panton; Messrs. Rich, Leonard Beedle, L. C. Remele, E. R. Robinson, T. Stickney & Son, Alvin Clark, Lucius Robinson, Shoreham; N. A. Saxton, O. C. Bacon, Waltham; A. A. Farnsworth, Brooksville; Henry Lane, C. D. Lane, James DeLong, Milo J. Ellsworth, Chas. Benedict, S. S. Rockwell, H. Sanford, Cornwall; J. G. Barker, Leicester; H. M. Graves, Salisbury.

HISTORY.

Flock commenced in the Fall of 1863 by Chilon Crane, father of the present owner, who purchased two ewes of Atwood and Robinson blood; they were sired by America (35), and out of the ewes bred by P. Elitharp.

These ewes and their progeny were bred to Eureka (58), Young Eureka (81), Chunk (95), Young Chunk (206), Sea Lion (83), and Stoga (355), were used. Since the flock came into the possession of the present owner, the following purchases have been made: Oct., 1874, fifty Atwood and Robinson ewes were purchased of C. D. Lane; November, 1874, ten ewes, of Beedle and Atwood blood, were purchased of H. E. Sanford; Dec., 1877, twelve ewe tegs were purchased of O. P. Lee.

RAMS USED.

In 1874, Bonaparte (176), Silver Horns (177), and Young Stoga (356); the stock of the first two is all sold; in 1875, Eureka, 3d (228), and Ironsides, 3d (357), were used; in 1876, Ironsides, 3d (357); and a few to Bismarck (221), Stub, 222), and Black Top (225).

———o———

FLOCK 59.

Owned by JULIUS J. CRANE, BRIDPORT, ADDISON CO., VT.

20 RAMS.

Marked with metallic labels in their ears, having thereon J. J. Crane and numbers 1 to 20.

68 EWES.

Labeled same as rams and numbers 1 to 68.

PEDIGREE.

Descended from importations from Spain through the flocks of D. Humphreys, Stephen Atwood, Conn.; A. Cock, Long Island; Wm. Jarvis, Weathersfield; Messrs. Hammond, E. D. Munger, H. R.Holden, Middlebury; Messrs.Rich, L. C. Remele, Leonard Beedle, E. R Robinson, E.A. Birchard, Shoreham; W. R. Sanford, Orwell; Prosper Elitharp, C. N. Hayward, D. F. Doty, H. C. Burwell, B. Myrick, G. Bruce, Bridport; R. P. Hall, Cornwall; O. C. Bacon, N. A. Saxton, Waltham.

EWES PURCHASED.

In 1862, one Atwood ewe, bred by P.Elitharp, and a pair of twin ewe lambs were purchased of C. N. Hayward; the lambs were sired by America (35); soon after this ten ewes were leased of D. F. Doty upon shares; a few Atwood ewes were also purchased of C. Abernathy, bred from Hammond stock; three more ewes were purchased of C N. Hayward, same blood as the first, and one sired by Eureka (58) out of an Atwood ewe, bred by Victor Wright; Feb., 1876, nineteen Atwood ewes were purchased of John T. Whitlock, that were bred by H. R. Holden; Sept. 28, 1877, one Atwood ewe was purchased of Geo. Bruce; she was sired by D. F. Doty's ram, and out of an ewe bred by O. C. Bascom.

RAMS USED.

The first ewe purchased was bred to Sweepstakes (32) and Ironsides (80); one of the twin ewes was bred to Gold Drop (64), and four times to Eureka (58); Young Eureka (81) was used on the Doty ewes, and was the stock ram of the flock for a number of years; Silver Horns (177) was used to two ewes in 1872, and the progeny was Ironsides (357) and Eureka, 3d (223); the latter has been the stock ram of the flock since 1873; Sea Lion (83) and Hamilton & Crane (175) were used to some extent, but the stock from them is all sold, none remaining in the flock.

——o——

FLOCK 60.

Owned by CHARLES E. CRANE, BRIDPORT, ADDISON CO., VT.

22 RAMS.

Marked with metallic labels in their ears, having thereon C. E. Crane and numbers 101 to 122.

68 EWES.

Labeled same as rams, and numbers 1 to 68.

PEDIGREE.

Descended from importations from Spain through the flocks of David Humphreys, Stephen Atwood, Connecticut; Andrew Cock, Long Island; Messrs. Hammond, Middlebury; N. Richards, Panton; P. Elitharp, C. N. Hayward, B. Myrick, H. C. Burwell, J. J. Crane, D. F. Doty, C. P. Crane, Bridport; Messrs. Rich, L. C. Remele, E. R. Robinson, L. Beedle, E. A. Birchard, Shoreham; W. R. Sanford, Orwell; O. C. Bacon, Waltham; Victor Wright, Weybridge; the blood of a small part of the flock pedigrees in part through the flocks of David Buffum, Wm. Bailey, Joseph I. Bailey, Rhode Island; S. T. Baker, Whiting; J. M. Ormsbee, D. & G. Cutting, Shoreham; J. O. Hamilton, Bridport.

EWES PURCHASED.

Flock commenced in 1863, by a purchase of four Atwood ewes of C. N. Hayward; in 1865, eight ewes were leased upon shares of George Bruce; five of these ewes were bred by Nelson Richards, two by O. C. Bacon, and one by C. N. Hayward; one Atwood ewe was also leased of P. Dakin, that was bred by H. B. Dodge; this ewe was sired by a ram bred by V. Wright, and out of ewe bred by O. C. Bacon.

RAMS USED.

The stock rams of C. N. Hayward, D. F. Doty and J. J. Crane, including Ironsides (80), Chunk (95), Doty Ram (134), Young Eureka (81), Young Stoga (356), Eureka 3d (223); also those of H. C. Burwell, including Bismarck (221), Stub (222), Black Top (225) and Acme (524); one of J. O. Hamilton's stock rams, Greasy (297), was used to a few ewes, and a ram sired by him was used to a few; there are sixteen ewes now in the flock sired by these rams; besides the rams named, rams bred within the flock have been used.

———o———

FLOCK 61.

Owned by the Estate of GERMAN CUTTING, EAST SHOREHAM, ADDISON Co., VT.

10 RAMS.

Marked with metallic labels in their ears, having thereon Cutting and Nos. 78 to 87.

78 EWES.

Labeled same as rams and numbers 1 to 78.

PEDIGREE.

Descended from importations from Spain through the flocks of David Buffum, Wm. Bailey, Joseph I. Bailey, Rhode Island; D. Humphreys, S. Atwood, Conn.; A. Cock, Long Island; Wm. Jarvis, Weathersfield; S. T. Baker, W. P. Wright, Whiting; R. P. Hall, E S. Stowell, R. J. Jones, E. Sanford, Cornwall; N. A. Saxton, Waltham, W. R. Sanford, Orwell; Messrs. Hammond, Middlebury; V. Wright, Weybridge; Messrs. Rich, L. C. Remele, E. R. Robinson, T. Stickney & Son, G. H. Hall, Shoreham; Baker & Smith, P. Elitharp, Bridport.

EWES PURCHASED.

The first purchase of ewes to found this flock was by D. & G. Cutting about 1841, and consisted of about eighty ewes of Rhode Island blood, of James M. Ormsbee, as is noted at page 58 of this work. About the year 1854 a few Atwood ewes were purchased of N. A. Saxton and E. Sanford, and a few that combined the Cock and Jarvis bloods with the Atwood of Wm. Wright. Oct. 55, 1862, a yearling Atwood ewe was purchased of N. A. Saxton, and Feb. 10, 1863, another ewe was purchased of the same.

RAMS USED.

Two rams of the same blood of the first purchase of ewes were purchased with them and were used at first; soon after the stock rams of Messrs. Hammond and W. R. Sanford, including Old Black (9), and others; also rams bred by Messrs. Hammond, including Wooster (16). (He sired about fifty lambs for Messrs. Cutting.) A ram was purchased of S. Atwood, Atwood Ram (25), and another of George Atwood, George Atwood Ram (26). Saxton Ram (27), was purchased of N. A. Saxton, Sept. 13, 1853. Young Wooster (28), Greasy (31), Young Sweepstakes (50), Monitor (141), Addison Chief (143), were all bred within the flock, and were all used more or less as the stock rams of the flock, Sweepstakes (32). Old Robinson Ram (38), Eureka (58), Wright's California (43), were all used to some extent; later Gold Drop (64), was used to a very few ewes. Later still the stock rams of E. S. Stowell, W. R. Sanford, T. Stickney & Son and rams bred by L. C. Remele have been used, Little Hall Ram (467) was purchased and used. Gen. Grant (474) was used in 1877, and Banker (471) also in the same year. Soon after this flock was commenced, it was divided by the brothers David and German Cutting, but the blood and breeding of the two flocks was substantially the same. Soon after the decease of David Cutting his flock came into the possession of his nephews, G. A. Cutting and H. M. Perry, and the other part of the flock, belonging to German Cutting, since his death has descended to his heirs.

FLOCK 62.

Owned by F. H. DEAN, WEST CORNWALL, ADDISON Co., VT.

20 RAMS.

Marked with metallic labels in their ears, having thereon F. H. Dean, and numbers 1 to 20.

180 EWES.

Labeled same as rams and numbers 21 to 200.

PEDIGREE.

Descended from importations from Spain through the flocks of D. Humphreys, S. Atwood, Conn.; A. Cock, Long Island; Wm. Jarvis, Weathersfield; Messrs. Hammond, S. W. Remele, David Hooker, Middlebury; A. A. Farnsworth, Brooksville; Wm. M. Gage, Ferrisburgh; A. J. Stowe, V. Wright, Weybridge; N. A. Saxton, Waltham; Messrs. Rich, Leonard Beedle, E. R. Robinson, L. C. Remele, R. N. Atwood, E. A. Birchard, Levi Wolcott, Shoreham; Charles Forbes, West Haven; Henry Root, J. S. Griswold, I. Dickinson, G. Root, Benson; W. R. Sanford, Orwell; Baker & Smith, P. Elitharp, B. Myrick, C. N. Hayward, H. C. Burwell, D. F. Doty, J. J. Crane, Bridport; C. B. Cook, H. Thorp, Charlotte; E. Meach, E. S. Rowley, Shelburn; E. S. Stowell, F. Hooker, W. H. Delong, R. P. Hall, R. J. Jones, Cornwall; L. P. Clark, Addison.

EWES PURCHASED.

Five ewe tegs were purchased of W. S. & E. Hammond about 1851. Some years after a purchase of two hundred Atwood ewes was made of Wm. M. Gage; a selection of sixty of these were retained in the flock. In 1863 ten Atwood ewe-tegs were purchased of L. P. Clark, and Aug. 27, 1864, ten ewe tegs that were Atwood blood bred direct from the flock of Messrs. Hammond. In 1874 twenty Robinson ewes were purchased of Levi Wolcott.

RAMS USED.

In 1851 an Atwood ram teg was purchased of W. S. & E. Hammond, and used for a few years. The stock rams of Messrs. Hammond and Col. E. S. Stowell have been used to some extent. Little Wrinkly (48) was purchased and used as the stock ram for many years; rams bred within the flock and sired by him have been used. Casius (257) was purchased in 1875 and has been the stock ram of the flock since. In 1876 Wrinkly (292) and W. H. Delong's 100 (360) were used to a part of the flock.

FLOCK 63.

Owned by WILLIAM H. DELONG, West Cornwall, Addison Co., Vt.

23 RAMS.

Marked with metallic labels in their ears, having thereon W. H. Delong and numbers 78 to 100.

71 EWES.

Labeled same as rams, with numbers 1 to 71.

PEDIGREE.

Descended from importations from Spain through the flocks of D. Humphreys, S. Atwood, Conn.; A. Cock, Long Island; Wm. Jarvis, Weathersfield; Messrs. Hammond, Middlebury; V. Wright, A. J. Stowe, Weybridge; N. A. Saxton, Waltham; L. P. Clark, Addison; W. R. Sanford, Orwell; Messrs. Rich, L. Beedle, L. C. Remele, E. R. Robinson, D. E. Robinson, E. A. Birchard, T. Stickney & Son, Shoreham; Baker & Smith, P. Elitharp, Bridport; Wm. M. Gage, Ferrisburgh; J. Q. Stickney, Whiting; S. Root, Hubbardton; L. Webster, Sudbury; R. P. Hall, E. S. Stowell, R. J. Jones, F. H. Dean, Cornwall.

EWES PURCHASED.

Flock commenced Jan. 11, 1862, by a purchase of seven ewes of Darwin E. Robinson. July 10, 1862, eleven ewes of the same blood were purchased of F. D. Douglas; the same year three Atwood ewes were purchased of F. H. Dean. April 16, 1873, thirty Atwood ewes were purchased of F. H. Dean; they were sired by Little Wrinkly (48).

RAMS USED.

Sweepstakes (32), Gold Drop (64), Comet (57), Golden Fleece (91), Eureka (58), Little Wrinkly (48), and Allright (169), have all been used to a few ewes. Delong's Remele Ram (361) was purchased and has been one of the stock rams of the flock since 1875. Gen Grant (72), Fortune (249), and W. H. Delong's 100 (360), have been bred within the flock and used as stock rams.

——o——

FLOCK 64.

Owned by HARRISON F. DEAN, West Cornwall, Addison Co., Vt.

2 RAMS.

Marked with metallic labels in their ears, having thereon H. F. Dean and numbers 34, 35.

33 EWES.

Labeled same as rams and numbers 1 to 33.

PEDIGREE.

Descended from importations from Spain through the flocks of Andrew Cock, Long Island; David Humphreys, Stephen Atwood, Conn.; William Jarvis, Weathersfield; Messrs. Hammond, Middlebury; Messrs. Rich, E. R. Robinson, Leonard Beedle, A. C. Harris, T. Stickney & Son, E. A. Birchard, C. E. Bush, Shoreham; J. Q. Stickney, Whiting; S. Root, Hubbardton; L. Webster, Sudbury; C. Merrill, Addison; C. B. Cook, Charlotte; W. R. Sanford, J. H. Thomas, Orwell; Baker & Smith, P. Elitharp, C. N. Hayward, D. F. Doty, B. Myrick, H. C. Burwell, J. J. Crane, Bridport; R. P. Hall, R J. Jones, Cornwall.

EWES PURCHASED.

The flock was commenced in the Spring of 1876, by a purchase of twenty Robinson and Stickney ewes of C. E. Bush; they were sired by Fremont, Jr. (215); the same season eight Atwood and Robinson ewes were purchased of J. J. Crane; they were sired by Eureka, 3d (223); Oct. 5, 1877, four ewes were purchased of C. E. Bush, same blood as the first.

RAMS USED.

In 1876, Allright (169); one yearling ewe and two rams were sired by him.

———o———

FLOCK 65.

Owned by DAVID F. DOTY, Bridport, Addison Co., Vt.

23 RAMS.

Marked with metallic labels in their ears, having thereon D. F. Doty and numbers 1 to 23.

57 EWES.

Labeled same as rams with numbers 1 to 57.

PEDIGREE.

Descended from the importations from Spain through the flocks of David Humphreys, Stephen Atwood, Conn.; William Jarvis, Weathersfield; Messrs. Hammond, Middlebury; Messrs. Rich, Leonard Beedle, E. R. Robinson, L. C. Remele; E. A. Birchard, A. Clark, Shoreham; N. Richards, Panton; Baker & Smith, P. Elitharp, C. N. Hayward, B. Myrick, H. C Burwell, J. J. Crane, Bridport; W. R. Sanford, Orwell.

EWES PURCHASED.

In 1853, seven Atwood ewes were purchased of P. Elitharp.

RAMS USED.

America (35), Ironsides (80), Sea Lion (83), Eureka (58), Young Eureka (81) and Eureka, 3d (223), have all been used more or less; Doty Ram (134) and others bred within the flock, have also been used.

——o——

FLOCK 66.

Owned by S. & S D. DOUD, New Haven, Addison Co., Vt.

5 RAMS.

Marked with metallic labels in their ears, having thereon S. & S. D. Doud and numbers 1 to 5.

35 EWES.

Labeled same as rams and numbers 6 to 40.

PEDIGREE.

Descended from the importation from Spain by Col. David Humphreys, through the flocks of Stephen Atwood, Conn.; Messrs. Hammond, Middlebury; A. J. Stowe, Weybridge; S. G. Holyoke, St. Albans.

EWES PURCHASED.

The flock was commenced Oct. 11, 1853, by a purchase of five Atwood ewe tegs of S. G. Holyoke; a year later four more Atwood ewes were purchased from the same flock.

RAMS USED.

These ewes were bred to rams owned or bred by Messrs. Hammond or those bred within the flock, except a few ewes to Little Wrinkly (48), and a few to an Atwood ram, bred by A, J Stowe; Eureka (58) was used to a few ewes, but the produce is all sold; of those bred within the flock, Doud's Wrinkly (363) was probably used longest of any.

——o——

FLOCK 67.

Owned by F. D. DOUGLAS, Whiting, Addison Co., Vt.

13 RAMS.

Marked with metallic labels in their ears, having thereon F. D. Douglas and numbers 1 to 13.

26 EWES.

Labeled same as rams and numbers 1 to 26.

PEDIGREE.

Descended from importations from Spain through the flocks of Andrew Cock, Long Island; D. Humphreys, S. Atwood, Conn.; David Buffum, William Bailey, Joseph l. Bailey, Rhode Island; Messrs. Rich, E. R. Robinson, L. Beedle, J. M. Ormsbee, D. & G. Cutting, L. C. Remele, Shoreham; Messrs. Hammond, Middlebury; Baker & Smith, P. Elitharp, Bridport; S. T. Baker, Whiting.

EWES PURCHASED.

About 1851, twenty-four ewes were purchased of J. Sheldon, of Fair Haven, Vt., that were bred by and purchased of J. Thurman Rich; about the same time twenty breeding ewes were purchased of E. R. Robinson. In 1860 thirty-two ewes were leased of Mrs. E. R. Robinson upon shares.

RAMS USED.

Robinson rams at first, but soon after the foundation of the flock Cutting blood was introduced by using Old Black (139); after that, Monitor (140) was bred within the flock and used a few years; after that, Rich and Robinson rams, principally bred within the flock, were used. No records have been kept to show dates or names of rams used, but care was taken to use none but pure Spanish Merinos.

——o——

FLOCK 68.

Owned by L. B. DODGE, Weybridge, Addison Co., Vt.

13 RAMS.

Marked with metallic labels in their ears, having thereon L. B. Dodge and numbers 25 to 37.

62 EWES.

Labeled same as rams, and numbers 1 to 24 and 38 to 75.

PEDIGREE.

Descended from importations from Spain, through the flocks of David Humphreys, Stephen Atwood, Conn.; Andrew Cock, Long Island; William Jarvis, Weathersfield; Messrs. Hammond, S. Piper, J. Piper, S. W. Remele, Middlebury; A. J. Stowe, V. Wright, S. Dodge, H. B. Dodge, Weybridge; R. P. Hall, E. S. Stowell, Cornwall; Messrs. Rich, E. R. Robinson, E. G. Farnham, A. Clark, L. C. Remele, Shoreham; W. R. Sanford, Orwell; D. Spear, Charlotte.

RAMS PURCHASED.

In 1872 six old Atwood ewes were purchased of S. Dodge & Son, that were bred by E. Hammond ; soon after a few Atwood ewes were purchased of A. J. Stowe, that were also bred by E. Hammond ; in 1876, the entire flock of S. Dodge & Son, consisting of fifty-six ewes and ewe tegs, was purchased ; this flock were bred by Messrs. S. Dodge & Son, from a few ewes that were purchased of E. Hammond, a few of A. J. Stowe, that were bred by E. Hammond, and a few ewes of D. Spear, that were bred from Victor Wright's flock ; on these ewes were used Atwood rams, bred by Messrs. Hammond and within the flock, except Sea Lion (83) was used to a few ewes ; and Keeler Ram (287) was also used to a number of ewes the last years they were owned by S. Dodge & Son.

RAMS USED BY L. B. DODGE.

The stock rams of S. Dodge & Son ; also the stock rams of A. J. Stowe, one of which was owned in company ; this ram was bred by S. W. Remele ; an Atwood and Robinson ram, bred by J. H. Sprague ; a few ewes were served two seasons by the French & Mason Ram (362); in 1872 a ram was used from the E. R. Clay flock (see Clay & Wing, Flock 46, for blood); after this, a ram sired by him, dam by French & Mason (362). With the ewes purchased in 1876, was also purchased a ram of the same blood of the ewes, that has been used since.

———o———

FLOCK 69.

Owned by LEMUEL S. DREW, Burlington, Chittenden Co., Vt.

9 RAMS.

Marked with metallic labels in their ears, having thereon L. S. D. and numbers 47 to 55.

46 EWES.

Labeled same as rams and numbers 1 to 46.

PEDIGREE.

Descended from the importations from Spain by Col. David Humphreys, through the flocks of Stephen Atwood, Conn.; Messrs. Hammond, Middlebury ; W. R. Sanford, Orwell ; Victor Wright, Weybridge ; S. G. Holyoke, St. Albans ; C. B. Cook, Henry Thorp, Charlotte ; H. Harrington, Keeler's Bay.

EWES PURCHASED.

In the Summer of 1877, six Atwood ewes were purchased of H. Harrington ; Oct. 13, 1877, thirty-three Atwood ewes were purchased of S. G. Holyoke & Son.

RAMS USED.

Fall, 1877, Fortune (369) was used, and the ram tegs from 47 to 55, and the ewe tegs from 40 to 46 were sired by him.

———o———

FLOCK 70.

Owned by A. F. ELLSWORTH, Whiting, Addison Co., Vt.

22 RAMS.

Marked with metallic labels in their ears, having thereon A. F. E. and numbers 29, 42 to 55, 57 to 63.

67 EWES.

Sixty-three labeled same as rams, and numbers 1 to 28, 30 to 41, 64 to 86, and four marked on labels J. H. T. and numbers 57, 63, 66, 76.

PEDIGREE.

Descended from importations from Spain through the flocks of David Humphreys, Stephen Atwood, Conn.; A. Cock, Long Island; Messrs. Hammond, W. R. Remele, Middlebury; C. I. Benedict, Arlington; Jesse Hinds, Brandon; W. R. Sanford, J. H. Thomas, Orwell; Messrs. Rich, L. C. Remele, E. R. Robinson, Leonard Beedle, Shoreham; Wm. M. Gage. Ferrisburgh; H. F. Dean, Cornwall; Baker & Smith, Prosper Elitharp, Bridport.

EWES PURCHASED.

In 1847, a number of Atwood ewes were purchased of J. Hinds; in 1858, five more Atwood ewes were purchased of F. H. Dean; in 1863, five Atwood ewes were added to the flock by a purchase from C. I. Benedict. of Arlington; in 1863, eight more Atwood ewes were purchased of the same; in March, 1874, ten Rich ewes were purchased of J. H. Thomas.

RAMS USED.

In 1857 a Rich ram was purchased of J. T. & V. Rich, and used one year; a few ewes were bred to the Sanford & Gibbs Ram (56); in 1863, nine ewes were served by Eureka (58), and two more by the same ram in 1867; in 1872, a ram was used bred by L. C. Remele; and in 1873, Ellsworth's Remele Ram (328) was purchased, and was the stock ram of the flock for a number of years after; Atwood rams from the flock of C. I. Benedict have been used a number of years; of the rams bred within the flock, Ellsworth's Eureka (327), A. F. Ellsworth's 58 (329), and Ellsworth's 59 (330), have been used extensively.

FLOCK 71.

Owned by MILO J. ELLSWORTH, MIDDLEBURY, ADDISON CO.,VT.

57 RAMS.

Marked with metallic labels in their ears, having thereon M. J. E., and numbers 1 to 27, 68 to 84, 197, 213 to 224.

172 EWES.

Labeled same as rams, and numbers 28 to 67, 85 to 185, 187 to 196, 198 to 200, 225 to 242.

PEDIGREE.

Descended from importations from Spain, through the flocks of A. Cock, Long Island; David Humphreys, Stephen Atwood, Conn.; Wm. Jarvis, Weathersfield ; Messrs. Hammond, W. R. Remele, Middlebury ; Baker & Smith, P. Elitharp, Bridport ; Victor Wright, Weybridge ; R. P. Hall, Charles I. Benedict, J. B. Hamblin, Lyman H.Payne, R.J.Jones, Cornwall ; C.B Cook, Charlotte ; Messrs. Rich, L. C. Remele, E. R. Robinson, Leonard Beedle, M. R. Atwood, T. Stickney & Son, Shoreham ; W. R. Sanford, Orwell.

EWES PURCHASED.

The flock was commenced in 1860 by a purchase of thirty Atwood ewes of Charles Benedict ; the Spring following, thirty more ewes were purchased of Charles Benedict ; these ewes were bred from ewes bred by P. Elitharp and C. B. Cook, and from rams bred by Messrs. Hammond and W. R. Remele ; about the same time one Atwood ewe was purchased of J. B. Hamblin, and about the same time another ewe was added to the flock, of the same blood, and from the same stock ; in 1876 five ewe tegs and one two-year-old ewe were purchased of L. H. Payne ; they were of Robinson blood ; the last-named was in lamb by Allright (169).

RAMS USED.

Those without the flock have been : Sanford & Gibbs (56), Sea Lion (83), Eureka (58), in 1873, a ram was purchased of M. R. Atwood ; this ram was used two years ; in 1875–6, Ellsworth's Remele Ram (328) was used ; in 1876–7, M. J. Ellsworth's 1 (504) and M. J. Ellsworth's 2 (558), were used.

———o———

FLOCK 72.

Owned by FRED H. FARRINGTON, BRANDON, VT.

35 RAMS.

Thirty-four marked with metallic labels in their ears, having thereon F. H. F. and numbers 38 to 44, 49, 51, 53 to 55, 59, 60, 64 to 76, 78 to 82, 93, 94, and one having on his label B. & F. and number 1.

67 EWES.

Sixty marked same as rams with numbers 1 to 37, 45 to 48, 50, 52, 56 to 58, 61 to 63, 83 to 90, 95, 96, 97.

Seven marked J. T. & V. R. and numbers 98, 100, 101, 109, 112, 113, 116.

PEDIGREE.

Descended from importations from Spain through the flocks of David Humphreys, Stephen Atwood, J. N. Blakeslee, John Nettleton, Conn.; A. Cock, Long Island; William Jarvis, Weathersfield; W. R. Sanford, H. G. Hibbard, Orwell; Messrs. Hammond, W. R. Remele, A. Chapman, Middlebury; Messrs. Rich, E. R. Robinson, Lucius Robinson, L. C. Remele, E. A. Birchard, T. Stickney & Son, Jasper Barnum, Geo. D. Bryant, E. G. Farnum, George H. Hall, S. L. Bissell, E. N. Bissell, Shoreham; R. P. Hall, John Towle, E. S. Stowell, E. Sanford, F. H. Dean, R. J. Jones, Cornwall; Ward M. Lincoln, Jesse Hinds, Lucius Merriam, Brandon; N. A. Saxton, Waltham; Victor Wright, A. J. Stowe, Weybridge; Ezra Meech, S. T. Rowley, Shelburn; Henry Thorp, C. B. Cook, Charotte; Baker & Smith, P. Elitharp, J. J. Crane, B. J. Myrick, F. D. Doty, H. C. Burwell, Bridport; Wm. Gage, Ferrisburgh; Julius G. Barker, Leicester; J. Q. Stickney, Whiting; Seneca Root, Hubbardton; L. Webster, Sudbury.

EWES PURCHASED.

Flock commenced April 30, 1873, by a purchase of ten ewes of E. N. Bissell, of Atwood, Rich and Blakeslee blood. Fall of 1875 two more were purchased from the same flock of the same blood. Oct. 15, 1876, six ewe tegs were purchased of F. H. Dean of Atwood and Robinson blood, they were sired by Cassius (257). Nov. 20. 1876, forty ewes were purchased of Lucius Merriam of Atwood and Robinson blood, of these nineteen are retained in the flock, the remainder were sold to Dennison Blackmer. Spring of 1877. seven ewes were purchased of E. N. Bissell that were bred by J. T. & V. Rich.

RAMS PURCHASED AND USED.

In 1873-4 E. N. Bissell's Stickney Ram (282) was used; five ewes of his get are now reserved. In 1874, Stoga (212) was used; four ewes are now retained sired by this ram. In 1876 Bissell's Stickney (282) was purchased; same year Cassius (257), Silver Horns, Jr. (401), Fremont, Jr. (215) and a ram belonging to E. Sanford were used. Dec. 16, 1876, Snowflake was purchased and he was used in 1877.

FLOCK 73.

Owned by E. G. FARNHAM, West Cornwall, Addison Co., Vt.

27 RAMS.

Marked with metallic labels in their ears, having thereon E. G. Farnham and numbers 1 to 22 and 136 to 140.

94 EWES.

Labeled same as rams and numbers 23 to 96 and 116 to 135.

PEDIGREE.

Descended from importations from Spain through the flocks of A. Cock, Long Island; D. Humphreys, S. Atwood, Conn.; Wm. Jarvis, Weathersfield; Messrs. Rich, L. Beedle, E. R. Robinson, D. E. Robinson, George Farnham, A. B. Treadway, G. D. Bryant, B. B. Tottingham, H. W. Jones, Kent Wright, A. C. Harris, E. A. Birchard, E. D. Bush, L. C. Remele, Shoreham; Baker & Smith, P. Elitharp, C. N. Hayward, W. W. Winchester, Bridport; Nelson Richards, Panton; C. B. Cook, Charlotte; Messrs. Hammond, D. Hooker, S. W. Remele, O. Severance, Middlebury; N. A. Saxton, Waltham; D. Hooker, Cornwall.

HISTORY.

In 1840, George Farnham, father of the present owner of the flock, purchased ten pure Cock ewes of E. R. Robinson; this was before the introduction of Atwood or Jarvis blood into the flock of E. R. Robinson. There were also subsequent purchases from the same flock; these ewes were bred pure to rams of the same blood until the estate of Geo. Farnham was settled in 1856, when the present owner came into possession of the flock. In 1856 and 1857 the sheep that fell to the share of D. E. Robinson in the estate of D. R. Robinson were leased upon shares by E. G. Farnham for one-half of the increase. Sept. 8, 1878, twenty Atwood and Robinson ewes were purchased of Kent Wright; they were bred from stock purchased of H. W. Jones, bred from the flocks of Mrs. E. R. Robinson, E. A. Birchard, E. G. Farnham, B. B. Tottingham and A. C. Harris; these last twenty ewes were sired by the stock rams of B. B. Tottingham.

RAMS USED.

For a number of years those received with the sheep of D. E. Robinson; after that two Atwood rams owned by L. Treadway, bred by E. Hammond; one of these was Sanford & Treadway (54); after this a ram bred within the flock, sired by one of the Treadway rams. A Rich ram bred by A. B. Treadway was used next; then rams bred within the flock until America (295) was purchased of G. D. Bryant; after that rams bred within the flock until the last two years Wrinkly (292) has been used.

FLOCK 74.

Owned by EDWIN FANNING, Richville (in Shoreham), Addison Co., Vt.

13 RAMS.

Marked with metallic labels in their ears, having thereon E. Fanning and numbers 25 to 37.

24 EWES.

Labeled same as rams and Nos. 1 to 24.

PEDIGREE.

Descended from importations from Spain through the flocks of A. Cock, Long Island; D. Humphreys, S. Atwood, Conn.; Messrs. Rich, L. C. Remele, E. R. Robinson, L. Beedle, T. Stickney, J. T. Stickney, Shoreham; R. P. Hall, John Towle, B. F. Fields, Cornwall; Baker & Smith, P. Elitharp, Bridport; Messrs. Hammond, Middlebury; V. Wright, Weybridge.

EWES PURCHASED.

In 1867 six ewes were purchased of L. C. Remele, and about the same time one ewe of J. T. & V. Rich and one from the Robinson flock.

RAMS USED.

The stock rams of John Towle, J. T. & V. Rich, and J. T. Stickney, besides rams bred within the flock.

———o———

FLOCK 75.

Owned by AUGUSTUS A. FARNSWORTH, Brooksville, Addison Co., Vt.

5 RAMS.

Marked with metallic labels in their ears, having thereon A. A. F. and numbers 116 to 120.

115 EWES

Labeled same as rams, and numbers 1 to 115.

PEDIGREE.

Descended from importations from Spain through the flocks of David Humphreys, Stephen Atwood, Conn.; Andrew Cock, Long Island; Wm. Jarvis, Weathersfield; Messrs. Hammond, O. Severance, S. W. Remele, D. Hooker, R. D. C. Robbins, Middlebury; R. P. Hall, E. S. Stowell, F. Hooker, Cornwall; Messrs. Rich, E. R. Robinson, L. C. Remele, E. A. Birchard, Shoreham ; Baker & Smith, P. Elitharp, B. Myrick, C. N. Hayward, J. J. Crane, H. C. Bissell, Bridport ; N. Richards, Panton ; Victor Wright, Weybridge ; F. D. Barton, N. A. Saxton, Waltham ; W. M. McIntyre, New Haven.

EWES PURCHASED.

The entire crop of yearling ewes of one year was purchased of Stephen Atwood, and the same of R. P. Hall ; these purchases formed the foundation of the flock ; Oct., 1876, forty ewes were purchased of C. Abernathy, of New Haven, that he had purchased of W. M. McIntyre ; they were bred pure from the flocks of E. Hammond and R. P. Hall by Prof. Robbins.

RAMS USED.

Rams from the flocks of Messrs. Hammond, E. S. Stowell, F. D. Barton, W. R. Remele, O. Severance and F. Hooker, have been used, besides a number of excellent quality raised within the flock ; Silver Mine 2d (84) was used to 75 ewes in 1865 ; David (118) was used two seasons, 1872, and in 1876 ; Phil Sheridan (172) was used one or two seasons ; Harry (343) was used in 1873 ; the Robinson strain of blood was introduced in 1875, by using Sam (368) ; the lambs of 1876 were sired by him and are the only sheep in the flock that have any but Atwood blood ; E. S. Stowell's 274 (500) was purchased and used in 1877.

---o---

FLOCK 76.

Owned by C. B. FISK, BROOKFIELD, ORANGE CO., VT.

50 EWES.

Marked with metallic labels in their ears, having thereon C. B. Fisk and numbers 1 to 50.

PEDIGREE.

Descended from importations from Spain by Col. David Humphreys, through the flocks of S. Atwood, Conn.; L. D. Gregory, V. Wright, Weybridge ; Messrs. Hammond, Middlebury ; E. S. Stowell, M. B. Williamson, S. Benton, R. P. Hall, Cornwall ; A. A. Farnsworth, Brooksville.

HISTORY.

This flock was formed by a purchase of nineteen Atwood ewes in the year 1862 or '63 of Stephen Benton by Aaron A. Fisk, father of the present owner of the flock.

RAMS USED.

At the time the ewes were purchased, a ram was purchased of M. B. Williamson, that was bred from the flocks of E. S. Stowell and Victor Wright; this ram and those bred within the the flock descended from him have been the only ones used up to the time the flock was offered for record, August, 1877.

——o——

FLOCK 77.

Owned by ORIN A. FIELD, WEST CORNWALL, ADDISON Co., VT.

9 RAMS.

Marked with metallic labels in their ears, having thereon O. A. Field and numbers 1, 4, 73 to 79.

44 EWES.

Labeled same as rams, and numbers 17 to 33,35 to 43, 45 to 50, 61 to 72.

PEDIGREE.

Descended from importations from Spain through the flocks of Andrew Cock, Long Island; David Humphreys, Conn.; William Jarvis, Weathersfield; Messrs. Rich. L. Beedle, L. C. Remele, E. R. Robinson, T. Stickney & Son, Levi Wolcott, R. N. Atwood, Shoreham; R. P. Hall, M. Bingham, W. H. Delong, Cornwall; H Cross, N. A. Saxton, Waltham; V. Wright, Weybridge; Messrs. Hammond, Middlebury; J. Q. Stickney, Whiting; Seneca Root, Hubbardton; L. Webster, Sudbury; Ward M. Lincoln, Brandon; J. Sheldon, Fair Haven; Charles Forbes, West Haven; H. Root, J. S. Griswold, J. Dickinson, Geo. Root, Benson; Baker & Smith, P. Elitharp, Bridport

EWES PURCHASED.

Sept. 5, 1866, twenty Atwood ewes were purchased of Merrill Bingham; a part of these were bred by H. Cross from R P. Hall stock, and a part were bred by V. Wright; Sept. 1, 1875, five ewes were purchased of L. C. Remele, of Atwood, Jarvis and Rich blood.

RAMS USED.

Fortune (249); next, a ram sired by Fortune (249); Fremont, Jr. (215) was used one year; Levi (552), was used to some extent; a ram sired by Fremont, Jr. (215), bred within the flock, has been used; also O. A. Fields' 1 (553).

30

FLOCK 78.

Owned by JAMES FORBES. Jr., Shoreham, Addison Co., Vr.

24 RAMS.

Marked with metallic labels in their ears, having thereon J. Forbes, Jr., and numbers 42 to 46, 57 to 76.

97 EWES.

Labeled same as rams, with numbers 1 to 97.

PEDIGREE.

Descended from importations from Spain through the flocks of David Humphreys, Stephen Atwood, J. N. Blakeslee, J. R. Nettleton, Conn.; David Buffum, William Bailey, Joseph I. Bailey, Rhode Island; Andrew Cock, Long Island; Wm. Jarvis, Weathersfield; Messrs. Rich, Leonard Beedle, G. H. Hall, L. C. Remele, E. R. Robinson, Darwin E. Robinson, E. A. Birchard, B. B. Tottingham, E. D. Bush, J. M. Ormsbee, F. D. Douglass (now of Whiting), D. & G. Cutting, J. N. North, T. Stickney & Son, Denny & Harris, L. Catlin, A. C. Harris, Shoreham; S. T. Baker, Whiting; Ward M. Lincoln, Brandon; C. S. Rumsey, Hubbardton; W. R. Sanford, O. S. Branch, J. T. Branch, D. B. & J. N. Buel, O. H. Bascom, H. T. Cutts, H. G. Hibbard, Orwell; R. P. Hall, John Towle, R. J. Jones, Cornwall; Messrs. Hammond, W. R. Remele, O. Severance, Middlebury; E. Porter, Rutland; Baker & Smith, Prosper Elitharp, C. N. Hayward, W. W. Winchester, Bridport; C. B. Cook, Charlotte; N. A. Saxton, Waltham; Nelson Richards, Panton; Victor Wright, Weybridge.

EWES PURCHASED.

Flock commenced Nov. 18, 1865, by a purchase of ten ewes of C. S. Rumsey (Atwood and Blakeslee blood); same Fall, a purchase of seven ewes of Orson Martin, of Benson, was made; they were bred by F. D. Douglass (of Rich, Jarvis and Atwood blood); in 1870, fifteen ewes were purchased of Lynde Catlin (of Robinson and Rhode Island blood); the next purchase was in 1874, seven ewes of Alanson Towner, Shoreham, that he had purchased of J. N. North (of Robinson blood); next a purchase of twenty-seven ewes was made of H. T. Cutts; they were purchased by Mr. Cutts from the estate of O. S. Branch, and O. H. Bascom; they were bred from the flock of W. R. Sanford (those from the flock of Mr. Bascom had a small amount of Jarvis blood); the next purchase was thirteen ewes and eight lambs, of J. T. Branch, of Atwood blood (through the flocks of W. R. Sanford and J. Towle); this last purchase was made August 15, 1877.

RAMS USED.

From 1865 to 1868, ram Echo (250); in 1868, one of T. Stickney & Son's stock rams to eighteen ewes; in 1869 and 1870, two rams, bred and owned by A. C. Harris; one of these rams was sired by Sweepstakes, the other by Hammond's Green Mountain; dams of both Robinson and Atwood blood; in 1871, a ram bred and owned by J. T. & V. Rich; in 1872, Major (181); in 1873, '74, '75, '76, ram Snowflake (277).

——o——

FLOCK 79.

Owned by A. E. FULLER, POMFRET, WINDSOR CO., VT.

10 RAMS.

Marked with metallic labels in their ears, having thereon A. E. Fuller and numbers 28, 32, 41, 44, 69 to 74.

61 EWES.

Labeled same as rams with numbers 2 to 5, 7 to 9, 11 to 22, 24, 26, 27, 29 to 32, 35 to 39, 43, 45 to 49, 51, 53 to 58, 60 to 63, 65, 66, 68, 75 to 84.

PEDIGREE.

Descended from the importations from Spain by Col. David Humphreys, through the flocks of Stephen Atwood, Conn.; Messrs. Hammond, W. R. Remele, Middlebury; W. R. Sanford, Orwell; R. P. Hall, E. S. Stowell, Cornwall; L. S. Drew, Burlington; A. E. Perkins, Pomfret.

EWES PURCHASED.

Flock commenced 1864, by a purchase of half of eleven Atwood ewes of P. Fuller, the father of the present owner of the flock, and B. W. Couch; these ewes were bred by W. R. Sanford; the same year, five Atwood ewes were purchased of A. E. Perkins; two of them were bred by W. R. Sanford, the other three by E. Hammond; in 1865, a part of ten ewes were purchased of P. Fuller and B. W. Couch, that they had purchased of E. Hammond; eight Atwood ewes were purchased of B. Chandler, that were bred by and purchased of W. R. Sanford; in 1868, twenty-five ewe tegs were purchased of W. R. Sanford.

RAMS USED.

Comet, Jr. (154), Wrinkly (150), Champion (170), Champion, Jr. (174), Stowell Ram (240), Green Mountain, Jr. (255), Perkins' Stowell Ram (256), Wool Mine (424), Black Top (425), Fortune (426), and others, bred within the flock, have successively been used.

FLOCK 80.

Owned by RECTOR GAGE, Addison, Addison Co., Vt.

16 RAMS.

Marked with metallic labels in their ears, having thereon R. Gage and numbers 24, 25, 66 to 79.

63 EWES.

Labeled same as rams, and numbers 1 to 23, 26 to 65.

PEDIGREE.

Descended from the importations from Spain by Col. David Humphreys, through the flocks of Stephen Atwood, Conn.; A. A. Farnsworth, Brooksville; R. P. Hall, Cornwall; Messrs. Hammond, W. R. Remele, Middlebury; W. R. Sanford, Orwell; V. Wright, Weybridge; N. A. Saxton, Waltham; H. Adams, W. G. Sprague, Vergennes; J. Smith, L. P. Clark, Addison; E Roberts, Ferrisburgh.

EWES PURCHASED.

In Jan., 1860, a number of Atwood ewes were purchased of the estate of Joseph Smith, that he bred from stock purchased of E. Roberts and H. Adams, who procured their stock of S. Atwood; in Oct., 1873, ten Atwood ewes were purchased of W. G. Sprague, that were bred from stock he purchased of A. A. Farnsworth; Oct., 1875, eight Atwood ewes were purchased of Mrs. Victor Wright, that were bred by Mr. Wright before his death.

RAMS USED.

Sweepstakes (32), Eureka (58), Little Wrinkly (48), Long Wool (205), Vigor (209), Chunkhead (182), Kilpatrick (71), Pat Henry (183), General (210), Perfect (242), R. Gage's 76 (251), have all been used to breed up the present flock, and a few other rams bred within the flock, and sired by some of the above, have been used to a limited extent.

———o———

FLOCK 81.

Owned by HENRY GIDDINGS, Fairfax, Franklin Co., Vt.

70 EWES.

Marked with metallic labels in their ears, having thereon H. Giddings and numbers 1 to 70.

PEDIGREE.

Descended from the importations from Spain by Col. David Humphreys, through the flocks of Stephen Atwood, Connecticut;

Messrs. Hammond, W. R. Remele, Middlebury ; Victor Wright, Weybridge ; W. R. Sanford, Orwell ; Henry Thorp, Charlotte ; R. P. Hall, Cornwall.

EWES PURCHASED.

Oct. 31, 1854, thirty-two Atwood ewes were purchased of W. S. & E. Hammond, by Chapman & Henry Giddings, which formed the foundation of the flock.

RAMS USED.

None but pure Atwood rams have been used, from the flocks of Messrs. Hammond, W. R. Remele, Victor Wright, Henry Thorp and those raised within the flock.

——o——

FLOCK 82.

Owned by GIDDINGS & OGILVIE, Hartford, Licking Co., Ohio.

13 RAMS.

Marked with metallic labels in their ears, having thereon D. Giddings, and numbers 19 to 23, 69, 72, 74 to 78, 80.

40 EWES.

Eighteen labeled same as rams and numbers 1 to 5, 7 to 10, 12, 13, 16, 18, 31, 56, 64, 70, 79 ; twenty-two marked on their labels H. Giddings and numbers 1 to 22.

PEDIGREE.

Descended from the importation from Spain by Col. David Humphreys through the flocks of S. Atwood, Conn.; Messrs. Hammond, W. R. Remele, Middlebury ; R. P. Hall, Cornwall ; V. Wright, Weybridge ; N. A. Saxton, Waltham ; L. P. Clark, Addison ; Henry Thorp, Charlotte ; Henry Giddings, Fairfax.

EWES PURCHASED.

A few ewes were purchased of V. Wright, before Mr. Wright's death, at a very high price ; three were purchased of L. P. Clark ; nine ewes were purchased of H. Thorp ; a few were purchased of H. W. Hammond.

RAMS USED.

Sweepstakes (32), Silver Mine (61), Gold Drop (64), and some of Messrs. Hammond's other stock rams. Black Top (90), Long Wool (112) and Pat Henry (183) have also been used, besides rams bred within the flock.

FLOCK 83.

Owned by ROLLIN GEASON, Benson, Rutland Co., Vt.

6 RAMS.

Marked with metallic labels in their ears, having thereon R. Gleason and numbers 1 to 6.

54 EWES.

Labeled same as rams and numbers 1 to 54.

PEDIGREE.

Descended from the importation from Spain by Col. David Humphreys through the flocks of S. Atwood, Conn.; Messrs. Hammond, W. R. Remele, S. W. Remele, D. Hooker, Middlebury; A.A. Farnsworth, Brooksville; R. P. Hall, John Towle, F. Hooker, Cornwall; N. A. Saxton, Waltham; V. Wright, Weybridge; W. R. Sanford, Orwell.

EWES PURCHASED.

Dec. 22, 1855, two ewes of W. R Sanford; Dec. 28, 1856, four ewes of the same; Oct 28, 1858, two more, and March 19, 1863, two more ewes from the same flock. The blood of all these ewes was Atwood and all were in lamb by the stock rams of Mr. Sanford when purchased. Dec. 5, 1861, five ewes were purchased of N. A. Saxton; Oct. 25, 1862, two ewes were purchased of the same; Jan. 23, 1863, one ewe; Dec. 31, 1866, one, and Dec., 1872, one. All were purchased from the same flock, all were Atwood blood and all except the last were in lamb by Mr. Saxton's stock rams, the last by one of W. R. Sanford's stock rams. Dec. 21, 1864, three ewes were purchased of R. P. Hall, and Feb., 1867, one more from the same flock; these, like the others, were Atwood blood and were in lamb by Mr. Hall's stock rams.

RAMS USED.

In 1856, W. R. Sanford's stock rams Old Greasy (18) and Young Matchless (17) were used, and also the same in 1857. In 1858 the same, and a few to stock rams of Messrs Hammond and W. R. Remele; in 1859, the stock rams of Messrs. Hammond and Green Mountain (46); 1860 and '61 the stock rams of N. A. Saxton, W. R. Sanford and Messrs. Hammond; in 1862, Major (94) and Sweepstakes (32); 1863, Major (94) and Silver Mine, 2d (84), also a few to Messrs. Hammond's stock rams; 1864, '65, '66, '67, '68 and '69, Silver Mine, 2d (84), Green Mountain (70), Golden Fleece (91), and a few to Messrs. Hammond's other stock rams. Red Leg (109) was used the last year to a few ewes. In 1870, W. R. Sanford's rams, and Silver Mine, 2d (84) were used; in 1871 and '72, Red Leg, 2d (229) was used; in 1873, N. A. Saxton's stock ram was purchased and he was used to most of the flock in 1873–4. A ram bred within the flock, sired by W. R. Sanford's stock ram by Comet, was used to a few ewes in 1874, and the same rams were used in 1875. Buell's Towle Ram (290) was used to a few ewes in 1874 and also in 1875. In 1876, one ewe was taken taken to Geo. H. Hall's Ben (339), otherwise same as in 1875; in 1877, one ewe to Wrinkly (292,) otherwise the same as in 1874-5.

———o———

FLOCK 84.

Owned by E. D. GRISWOLD, Orwell, Addison Co., Vt.

50 EWES.

Marked with metallic labels in their ears, having thereon E. D. Griswold, and numbers 1 to 50.

PEDIGREE.

Descended from importations from Spain by Col. David Humphreys through the flocks of S. Atwood, Conn.; Messrs. Hammond, W. R. Remele, Middlebury; Victor Wright, Weybridge; W. R. Sanford, Orwell.

EWES PURCHASED.

November, 1859, ten Atwood ewes were purchased of W. R. Sanford in company with C. H. Conkey; of these E. D. Griswold had five.

RAMS USED.

The present flock has been bred up by using Atwood rams from the flock of W. R. Sanford and none others.

———o———

FLOCK 85.

Owned by DARWIN E. GROVENOR, Bridport, Addison Co., Vt.

25 RAMS.

Marked with metallic labels in their ears, twenty-one having thereon D. E. Grovenor, and numbers 1, 2, 38 to 56; four having on labels J. O. Hamilton and numbers 71, 73, 75, 91.

46 EWES.

Labeled same as rams, and numbers 3 to 37, and 57 to 67.

PEDIGREE.

Descended from importations from Spain through the flocks of D. Humphreys, S. Atwood, Conn.; A. Cock, Long Island; David Buffum, Wm. Bailey, J. I. Bailey, Rhode Island; Wm. Jarvis,

Weathersfield; E. Hammond, W. R. Remele, Middlebury; W. R. Sanford, Orwell; Messrs. Rich, E. R. Robinson, L. C. Remele, E. A. Birchard, J. M. Ormsbee, D. & G. Cutting, Shoreham; Baker & Smith, P. Elitharp, C. N. Hayward, J. J. Crane, D. F. Doty, J. Kellogg, J. O. Hamilton, H. N. Solace, Bridport; S. T. Baker, Whiting; N. A. Saxton, Waltham; N. Richards, Panton.

EWES PURCHASED.

Three old ewes and one year-old ewe were purchased of J. Howe bred by J. J. Crane; Nov., 1877, thirty-one ewes were purchased of H. N. Solace, bred from the flocks of D. F. Doty, J. J. Crane and the Robinson flock, through that of J. Kellogg, by using rams from the flocks of D. F. Doty, J. J. Crane and J. O. Hamilton.

RAMS PURCHASED OR USED.

A ram (No. 1) was purchased of Geo. H. Smith that was sired by Eureka, 3d (223), and out of an Atwood and Robinson ewe. This ram was used in 1877 and sired the tegs in rams from 36 to 56, and in ewes 57 to 67. One ram was bred from the Crane ewes and sired by a ram of Robinson and Atwood blood belonging to E. M. Wheeler; in 1877, four rams of Cutting and Robinson blood were purchased of J. O. Hamilton; their numbers are 71, 73, 75, 91.

——o——

FLOCK 86.

Owned by HENRY HARRINGTON, Keeler's Bay, Grand Isle Co., Vt.

8 RAMS.

Marked with metallic labels in their ears, having thereon H. H. and numbers 52, 54 to 60.

41 EWES.

Labeled same as rams and numbers 11 to 15, 18, 19, 21 to 32, 35 to 51, 53, 61 to 65.

PEDIGREE.

Descended from the importation from Spain by Col. David Humphreys, through the flocks of S. Atwood, Conn.; C. B. Cook, H. Thorp, Charlotte; Messrs. Hammond, Middlebury; V. Wright, Weybridge; H. N. Newell, Shelburn.

EWES PURCHASED.

Nov. 4, 1867, ten Atwood ewes were purchased of C. B. Cook,

RAMS USED.

At the same time of the purchase of the ewes an Atwood ram was purchased of Henry Thorp; in 1869, another Atwood ram was purchased of Mr. Thorp; an Atwood ram was also purchased of H. N. Newell; one was used belonging to Mr. Hall, of Grand Isle, that was bred and sold by Henry Thorp; these have all been used, besides some bred within the flock; South Hero (571) has been the stock ram of the flock for two seasons past.

——o——

FLOCK 87.

Owned by J. O. HAMILTON, BRIDPORT, ADDISON CO., VT.

22 RAMS.

Marked with metallic labels in their ears, having thereon J. O. Hamilton, and numbers 51, 53, 55, 57, 59, 61, 63, 65, 67, 69, 71, 73, 75, 77, 79, 81, 83, 85, 87, 89, 91, 93.

70 EWES.

Labeled same as rams, with numbers 1 to 50, 52, 54, 56, 58, 60, 62, 64, 66, 68, 70, 72, 74, 76, 78, 80, 82, 84, 86, 88, 90.

PEDIGREE.

Descended from importations from Spain through the flocks of A. Cock, Long Island; D. Humphreys, S. Atwood, Conn.; David Buffum, Wm. Bailey, J. I. Bailey, Rhode Island; Wm. Jarvis, Weathersfield; Messrs. Rich, E. R. Robinson, L. C. Remele, E. A. Birchard, J. M. Ormsbee, D. & G. Cutting, Shoreham; S. T. Baker, Whiting; N. Richards, Panton; Baker & Smith, P. Elitharp, Bridport.

EWES PURCHASED.

In 1861 or 1862, twenty-three ewes were purchased of David Cutting, of Cutting blood.

RAMS USED.

Ironsides (80) was used two years; from that date to 1872, rams bred within the flock were used; in 1872, the Old Dea. James Ram (52) was used, and also for three years after that, Greasy (297) and J. O. Hamilton's 51 (517). J. O. Hamilton's 53 (518) is now the stock ram of the flock.

31

FLOCK 88.

Owned by MRS. C. N. HAYWARD, Bridport, Addison Co., Vt.

13 RAMS.

Marked with metallic labels in their ears, having thereon Mrs. S. E H., and numbers 15, 17, 19, 21, 36, 37, 38, 40, 43, 44, 50, 52, 77.

57 EWES.

Labeled same as rams, and numbers 1 to 9, 11 to 14, 16, 18, 22 to 33, 39, 45, 47, 48, 49, 51, 53 to 76.

PEDIGREE.

Descended from importations from Spain through the flocks of D. Humphreys, S. Atwood, Conn.; A. Cock, Long Island; Wm. Jarvis, Weathersfield; Messrs. Hammond. W. R. Remele, Middlebury; V. Wright, Weybridge; N. Richards, Panton; N. A. Saxton, Waltham; Messrs. Rich, L. C. Remele, L. Beedle, E. R. Robinson, Shoreham; Baker & Smith, P. Elitharp, B. Myrick, H. C. Burwell, D. F. Doty, J. J. Crane, Bridport; C. B. Cook, Charlotte.

EWES PURCHASED.

In the Fall of 1874, a purchase of ten Atwood ewes and forty or more ewe tegs that had a small portion of Cock and Jarvis blood in them, of Prosper Elitharp; a few more were purchased of C. B. Cook; a few ewes were also purchased, at different times, of V. Wright; March 29th, 1861, three ewes were purchased from this flock.

RAMS USED.

America (35) was used to a great extent; a ram teg was purchased in 1860 of N. A. Saxton; Ironsides (80) was used for a number of years; Young Ironsides (82) was used to some extent; Chunk (95) was the stock ram of the flock for a number of years; Young Chunk (206) was used to some extent; these were all Atwood rams; rams bred by H. C. Burwell, having a small portion of Robinson blood, have been used for a few years past.

——c——

FLOCK 89.

Owned by J. B. HAMBLIN, Cornwall, Addison Co., Vt.

38 RAMS.

Marked with metallic labels in their ears, having thereon J. B. Hamblin and numbers 1 to 13, 80 to 89, 91 to 105.

79 EWES.

Labeled same as rams, seventy-eight with numbers 14 to 79, 106 to 117, and one marked on label W. McCauley and number 43

PEDIGREE.

Descended from importations from Spain through the flocks of D. Humphreys, S. Atwood, Conn.; A. Cock, Long Island; Wm. Jarvis, Weathersfield; Messrs. Hammond, W. R. Remele, Middlebury; V. Wright, L. D. Gregory, Weybridge; Messrs. Rich, L. Beedle, L. C. Remele, E. R. Robinson, E. A. Birchard, A. Clark, Shoreham; R. P. Hall, J. Ellsworth, M. J. Ellsworth, S. S. Gibbs, E. Sanford, R. R. Wright, B. S. Fields, Henry Lane, Rollin Lane, R. J. Jones, S. S. Rockwell, Cornwall; W. M. Lincoln, J. Hinds, Brandon; A. Hull, Wallingford; Baker & Smith, C. A. Hayward, P. Elitharp, Bridport; W. R. Sanford, Orwell; O. Smith, Pittsford; C. B. Cock, Charlotte; N. Richards, Panton; N. A. Saxton, Waltham.

EWES PURCHASED.

The first ewes purchased were sixteen of C. W. Foot that were bred by R. J. Jones and certified by him to be pure blooded Spanish merino. A few years later twenty or twenty-two Atwood ewes were purchased of A. W. Dana that he purchased of Orlin Smith and bred from stock from Messrs. Hammond, J. Hinds and W. M. Lincoln; these ewes were purchased in company with C. Benedict and equally divided with him. One ewe was purchased of Wm. McCauley, of Atwood and Cock blood.

RAMS USED.

A ram was purchased of Jesse Ellsworth and used several years; he was sired by an Atwood ram bred by R. P. Hall and an ewe purchased of A. Hull; a Robinson ram purchased of S. S. Gibbs by M. J. Ellsworth was next used, and other rams bred by and belonging to M. J. Ellsworth were used. Sea Lion (83), Eureka (58), Old Greasy (74) and Field's Eureka (450) were all used to a considerable extent. A ram bred by E. Sanford was purchased in company with M. J. Ellsworth and used a few seasons. A ram sired by one of V. Wright's stock rams and out of an Atwood ewe purchased of L. D. Gregory was purchased of R. R. Wright and used. Hamblin's Lane Ram (502) was used two years. M. J. Ellsworth's (504) was used for a part of the flock in 1876. J. B. Hamblin's 1 (503) has been the stock ram of the flock for three or four years past.

——o——

FLOCK 90.

Owned by O. & E. S. HALL, East Randolph, Orange Co., Vt.

28 RAMS.

Marked with metallic labels in their ears, having thereon O. & E. S. Hall and numbers 38 to 45, 76, 84 to 102.

90 EWES.

Labeled same as rams, and numbers 1 to 37, 46 to 75, 103 to 125.

PEDIGREE.

Descended from importations from Spain by Col. David Humphreys through the flocks of S. Atwood, Conn.; Messrs. Hammond, Middlebury.

EWES PURCHASED.

In the spring of 1867 a few ewes were purchased of E. Hammond & Son, and a few after that, until in 1870 ten ewes were purchased of E. Hammond & Son. In the fall of 1871 fourteen ewes were purchased of Geo. Hammond. In all, these purchases numbered fifty-three sheep. (The junior partner of this firm, E. S. Hall, was Messrs. Hammond's shepherd from Oct. 15, 1867, to Feb. 26, 1870, and made most of these purchases.)

RAMS USED.

In the fall of 1866 a ram teg was purchased of Messrs. Hammond; he was sired by Green Mountain (70); his dam was an ewe Mr. Hammond called Black Rose. This teg was called Giant and was used a number of years. The fall of 1870 Bull Dog (104) was hired and used and the following spring he was purchased and used every season until he died. Hall's Gold Drop (414) and Queen Ram (415) have been the stock rams of the flock for four or five years.

———o———

FLOCK 91.

Owned by GEO. H. HALL, Shoreham, Addison Co., Vt.

1 RAM.

Marked with metallic label in his ear, having thereon Geo. H. Hall and number 90.

48 EWES.

Labeled same as ram and Nos. 1 to 4, 6, 9 to 13, 15 to 27, 33, 34, 36, 37, 40 to 43, 47, 51, 52, 57, 165 to 169, 175 to 182.

PEDIGREE.

Descended from importations from Spain by Col. David Humphreys through the flocks of S Atwood, Conn.; Messrs. Hammond, W. R. Remele, E. R. Clay, Middlebury; R. P. Hall, Henry Lane, R. J. Jones, Cornwall; W. R Sanford, Orwell; C. B. Cook, Charlotte; N. A. Saxton, Waltham; V. Wright, Wales Bros., Weybridge; L P. Clark, Addison; J. Hinds, E. D. Hinds, Brandon; P. Elitharp, C. N. Hayward, Bridport.

EWES PURCHASED.

Eleven Atwood ewes were purchased of R. J. Jones in March, 1865; a part were bred by R. P. Hall, the rest were bred from ewes from the flock of Mr. Hall and C B. Cook by using Atwood rams; these ewes were in lamb by Seville (75). In Oct., 1876, eight Atwood ewes were purchased of Wales Bros. from stock bred by E. R. Clay, by rams bred by J. Hinds, L. P. Clark, C. N. Hayward and E. R. Clay.

RAMS USED.

Matchless (121), Fashion (180), Major (181) and Ben (339), all bred within the flock, have been used successively as stock rams, the last being the stock ram at the present time.

------o------

FLOCK 92.

Owned by JOHN H. HAZEN, HARTFORD, WINDSOR Co., VT.

5 RAMS.

Marked with metallic labels in their ears, having thereon J. H. Hazen and numbers 66 to 70.

65 EWES.

Labeled same as rams and numbers 1 to 65. The first 39 of these are pure Atwoods direct from the Humphreys importation; the remainder of them have a small strain of Robinson blood.

PEDIGREE.

Descended from importations from Spain through the flocks of David Humphreys, Stephen Atwood, Conn.; A. Cock, Long Island; Wm. Jarvis, Weathersfield; Messrs. Hammond, W. R. Remele, Middlebury; W. R. Sanford, Orwell; Messrs. Rich, E. R. Robinson, L. C. Remele, E. A. Birchard, Rollin Birchard, L. Beedle, Shoreham; E. S. Stowell, Cornwall; Baker & Smith, P. Elitharp, Bridport; A. Lawrence, Hinesburgh; J. Arey, Salisbury, N. H.; A. E. Fuller, A. E. Perkins, Peter Fuller, Pomfret; Nathan Cushing, Woodstock; A. H. Sperry, Cornwall; V. Wright. Weybridge.

EWES PURCHASED.

Flock commenced some thirty years ago by a purchase of Jarvis ewes from the flock of Nathan Cushing. In 1859 a few Atwood ewes were purchased of Peter Fuller, bred from the flocks of W. R. Sanford and E. Hammond; soon after this, five ewes were purchased of A. E. Perkins, bred from the same flocks, and after this a few of A. E. Fuller of the same blood, bred from the same flocks. In the fall of 1873 eighteen or twenty ewes were purchased of J. Arey; twelve of these were pure Atwoods, and the remainder had a strain of Robinson blood.

RAMS USED.

At first a Jarvis ram from the Cushing flock three or four years, then two rams bred by Messrs. Hammond and W. R. Sanford that were purchased or hired of Peter and A. E. Fuller. One ram was purchased of Almon Lawrence, bred from the Hammond flock. Stowell's Ram (240) is the present stock ram of the flock.

———o———

FLOCK 93.

Owned by E. J. & E. W. HARDY, Osceola, Mich.

36 RAMS.

Marked with metallic labels in their ears, having thereon E. J. & E. W. H. and numbers 131 to 165, 170.

167 EWES.

Labeled same as rams, and numbers 1 to 130, 166 to 169, 171 to 203.

PEDIGREE.

Descended from importations from Spain through the flocks of A. Cock, Long Island; D. Humphreys, S. Atwood, Conn.; Wm. Jarvis, Weathersfield; Messrs. Rich, L. Beedle, E. R. Robinson, L. C. Remele, E. A. Birchard, H. B. Wright, Dewey & Harris. E D. Bush, E. G. Farnham, B. B. Tottingham, James Forbes, Jr., F. & L. E. Moore, T. Stickney & Son, Darwin E. Robinson, F. D. Douglas (now of Whiting), A. B. Treadway, E. Barnum, H. W. Jones, Shoreham; Messrs. Hammond. W. R. Remele, O. Severance, Middlebury; W. R. Sanford, L. Wilcox, O. S. Branch, J. T. Branch, D. B. & J. N. Buell, O. H. Bascom, H. T. Cutts, Orwell; R. P. Hall, John Towle, Cornwall; Baker & Smith, P. Elitharp, C. N. Hayward, W. W. Winchester, Bridport; C. B. Cook, Charlotte; N. A. Saxton, Waltham; N. Richards, Panton; V. Wright, Weybridge; J. Q. Stickney, Whiting; Seneca Root, C. S. Rumsey, Hubbardton; L. Webster, Sudbury; William Ball, Hamburg, Mich.

EWES PURCHASED.

In 1873, fifty of Atwood, Robinson and Rich blood, were purchased of Wm. Ball. In 1874, in company with Wm. Ball, seventeen ewe tegs were purchased of E. D. Bush, eight of which were retained in this flock; also at the same time twenty ewe tegs were purchased from the flocks of H. W. Jones, also thirty-one of J. Forbes, Jr., and twenty-six of B. B. Tottingham; all these combined the blood of the Atwood, Cock and Jarvis flocks.

RAMS USED.

Don Pedro (276), Maximilian (285), Little Wrinkly (563) and Sammy (564) have been used and have sired the younger part of the flock.

———o———

FLOCK 94.

Owned by GEORGE HAMMOND, MIDDLEBURY, ADDISON Co., VT.

1 RAM.

Marked with metallic label in his ear, having thereon G. Hammond, and number 200. This ram is Rarus (569).

118 EWES.

Sixty-six labelled same as ram and numbers 1 to 5, 7, 9 to 11, 13 to 19, 21 to 29, 31, 32, 35 to 38, 40 to 42, 45. 46, 48, 50 to 56, 58 to 67, 69 to 72, 74, 77, 78, 79, 89, 98. 125, 128. Fifty-two with labels marked L. J. O. and numbers 4, 5, 8, 9, 10, 14 to 17, 19, 20, 27, 29, 30, 32, 35 to 38, 43, 48, 49, 51, 52, 54, 60 to 62, 66, 68, 70, 71, 74, 75, 76, 78, 82. 86, 87, 124, 131, 137, 142, 143, 144, 153, 166, 171, 176, 182, 186, 410.

PEDIGREE.

Descended from the importations from Spain by Col. David Humphreys through the flocks of S. Atwood, Conn.; Messrs. Hammond, W. R. Remele. Middlebury; W. R. Sanford, Orwell; R. P. Hall, R. J. Jones. Cornwall; V. Wright, Weybridge; H. Thorp, C. B. Cook, Charlotte; D. E. Hill, H. C. Burwell, Bridport; L. J. Orcutt, Cummington, Mass.

EWES PURCHASED.

This flock is made up of the old Hammond and Sanford flocks, repurchased by George Hammond of L. J. Orcutt, late in 1877.

RAMS USED.

The rams used by Mr. Orcutt and Mr. Hammond have been Acme (524), Rarus (569), and H. Thorp's No. 1 (378) and one or two bred in the old Hammond flock.

FLOCK 95.

Owned by G. W. HERVEY, Unionport, Jefferson Co., Ohio.

2 RAMS.

Marked with metallic labels in their ears, having thereon G. W. Hervey, and numbers 56, 57.

13 EWES.

Twelve labeled same as rams, and numbers 44 to 55 and one labeled L. P. Clark and number 27.

PEDIGREE.

Descended from importations from Spain through the flocks of D. Humphreys, S. Atwood, Conn.; A. Cock, Long Island; Wm. Jarvis, Weathersfield; Messrs. Hammond. W. R. Remele, Middlebury; V. Wright, Weybridge; W. R. Sanford, Orwell; N. A. Saxton, Waltham; L. P. Clark, Wm. Hanks, Addison; Messrs. Rich, L. C. Remele, E. R. Robinson, Shoreham; C. B. Cook, Charlotte; Alex. McAlmont, W. L. Archer, Pa.; W. G. Markham, W. Cole, W. C. Hatch, S. B. Lusk, Western New York; J. E. Parker, C. K. Williams, Whiting; F. H. Dean, R. P. Hall, S. S. Gibbs, E. S. Dana. R. J. Jones, Cornwall.

EWES PURCHASED.

March 28, 1876, three Atwood ewes were purchased of Alexander McAlmont; they were bred from the flock of W. L. Archer. April 28. 1876, seven ewes were purchased of J. H. Close, of St. Clairsville, Ohio; two were Atwood ewes bred by L. P. Clark, two were Atwood and Robinson ewes bred by Wm. G. Markham, and three were Atwood ewes bred by Wm. Hanks from stock from W. R. Sanford. In the fall of 1877, Atwood ewe 27, bred by L. P. Clark, was purchased of J. H. Close.

RAMS USED.

Brick (371) sired rams 56 and 57. Symmetry (372) sired ewes 51 to 55.

——o——

FLOCK 96.

Owned by A. E. HITCHCOCK & SON, East Shoreham, Addison Co., Vt.

5 RAMS.

Marked with metallic labels in their ears, having thereon A. E H. & Son, and numbers 1 to 5.

45 EWES.

Labeled same as rams, with numbers 6 to 50.

PEDIGREE.

Descended from importations from Spain through the flocks of D. Humphreys, S. Atwood, Conn.; A. Cock, Long Island; Wm. Jarvis, Weathersfield; Messrs. Rich, L. C. Remele, E. R. Robinson, Earl R. and F. Delano, T. Stickney & Son, D. Cutting, Shoreham; R. P. Hall, John Towle, S. S. Gibbs, Cornwall; N. A. Saxton, Waltham; C. B. Cook, Charlotte; Baker & Smith, P. Elitharp, Bridport; Messrs. Hammond, W. R. Remele, Middlebury; W. R. Sanford, Orwell; J. Hinds, Brandon.

EWES PURCHASED.

August 29, 1863, twenty-five Atwood ewes were purchased of Jesse Hinds. In Oct., 1873, twenty-one ewe tegs were purchased of Earl R. and F. Delano of Atwood and Stickney blood bred from ewes purchased of T. Stickney, and rams from the flocks of S. Atwood, E. R. Robinson, S. S. Gibbs, John Towle; an Atwood ram bred by D. Cutting, Young Sweepstakes (50), and one ewe to Eureka (58).

RAMS USED BY MESSRS. HITCHCOCK.

The first three years Young Sweepstakes (50) was used. From 1863 to 1872 rams bred within the flock were used. In 1872 a Cutting ram was used, but none of the stock is included in the flock registered. In 1873 and '74 a ram of Atwood and Stickney blood, bred by Messrs. Delano, was used; in 1875, one bred by T. Stickney & Son; and in 1876 one bred by J. T. & V. Rich.

———o———

FLOCK 97.

Owned by H. G. HIBBARD, Orwell, Addison Co., Vt.

7 RAMS.

Marked with metallic labels in their ears, having thereon H. G. Hibbard, and numbers 1 to 7.

21 EWES.

Labeled same as rams with numbers 8 to 28.

PEDIGREE.

Descended from importations from Spain through the flocks of D. Humphreys, S. Atwood, Conn.; A. Cock, Long Island; Wm.

Jarvis, Weathersfield; Messrs. Hammond. W. R. Remele, Middlebury; Messrs. Rich, T. Stickney, E. R. Robinson, L. C. Remele, E. A. Birchard, B. B. Tottingham, Shoreham; W. R. Sanford, Orwell; R. P. Hall, John Towle, Cornwall; A. H. Hubbard, Whiting; W. M. Lincoln, Brandon; E. Porter, Rutland; Baker & Smith, P. Elitharp, Bridport; C. B. Cook, Charlotte.

EWES PURCHASED.

In 1858, eight ewes were purchased of E. Porter, they were descended from Atwood stock, bred by Ward M. Lincoln.

RAMS USED.

A ram of Atwood blood, and bred from the same stock, was purchased with the ewes, and he was used the first year; in 1859, York State (145) was used; after this, rams bred within the flock were used; also, to a limited extent, the stock rams of T. Stickney & Son, W. R. Sanford, B. B. Tottingham, F. H. Dean, Messrs. Hammond, A. H. Hubbard and J. Towle; Gold Drop (64), Little Wrinkly (48), Towle Ram (164), Hubbard (165) and Rich's stock ram (462) being among those used; of those bred within the flock Woolly (213), and Hibbard (214) have been used most.

———o———

FLOCK 98.

Owned by E. D. HINDS, Brandon, Rutland Co., Vt.

11 RAMS.

Marked with metallic labels in ears, having thereon J. Hinds, and numbers 51 to 54, 57 to 63.

76 EWES.

Labeled same as rams with numbers 64 to 98, 102, 122, 151, 152, 153, 155 to 169, 171, 179, 180, 202, 203, 207 to 210, 214, 225, 228, 230, 232, 236, 237, 242, 270, 271, 274, 295.

PEDIGREE.

Descended from importations from Spain through the flocks of David Humphreys, Stephen Atwood, Conn.; Andrew Cock, Long Island; William Jarvis, Weathersfield; Messrs. Hammond, W. R. Remele, S. W. Remele, D. Hooker, Middlebury; A. A. Farnsworth, Brooksville; Victor Wright, Weybridge; Leonard Beedle, Messrs. Rich, L. C. Remele, T. Stickney & Son, E R. Robinson, J. T. Stickney, Shoreham; W. R. Sanford, Orwell; C. B. Cook, Charlotte; R. P. Hall, John Towle, F. Hooker, Cornwall; Baker & Smith, Prosper Elitharp, Bridport.

EWES PURCHASED.

Flock commenced in 1850 by Jesse Hinds, father of the present owner, by a purchase of twenty-five Atwood ewes of W. S. & E. Hammond. A few other ewes were also purchased of Messrs. Hammond but the time cannot be given accurately.

RAMS USED.

The stock rams of Messrs. Hammond, Young Matchless (17), Old Greasy (18) and Long Wool (21) were used to the ewes purchased of Messrs. H., and also rams purchased from the same flock; later, Sweepstakes (32), Gold Drop (64) and Silver Mine (61) were used. In 1858 Green Mountain (46) was purchased and used a number of years. In 1867 Green Mountain (70) was used. In 1873 the Rich strain of blood was introduced by purchasing and using a ram, bred by L. C. Remele, for three years; during the same years Wrinkly (292); a ram bred by H. G. Hibbard was used in 1876. In 1877 Rarus (569) was used.

——o——

FLOCK 99.

Owned by FRANKLIN HOOKER, CORNWALL, ADDISON Co., VT.

20 RAMS.

Marked with metallic labels in their ears, having thereon F. Hooker and numbers 36 to 53, 55, 63.

40 EWES.

Labeled same as rams and numbers 1 to 35 and 72 to 76.

PEDIGREE.

Descended from importations from Spain through the flocks of D. Humphreys, S. Atwood, Conn.; A. Cock, Long Island; David Buffum, Wm. Bailey, J. I. Bailey, Rhode Island; Wm. Jarvis, Weathersfield; Messrs Hammond, W. R. Remele, S. W. Remele, D. Hooker, Middlebury; A. A. Farnsworth, Brookeville; N. A. Saxton, Waltham; C. B. Cook, Charlotte; R. P. Hall, Cornwall; Messrs. Rich, L. C. Remele, E. R. Robinson, L. Beedle, E. G. Farnham, G. D. Bryant, Messrs. Cutting, J. M. Ormsbee, Shoreham; S. T. Baker, Whiting; H. M. Graves, Salisbury; J. Hinds & Son, Brandon; Baker & Smith, P. Elitharp, Bridport; I. G. Barker, Leicester.

EWES PURCHASED.

In January, 1864, eleven Atwood ewe tegs and two yearling Atwood ewes were purchased of D. Hooker; they were bred from the flocks of E. Hammond, N. A. Saxton and A. A. Farnsworth.

Feb. 9, 1865, five Atwood ewes were purchased of S. W. Remele ; at the same time E. D. Searl purchased of S. W. Remele four Atwood ewes and soon after these ewes were purchased of Mr. Searl. In the fall of 1875 four old Atwood ewes were purchased of E. D. Hinds. In Feb., 1877, the Rich blood was introduced into the flock by a purchase of three ewes of J. G. Barker and they and their lambs are of this blood. There are also four ewes in the flock that are descended from Cutting stock by a purchase by David Hooker ; the remainder of the flock are Atwood blood.

RAMS USED.

The first ewes were bred to the stock rams of R. P. Hall, and Little Ram (279); after these to rams bred within the flock for some years. Little Wrinkly (48) was used in 1868. Wrinkly (292) has been the stock ram of the flock since 1869.

———o———

FLOCK 100.

Owned by S. G. HOLYOKE & SON, St. Albans, Vt.

19 RAMS.

Marked with metallic labels in their ears, having thereon S. G. H. & Son, and numbers 81 to 99.

80 EWES.

Marked same as rams with numbers 1 to 80.

PEDIGREE.

Descended from the Humphreys importation from Spain through the flocks of Stephen Atwood, Conn.; Messrs. Hammond, Middlebury ; W. R. Sanford, Orwell ; V. Wright, Weybridge.

EWES PURCHASED.

Flock commenced in 1845 by a purchase of one ewe of W. S. & E. Hammond. In 1846 eleven ewes were purchased of the same parties.

RAMS USED OR PURCHASED.

The stock rams of Messrs. W. S & E. Hammond were used until 1850, when a ram was purchased (Holyoke & Hunt's Atwood ram (367), and used until 1854 to a part of the ewes, the remainder to Messrs. Hammond's stock rams. Kilpatrick (71) was used to a part of the ewes for two years, 1865 and '66. The ram Fortune (369) was purchased of Geo. Hammond in 1876 and has since been the stock ram of the flock.

FLOCK 101.

Owned by JEROME HOLDEN, Westminster West, Windham Co., Vt.

10 RAMS.

Marked with metallic labels in their ears, having thereon J. Holden and numbers 96 to 105.

55 EWES.

Labeled same as rams and numbers 51 to 95, 106 to 115.

PEDIGREE.

Descended from the importations from Spain by Col. David Humphreys through the flocks of S. Atwood, Conn.; Messrs. Hammond, W. R. Remele, Middlebury; N. A. Saxton, Waltham; R. P. Hall, John Towle, E. S. Stowell, Henry Lane, Cornwall; C. B. Cook, Charlotte; J. D. Wheat, E. P. Washburn, Putney.

EWES PURCHASED.

Nov. 11, 1865, three Atwood ewes were purchased of N. A. Saxton. August 11, 1866, two Atwood ewes were purchased of J. Towle. Sept. 21, 1867, three Atwood ewes were purchased of E. Hammond & Son; these last were in lamb by Green Mountain (70). March 2, 1878, ten Atwood ewes were purchased of E. P. Washburn that were bred from ewes purchased of H. Lane and from Atwood rams bred by H. Lane, J. Holden and J. D. Wheat.

RAMS USED.

Golden Fleece (91), Bull Dog (104), (was hired and used one season) Victor (159), were used. Of those bred within the flock, Maximilian (179), Greasy (237), and Rocky Mountain (238), have all been used as stock rams.

———o———

FLOCK 102.

Owned by W. W. HOLMES, Short Creek, Harrison Co., Ohio.

1 RAM.

Marked with metallic label in his ear, having thereon W. W. Holmes, and number 1.

60 EWES.

Marked same as ram, with numbers 2 to 61.

PEDIGREE.

Descended from importations from Spain through the flocks of D. Humphreys, S. Atwood, Conn.; A. Cock, Long Island; Wm. Jarvis, Weathersfield ; Messrs. Hammond, W. R. Remele, S. W. Remele, Middlebury ; Baker & Smith, P. Elitharp, O. N. Hayward, H. C. Burwell, D. F. Doty, J. J. Crane, C. P. Crane, E. Grovenor, B. Myrick, E. Fitch, F. Wing, C. E. Crane, L. S Burwell, Bridport ; Messrs. Rich, E. R. Robinson, L. Beedle, L. C. Remele, E. A. Birchard, A. Smith, Shoreham ; R. P. Hall, J. Towle, M. J. Ellsworth, R. Lane, C. Benedict, B. Fields, J. B. Hamblin, E. Sanford, R. R. Wright, Cornwall ; C. B. Cook, Charlotte ; E. R. Sanford, Orwell ; V. Wright, L. D. Gregory, Weybridge ; O. Smith, Pittsford ; Wm. Lincoln, J. Hinds, Brandon.

EWES PURCHASED.

In the Spring of 1877, sixty ewes were purchased of J. B. Cherbino that were bred by H. C. Burwell, E Grovenor, D. F. Doty, J. B. Hamblin, S. W. Remele, L. S. Burwell ; the Grovenor ewes were descended from the Hayward, Elitharp and J. J. Crane flocks. A part of these ewes were in lamb by (222) and a part by the stock rams of the flocks from which they were purchased, eleven in all, but the larger part were young ewes not in lamb.

RAM USED.

At the time of the purchase of ewes the ram Constitution (366) was purchased through the same parties, and he has been the stock ram of the flock since.

——o——

FLOCK 103.

Owned by FRANK S. HOLLEY, CORNWALL, ADDISON Co., VT.

2 RAMS.

Marked with metallic labels in their ears, having thereon F. S. Holley and numbers 58, 59.

57 EWES.

Labeled same as rams, with numbers 1 to 57.

PEDIGREE.

Descended from importations from Spain through the flocks of D. Humphreys, S. Atwood, Conn.; A. Cock, Long Island ; Wm. Jarvis, Weathersfield ; Messrs. Hammond, W. R. Remele, Middlebury ; R. P. Hall, E. Sanford, C. D. Lane, Chas. Benedict, F. H. Dean, Cornwall ; Messrs. Rich, E. R. Robinson, T. Stickney & Son, L. C. Remele, Shoreham ; Baker & Smith, P. Elitharp, Bridport ; J. Q. Stickney, Whiting ; S. Root, Hubbardton ; L. Webster, Sudbury ; C. B. Cook, Charlotte ; N. A. Saxton, Waltham.

EWES PURCHASED.

Flock commenced in 1850 by a purchase of twenty Atwood ewes of R. P. Hall and W. R. Remele by Truman B. Holley, father of the present owner.

RAMS USED.

For the first three years a ram from the same flock as the ewes; the next two years a ram bred within the flock, sired by the first named; for four or five years after, a French ram was used, but the produce has all been sold off; none of that blood remains in the flock. In 1861 and '62, a ram of Atwood and Robinson blood, bred by C. D. Lane, was used; in the last year an Atwood ram from the flock of C. Benedict was used to a part of the flock and the same ram was used in 1863; in 1864 an Atwood and Robinson ram, bred by E. Sanford, was used; in 1865, and for three years after, a ram, bred within the flock, sired by the Benedict ram, was used; the next two years Bull Dog (392) was used; in 1871 a ram bred by J. Q. Stickney was used; for the next three years a ram bred by C. Benedict was used; since that time a ram bred by J. Q. Stickney has been used.

———o———

FLOCK 104.

Owned by A. H. HUBBARD, Whiting, Addison Co., Vt.

10 RAMS

Marked with metallic labels in their ears, having thereon A. H. Hubbard, and Nos. 51 to 60.

50 EWES.

Labeled same as rams and numbers 1 to 50.

PEDIGREE.

Descended from importations from Spain through the flocks of D. Humphreys, S. Atwood, Conn.; A. Cock, Long Island; D. Buffum, Wm. Bailey, Joseph I. Bailey, Rhode Island; Wm. Jarvis, Weathersfield; Messrs. Hammond, W. R. Remele, S. W. Remele, D. Hooker, Middlebury; R. P. Hall, J. Towle, F. Hooker, S. S. Gibbs, Cornwall; N. A. Saxton, Waltham; A. A. Farnsworth, Brooksville; Messrs. Rich, T. Stickney & Son, E. R. Robinson, L. C. Remele, L. Beedle, J. T. Stickney, Messrs. Cutting, J. M. Ormsbee, Shoreham; W. R. Sanford, Horace Bush, Orwell; C. B. Cook, Charlotte; Baker & Smith, P. Elitharp, Bridport; S. T. Baker, A. M. Baldwin, J. Q. Stickney, Joel Barlow, Whiting; Seneca Root, Hubbardton; L. Webster, Sudbury.

EWES PURCHASED.

About the year 1860 ten Atwood ewes were purchased of R. P. Hall and about the same time a few ewes of D. Cutting, and a few of Rich, Jarvis and Atwood blood of L. C. Remele; soon after a few Atwood ewes were purchased of F. Hooker; in 1875, eight Rich and Atwood ewes were purchased of H. Bush, and four or five Robinson ewes were purchased of Joel Barlow, who bred his flock direct from the Robinson.

RAMS USED.

For a number of years after the flock was founded, Atwood rams from the flocks of R. P. Hall and F. Hooker, including one purchased of F. Hooker and Towle Ram (164). Other rams bred within the flock have been used; Gibbs' Ram (171) a little for two years; a ram rented of J. T. Stickney, Fremont, Jr. (215), Hubbard's Rich Ram (269), Wrinkly (292) and Baldwin (449). Other rams bred within the flock have been used, among which was Hubbard Ram (165) which was used more than any other.

———o———

FLOCK 105.

Owned by CURTIS H. JAMES, Cornwall, Addison Co., Vt.

5 RAMS.

Marked with metallic labels in their ears, three having thereon C. H. James and numbers 76, 77, 78, and two having thereon C. P. Crane and numbers 3 and 29.

36 EWES.

Labeled same as rams, C. H. James, and numbers 1 to 36.

PEDIGREE.

Descended from importations from Spain through the flocks of D. Humphreys, S. Atwood, Conn.; A. Cock, Long Island; D. Buffum, Wm. Bailey, J. I. Bailey, Rhode Island; Wm. Jarvis, Weathersfield; Messrs. Hammond, W. R Remele, U. D. Twitchell, Middlebury; W. R. Sanford, Orwell; Messrs. Rich, L. Beedle, E. R. Robinson, T. Stickney & Son, J. T. Stickney, E. A. Birchard, D. & G. Cutting, J. M. Ormsbee, Shoreham; Baker & Smith, P. Elitharp, J. O. Hamilton, C. N. Hayward, J. J. Crane, C. P. Crane, D. F. Doty, B. Myrick, H. C. Burwell, Bridport; N. A. Saxton, Waltham; N. Richards, Panton; S. T. Baker, Whiting; R. P. Hall, C. Benedict, M. J. Ellsworth, C. D. Lane, Cornwall; V. Wright, Weybridge.

EWES PURCHASED.

In 1871 twenty-five Atwood ewes were purchased of U. D. Twitchell; in the fall of 1874 ten more of the same blood were purchased from the same flock. In 1876 seventeen ewes (Cutting and Robinson blood) were purchased of J. O. Hamilton.

RAMS USED.

The first ewes were bred to the Dea. James Ram (52), but the produce has all been sold; afterwards a ram bred by J. T. Stickney was purchased. Three years since two rams of Atwood and Robinson blood were purchased of M. J. Ellsworth, and in 1877, in company with S. James, two rams were purchased that were bred by C. P. Crane. All these rams have been used as stock rams. Greasey (297) was also purchased and used.

———o———

FLOCK 106.

Owned by JOHN A. JAMES, MIDDLEBURY, ADDISON CO., VT.

6 RAMS.

Marked with metallic labels in their ears, having thereon J. A. James, and numbers 51 to 56.

24 EWES.

Labeled same as rams and numbers 1 to 24.

PEDIGREE.

Descended from importations from Spain through the flocks of A. Cock, Long Island; D. Humphreys, S. Atwood, Conn.; Wm. Jarvis, Weathersfield; Messrs. Rich, E. R. Robinson, L. C. Remele, L. Beedle, A. Clark, E. A. Birchard, Shoreham; Baker & Smith, P. Elitharp, Bridport; C. B. Cook, Charlotte; N. A. Saxton, F. D. Barton, Waltham; A. A. Farnsworth, Brooksville; R. P. Hall, E. S. Stowell, M. B. Williamson, S. Benton, Cornwall; Messrs Hammond, W. R. Remele, Middlebury; V. Wright, L. D. Gregory, Weybridge; C. G. Seager, C. P. Morrison, Addison; C. B. Fisk, Brookfield.

EWES PURCHASED.

In April, 1877, fifteen Robinson ewes, bred by C. P. Morrison, were purchased of Cherbino & Williamson; in Oct. following, six Atwood ewes were purchased of C. B. Fisk.

RAMS USED.

Two rams belonging to Samuel James and C. H. James were used, the ram tegs 51 and 56 and the ewe tegs 22, 23 and 24 were sired by them.   33

FLOCK 107.

Owned by SAMUEL JEWETT, Independence, Jackson Co.,Mo.

13 RAMS.

Nine marked with metallic labels in their ears, having thereon S. Jewett and numbers 315, 316, 317, 323, 329, 331 to 334; four marked on labels J. J. Crane and numbers 6 to 9.

95 EWES.

Seventy-six marked same as rams with numbers 2, 6, 11, 16, 17, 26, 28, 38, 43, 101, 102, 103, 107, 108, 109, 111, 113, 119, 122, 123, 124, 129, 130, 131, 132, 138, 142, 143, 144, 146, 147, 148, 150, 187, 191, 193 to 197, 201, 210, 225, 229, 247, 249, 255, 257, 301 to 315, 318 to 322, 324 to 328, 330, 336, 337; nineteen marked on labels J. J. Crane, and numbers 33 to 51.

PEDIGREE.

Descended from importations from Spain through the flocks of D. Humphreys, S. Atwood, Conn.; A. Cock, Long Island; Wm. Jarvis, Weathersfield; Messrs. Hammond, W. R. Remele, E. R. Clay, Middlebury; V. Wright, J. B. Cherbino, Weybridge; R. P. Hall, J. Towle, S. S. Rockwell, Wm. McCauley, C. D. Lane, M. J. Ellsworth, F. H. Dean, C. Benedict, H. Lane, S. S. Gibbs, Cornwall; Messrs. Rich, L. C. Remele, L. Beedle, E. R. Robinson, T. Stickney & Son, J. T. Stickney, Shoreham; Baker & Smith, P. Elitharp, O. N. Hayward, D. F. Doty, J. J. Crane, B. W. Crane, B. Myrick, Geo. N. Payne, Bridport; A. F. Ellsworth, J. Q. Stickney, Whiting; W. R. Sanford, Orwell; C. B. Cook, Charlotte; N. A. Saxton, Waltham; N. Richards, Panton; J. Hinds, Brandon; Wm. M. Gage, Ferrisburgh; C. I. Benedict, Arlington; S. Root, Hubbardton; L. Webster, Sudbury.

EWES PURCHASED.

Nineteen Atwood and Robinson ewes were purchased of J. J. Crane; fourteen Robinson ewes were purchased of Geo. N. Payne; at one time ten and at another eight ewes that were bred by Wm. McCauley, from ewes he purchased of S. S. Rockwell; a part were sired by Little Wrinkly (48), the rest by Gold Dust (334); fourteen ewes were purchased of J. B. Cherbino that he bred from ewes purchased of C D. Lane; they were sired by Drake Ram (235) and Jim (234). One ewe was purchased of A. F. Ellsworth; she was sired by a ram bred by J. T. Stickney, he by Fremont, Jr. (215). Two Atwood ewes were purchased of B. W. Crane; they were the dams of Golden Fleece (432) and Buck Mountain (433). These purchases were made between the years 1868 and 1876, inclusive.

RAMS USED.

Dea. James Ram (52) to a few ewes; he sired Buck Mountain (433) that was used as the stock ram of the flock until the Doty Ram (134) was taken West, when he was used to a portion of the flock; he sired Golden Fleece (432) that was used as a stock ram for five years. Constitution (434) was purchased in 1876, and four rams of J. J. Crane; one of these last has been used as a stock ram to some extent. Matchless, bred within the flock, sired by Golden Fleece (432), has been used.

——o——

FLOCK 108.

Owned by E. A. JENNINGS, WEST CORNWALL, ADDISON CO., VT.

8 RAMS.

Marked with metallic labels in their ears, having thereon E. A. Jennings, and numbers 41 to 48.

40 EWES.

Labeled same as rams, and numbered 1 to 40.

PEDIGREE.

Descended from importations from Spain through the flocks of A. Cock, Long Island; D. Humphreys, S. Atwood, Conn.; David Buffum, Wm. Bailey, J. I. Bailey, Rhode Island; Wm. Jarvis, Weathersfield; Messrs. Rich, E. R. Robinson, Lucius Robinson, T. Stickney & Son, L. Beedle, E. B. Douglas, T. Delano, J. M. Delano, L. C. Remele, J. M. Ormsbee, D. & G. Cutting, J. Forbes, Jr., E. A. Birchard, E. G. Farnham, H. Birchard, J. T. Stickney, Shoreham; Messrs. Hammond, W. R. Remele, D. Hooker, S. W. Remele, Middlebury; R. P. Hall, F. Hooker, Cornwall; F. D. Douglas, S. T. Baker, Whiting; W. R. Sanford, Orwell; N. A. Saxton, Waltham; A. A. Farnsworth, Brooksville.

EWES PURCHASED.

About fifty ewes, Robinson blood, were purchased of T. & J. M. Delano in two purchases nearly twenty-five years since. In March, 1876, in company with I. G. Wooster, fifty-one Robinson and Cutting sheep were purchased of E. B. Douglas; these sheep had been bred by Mr. Douglas from the Robinson flock, principally from ewes purchased of J. M. Delano, now of Ticonderoga, N. Y., who bred them from ewes purchased of E. R. and Lucius Robinson, and by using rams from those flocks and from D. & G. Cutting; Mr. Douglas also had a few ewes that were bred by E. G. Farnham. The young sheep in the flock of Mr. Douglas were sired by Snowflake (277). Of the ewes bought of Mr. Douglas twelve were retained in the flock of Mr. Jennings.

RAMS USED.

Mr. Jennings used to the first ewes, Old Robinson (38), Tottingham (40), Eureka (58), Jenning's Ram (89), Wrinkly (292) and Delong's Remele Ram (361), besides those raised within the flock from these rams.

——o——

FLOCK 109.

Owned by CALVIN H. KETCHUM, Whiting, Addison Co., Vt.

8 RAMS.

Marked with metallic labels in their ears, having thereon C. H. Ketchum, and numbers 51 to 58.

42 EWES.

Labeled same as rams, and numbers 1 to 35, 59, 60, 62, 67 to 70.

PEDIGREE.

Descended from importations from Spain through the flocks of A. Cock, Long Island; D. Humphreys, S. Atwood, J. N. Blakeslee, J. Nettleton, Conn.; Wm. Jarvis, Weathersfield; Messrs. Rich, L. C. Remele, E. R. Robinson, T. Stickney & Son. S. L. Bissell, E. N. Bissell, E. A. Birchard, Shoreham; W. M. Lincoln, Brandon; W. R. Sanford, H. G. Hibbard, Orwell; R. P. Hall, B. S. Fields, F. H. Dean, J. Towle, Cornwall; Messrs. Hammond, W. R. Remele, Middlebury; E. Porter, Rutland; W. P. Wright, E. Rich, A. F. Ellsworth, J. Q. Stickney, Whiting; S. Root, Hubbardton; L. Webster, Sudbury; Baker & Smith, P. Elitharp, Bridport; C. B. Cook, Charlotte.

EWES PURCHASED.

In November, 1873, seven Robinson and Stickney ewes were purchased of W. P. Wright; Dec. 1, 1874, three more from the same flock; Dec. 2d, 1875, four, and Oct. 4, 1876, six more, all from the same flock and of the same blood. In December, 1874, fifteen Stickney ewes were purchased of T. Stickney & Son. Nov. 30, 1876, six ewes were purchased of E. N. Bissell; these ewes combined the blood of the Atwood and Rich flocks, with a trace of Blakeslee blood.

RAMS USED.

Rams from the flock of T. Stickney & Son, Co. Ram (453), Ellsworth's Remele (328), C. H. Ketchum's 51 (454), Hibbard (214), and one bred within the flock sired by Co. Ram (453).

FLOCK 110.

Owned by D. C. KETCHUM, Sudbury, Rutland Co., Vt.

23 EWES.

Marked with metallic labels in their ears, having thereon R. J. Smith and numbers 69 to 91.

PEDIGREE.

Descended from the importation from Spain by Col. David Humphreys through the flocks of S. Atwood, Conn.; Messrs. Hammond, W. R. Remele, Middlebury; V. Wright, Weybridge; W. R. Sanford, Orwell; R. J. Smith, Sudbury.

EWES PURCHASED.

In 1878, twenty-three Atwood ewes were purchased of R. J. Smith and these constitute the entire flock at time of registration.

—o—

FLOCK 111.

Owned by O. F. KITCHELL, Bridport, Addison Co., Vt.

12 RAMS.

Marked with metallic labels in their ears, having thereon O. F. K. and numbers 6, 10 to 16, 29 to 32.

23 EWES.

Labeled same as rams, with numbers 1 to 5, 7 to 9, 17 to 28, 33 to 35.

PEDIGREE.

Descended from importations from Spain through the flocks of D. Humphreys, S. Atwood, Conn.; A. Cock, Long Island; Wm. Jarvis, Weathersfield; Messrs. Hammond, W. R. Remele, O. Severance, Middlebury; V. Wright, Weybridge; R. P. Hall, F. H. Dean, E. S. Stowell, Cornwall; Baker & Smith, P. Elitharp, C. N. Hayward, H. C. Burwell, B. Myrick, D. F. Doty, J. J. Crane, W. W. Winchester, M. K. Barbour, Bridport; Messrs. Rich, E. R. Robinson, L. C. Remele, E. A. Birchard, L. Beedle, Shoreham.

EWES PURCHASED.

W. W. Winchester purchased of O. Severance a few Atwood ewes that were descended from the flocks of Hammond, Stowell, F. H. Dean and Victor Wright. These ewes Mr. Winchester kept a few years, then sold them to D. C. Barbour of Bridport and he sold them to O. F. Kitchell. The rams used during the time after the purchase by Mr. Winchester to the time of registration were from the flock of C. N. Hayward, and Charlie (253).

RAMS USED.

In 1877 the stock rams of H. C. Burwell and one owned by M. K. Barbour were used and the tegs, 29 to 32, in rams, and 33 to 35 were sired by them.

———o———

FLOCK 112.

Owned by J. W. KNAPP, RICHVILLE (in Shoreham), ADDISON Co., VT.

7 RAMS.

Marked with metallic labels in their ears, having thereon J. W. Knapp, and numbers 1 to 7.

38 EWES.

Labeled same as rams, and numbers 8 to 45.

PEDIGREE.

Descended from importations from Spain through the flocks of A. Cock, Long Island; D. Humphreys, S. Atwood, Conn.; D. Buffum, Wm. Bailey, J. I. Bailey, Rhode Island; Wm. Jarvis, Weathersfield; Messrs. Rich, L. C. Remele, E. R. Robinson, T. Stickney, J. T. Stickney, D. & G. Cutting, J. M. Ormsbee, R. N. Atwood, E. Fanning, M. R. Atwood, L. Beedle, Shoreham; Messrs. Hammond, W. R. Remele, Middlebury; N. A. Saxton, Waltham; W. R. Sanford, Orwell; A. F. Ellsworth, A. H. Hubbard, W. P. Wright, E. Rich, S. T. Baker, Whiting; R. P. Hall, B. S. Field, F. H. Dean, Cornwall; Wm. Gage, Ferrisburgh; C. B. Cook, Charlotte; Baker & Smith, P. Elitharp, Bridport; V. Wright, Weybridge.

EWES PURCHASED.

In August, 1864, four Stickney, Atwood and Rich ewes were purchased of W. P. Wright, and in December, 1867, another ewe was purchased from the same flock of like blood. About the same time of the last purchase two Atwood and Rich ewes were purchased of L. C. Remele, and five ewes bred from the flock of D. & G. Cutting.

RAMS USED.

Rams from the flocks of R. P. Hall, R. N. Atwood, G. Cutting, E. Fanning, and J. T. Stickney have been used to breed up the present flock.

FLOCK 113.

Owned by CHARLES D. LANE, CORNWALL, ADDISON CO., VT.

6 RAMS.

Marked with metallic labels in their ears, having thereon C. D. Lane and numbers 58, 80, 180, 198, 271, 274.

152 EWES.

Labeled same as rams and numbers 1 to 6, 8 to 10. 12 to 26, 28, 30 to 36, 37, 39, 40, 42 to 53, 55, 57, 60 to 62, 64 to 66, 68 to 72, 74, 76, 78, 79, 81, 83, 84, 86 to 89, 91, 92, 95, 96, 103, 104, 106, 111, 112, 114, 115, 118, 120, 122, 126, 130, 131, 134, 135, 137, 138, 140. 141, 143, 148, 149, 154 to 156, 159 to 161, 163 to 166, 168, 170, 172 to 177, 182 to 186, 189 to 194, 196, 197, 199, 200, 251 to 270, 272, 273.

PEDIGREE.

Descended from importations from Spain through the flocks of D. Humphreys, S. Atwood, Conn.; A. Cock, Effingham Lawrence, Long Island ; Wm. Jarvis, Weathersfield ; Messrs. Rich, Leonard Beedle, Zebulon Frost, Abram Frost, E. R. Robinson, Lucius Robinson, T. Stickney & Son, L. C. Remele, H. W. Walker, J. Delong, Shoreham ; Messrs. Hammond, W. R. Remele, U. D. Twitchell, E. R. Clay, Cherbino & Williamson, Middlebury ; A. A. Farnsworth, Brooksville ; O. C. Bacon, N. A. Saxton, Waltham ; R. P. Hall, E. S. Stowell, S. S. Gibbs, Henry Lane, E. Sanford. Milo J. Ellsworth, Charles Benedict, Cornwall ; C. B. Cook, Charlotte ; Baker & Smith, P. Elitharp, C. N. Hayward, B. Myrick, H. C. Burwell, D. F. Doty, J. J. Crane, Bridport ; N. Richards, Panton ; W. R. Sanford, O. H. Bascom, W. O. Bascom, H. D. Bascom, Orwell.

EWES PURCHASED.

In the Spring of 1857 twenty-six ewe togs were purchased of J. Delong, that he had bred direct from the flock of E. R. Robinson, by leasing a part of the flock upon shares, after the death of Mr. Robinson. The fall after, nine ewes were purchased of Lucius Robinson. In 1871, fourteen Atwood ewes were purchased of U. D. Twitchell. In 1872, forty Atwood and Robinson ewes were purchased of Charles Benedict ; the same year, forty ewes were purchased of O. H. Bascom, that were bred from the flock of W. R. Sanford, also twenty ewes of W. O. Bascom, bred from the same flock. In 1873, fifty-five Stickney ewes were purchased of Cherbino & Williamson, that they had purchased from Messrs. Stickney ; they were in lamb by Bonaparte (176) and Silver Horns (177). In August, 1876, twenty-five ewes were purchased of H. D. Bascom that were bred from the flock of W. R. Sanford. The fall of 1876. sixteen ewes were purchased of Miner Jones that were bred by H. W. Walker.

RAMS USED.

The Lute Robinson Ram (39) was first used ; next the Sanford & Treadway ram (54); a few years after the Lane ram (120). In 1868, Jim (234) was purchased and soon after the Drake ram (235). In 1870, Buck Mountain (247) was purchased. In 1875, Bonaparte (176) and Silver Horns (177) were used. Besides the rams enumerated, some bred within the flock have been used, among others C. D. Lane's 274 (501).

———o———

FLOCK 114.

Owned by HENRY LANE, Cornwall, Addison Co., Vt.

2 RAMS.

Marked with metallic labels in their ears, having thereon Henry Lane and numbers 41, 45.

58 EWES.

Labeled same as rams with numbers 1 to 40, 46 to 63.

PEDIGREE.

Descended from the importation from Spain by Col. David Humphreys through the flocks of S. Atwood, Conn.; Messrs. Hammond, W. R. Remele, Middlebury; R. P. Hall, John Towle, H. J. Manchester, E. S. Stowell, R. J. Jones, Cornwall ; C. B. Cook, Charlotte ; N. A. Saxton, Waltham.

EWES PURCHASED.

In January, 1858, thirty Atwood and Blakeslee ewes were purchased of C. B. Cook ; a part were pure Atwood ; in 1860 all these sheep and their increase were sold off except ten pure Atwood ewes ; Aug. 28, 1858, ten yearling Atwood ewes were purchased of N. A. Saxton ; in June, 1874, ten Atwood ewes were purchased of H. J. Manchester ; they were a selection from a number he purchased of John Towle.

RAMS USED.

With the ewes purchased of N. A. Saxton were purchased two yearling Atwood rams and used to the ewes purchased of C. B. Cook ; Cross Ram (42) was used to the Saxton ewes; Sanford Lane (60) was next used for two years; Sweepstakes (32) was used two or three years to a part of the flock ; from him was raised a ram that was used two years. Little Wrinkly (48) was used to forty ewes one year ; for some years after, rams bred within the flock were used ; for three years past H. Lane's Jones Ram (431) has been used.

FLOCK 115.

Owned by ROLLIN LANE, CORNWALL, ADDISON CO., VT.

32 RAMS.

Marked with metallic labels in their ears, having thereon R. Lane, and numbers 13, 14, 87 to 94, 130 to 151.

115 EWES.

Labeled same as rams, one hundred and ten having on their labels R. Lane and numbers 9 to 11, 15 to 86, 95 to 129; five having on their labels O. H. Bascom and numbers 10, 52, 53, 63, 74.

PEDIGREE.

Descended from importations from Spain through the flocks of D. Humphreys, S. Atwood, Conn.; A. Cock, Long Island; D. Buffum, W. Bailey, J. I. Bailey, Rhode Island; Wm. Jarvis, Weathersfield; Messrs. Hammond, W. R. Remele, A. Chapman, J. G. Wellington, S. W. Remele, D. Hooker, Middlebury; Messrs. Rich, L. C. Remele, E. R. Robinson, T. Stickney & Son, J. M. Ormsbee, D. & G. Cutting, E. A. Birchard, E. G. Farnham, Shoreham; R. P. Hall, F. H. Dean, H. C. Bingham, Henry Lane, A. J. Wooster, G. F. Casey, S. Benton, E. Sanford, E. S. Stowell, M. L. Keeler, Charles Benedict, M. J. Ellsworth, R. J. Jones, Cornwall; Baker & Smith, P. Elitharp, C. N. Hayward, H. C. Burwell, D. F. Doty, J. J. Crane, A. Hamilton, J. O. Hamilton, B. Myrick, Bridport; W. R. Sanford, O. H. Bascom, G. H. Bush, J. O. Wright, H. G. Hibbard, Orwell; M. Sheldon, Salisbury; E. S. Casey, S. T. Baker, Whiting; O. Smith, Pittsford; J. Hinds, W. M. Lincoln, Brandon; N. A. Saxton, Waltham; C. B. Cook, Charlotte; V. Wright, L. D. Gregory, A. W. Dana, Weybridge; N. Cushing, Woodstock; D. Davis, Windsor; Henry Thorp, C. B. Cook, Charlotte; E. Meach, E. S. Rowley, Shelburn; A. A. Farnsworth, Brooksville.

EWES PURCHASED.

Ten ewes were purchased of J. M. Parker, of Cornwall; eight of them were Atwood ewes bred by F. H. Dean, and two were Robinson ewes bred by E. G. Farnham; seventeen ewes were purchased of G. D. Bush, ten of them bred by O. H. Bascom from the flock of W. R. Sanford; seven of them were Atwood ewes bred by G. D. Bush; seven ewes were purchased of H. C. Bingham: two were bred by V. Wright: one was from a V. Wright ewe and sired by Eureka (58), and four were sired by Keeler Ram (287); seven Atwood ewe tegs were purchased of F. H. Dean; one ewe was purchased of M. J. Wright that was bred by A. Hamilton from an ewe bred by P. Elitharp and sired by Old Dea, James Ram (52) ; Sept. 11, 1878, 109 sheep were purchased of I. L. Eells that were bred by S. Benton from stock he purchased of L. D. Gregory ; twenty-two ewes and ewe tegs were purchased of Moses Sheldon bred from ewes bred by J. G. Wellington and Mrs. Victor Wright ; a part of the tegs were sired by Brooksville (354) and a part by Dunmore (346).

RAMS USED.

A ram teg was purchased of M. J. Keeler in 1875 that was used to some extent, but his get are all sold. (This ram was called Sheridan Wrinkly.) Three ram tegs were purchased of F. H. Dean in Jan., 1877, and two of them called Tiger and Little Wrinkly were used in 1877, and sired the tegs now in the flock in ewes from 108 to 129 and in rams from 130 to 151.

——o——

FLOCK 116.

Owned by HARRISON S. LANGDON, New Haven, Addison Co., Vt.

8 RAMS

Marked with metallic labels in their ears, having thereon H. S. L., and Nos. 14, 16, 19, 23, 26, 21, 22, 27.

21 EWES.

Labeled same as rams and numbers 0, 1, 2, 3, 4, 5, 6, 8, 10, 11, 12, 13, 15, 17, 18, 20, 24, 25, 28, 29, 30.

PEDIGREE.

Descended from importations from Spain through the flocks of David Humphreys, Stephen Atwood, Conn.; Andrew Cock, Long Island ; William Jarvis, Weathersfield ; Messrs. Hammond, B. Preston, W. R. Remele, Middlebury ; Victor Wright, Weybridge ; Messrs. Rich, E. A. Birchard, L. Beedle, L. C. Remele, E. R. Robinson, Alvin Clark, Shoreham ; N. A. Saxton, Waltham ; N. Richards, Panton ; E. S. Stowell, R. P. Hall, C. D. Lane, Hiram Peck, M. J. Ellsworth, Cornwall ; Baker & Smith, P. Elitharp, C. N. Hayward, H. C. Burwell, B. J. Myrick, J. J. Crane, D. F. Doty, Bridport ; James Ward, Lincoln ; C. B. Cook, Charlotte ; Seth Langdon, New Haven ; N. T. Sprague, Brandon.

EWES PURCHASED.

Flock commenced by a purchase of a few Atwood ewes of Seth Langdon, father of the present owner ; they had been bred from ewes and rams purchased of E. Hammond ; one ewe of Robinson and Atwood blood was also purchased of N. T. Sprague ; two ewes of Atwood blood were also purchased of Buel Preston ; April 16, 1877 three Atwood ewes were purchased of James Ward ; one was bred by H. W. Hammond. the other two were bred from rams and ewes bred by H. W. Hammond ; one of these ewes was sired by Green Mountain (70) and the other two were sired by Kearsarge (55).

RAMS USED.

Mason's Sweepstakes (429) sired ewes 0, 1, 5, 8; one of V· Wright's stock rams at the time of his death was leased of Geo. Hammond's administrator ; he sired ewes 2 and 3 ; Langdon's Cherbino Ram (430) sired ewes 4 and 6 ; numbers 10, 11, 12 are the Ward ewes. In 1876 and 1877 Bismark (221) was used, he sired rams 14, 16, 19, 23, 26, 21, 22, 27 and ewes 13, 15, 17, 20, 24, 25, 28, 29, 30.

—— o ——

FLOCK 117.

Owned by OTIS P. LEE, MIDDLEBURY, ADDISON Co., VT.

2 RAMS.

Marked with metallic labels in their ears, having thereon O. P. Lee and numbers 69, 70.

68 EWES.

Labeled same as rams, and numbers 1 to 68.

PEDIGREE.

Descended from importations from Spain through the flocks of D. Humphreys, S. Atwood, Conn.; A. Cock, Long Island ; Wm. Jarvis, Weathersfield ; Messrs. Rich, L. Beedle, E. R. Robinson, L. C. Remele, E. A. Birchard, E. G. Farnham, A. Clark, Geo. D. Bryant, Shoreham ; C. B. Cook, Charlotte ; Baker & Smith, P. Elitharp, C. N. Hayward, D. F. Doty, J. J. Crane, Bridport ; Messrs. Hammond, W. R. Remele, S. W. Remele, D. Hooker, Middlebury ; R. P. Hall, F. Hooker, Cornwall ; A. A Farnsworth, Brooksville ; J. G. Barker, Leicester; H. M. Graves, Salisbury ; N. A. Saxton, Waltham ; V. Wright, Weybridge.

EWES PURCHASED.

Flock commenced March, 1873, by a purchase of ten Atwood ewes of P. Elitharp, and in the fall of the same year, twenty Atwood and Robinson ewes were purchased of H. P. Elitharp, that were bred by Julius G. Barker.

RAMS USED.

These ewes were bred in 1873, '74 and '75 to Bonaparte (176), Silver Horns (177), and Phil Sheridan (172); in 1876 Bonny (428) was used; in 1877, Brooksville (354) was used.

———o———

FLOCK 118.

Owned by W. H. & T. P. D. MATHEWS, Cornwall, Addison Co., Vt.

1 RAM.

Marked with metallic label in his ear, having thereon T. H. & T. P. D. M. and number 20.

19 EWES.

Labelled same as ram and numbers 2 to 19, 21.

PEDIGREE OF EWES.

Descended from the importations from Spain by Col. David Humphreys through the flocks of S. Atwood, Conn.; Messrs. Hammond, W. R. Remele, Middlebury; R. P. Hall, S. Benton, E. S. Stowell, R. J. Jones, Cornwall; A. A. Farnsworth, Brooksville; T. D. Barton, N. A. Saxton, Waltham; V. Wright, L. D. Gregory, Weybridge.

The ram is Young Bonny (438) and his pedigree can be traced through his ancestors by Bonaparte (176) and the flock of S. S. Rockwell and H. E. Sanford.

EWES PURCHASED.

All the ewes recorded in this flock were bred by S. Benton and sold to I. L. Eells by him to R Lane, and by him to Messrs. Matherws in 1877. Ram used, Young Bonny (438) in 1877.

———o———

FLOCK 119.

Owned by ORSON C. MARTIN, Benson, Rutland Co., Vt.

9 RAMS.

Marked with metallic labels in their ears, six having thereon O. C. M. and numbers 17 to 22, and two having on their labels H. G. Hibbard and numbers 3 and 7.

16 EWES.

Labeled same as first six rams, with numbers 1 to 16.

PEDIGREE.

Descended from importations from Spain through the flocks of D. Humphreys, S. Atwood, Conn.; A. Cock, Long Island; D. Buffum, W. Bailey, J. I. Bailey, Rhode Island; Wm. Jarvis, Weathersfield; Messrs. Rich, L. Beedle, J. C. Remele, E. R. Robinson, E. D. Bush, E A. Birchard, A. Clark, D. C. Smith, J. Forbes, Jr., M. Orvis, D. & G. Cutting, J. M. Ormsbee, Shoreham; S. T. Baker, Whiting; Baker & Smith, P. Elitharp, J. O. Hamilton, Bridport; Messrs. Hammond, Middlebury; N. Richards, Panton; W. R. Sanford, Orwell; W. M. Lincoln, Brandon; E. Porter, Rutland.

EWES PURCHASED.

In Sept., 1877, twelve ewes, mostly Robinson, with a trace of Cutting blood, were purchased of H. Jones of Shoreham. Mr. Jones purchased them a short time before of Myron Orvis and he bred them from the flocks of D. C. Smith, Alvin Clark and E. D. Bush, with one ewe from the flock of J. O. Hamilton.

RAM USED.

H. G. Hibbard's 3 (572) was purchased and used in 1877.

———o———

FLOCK 120.

Owned by PETER MARTIN, Rush, Munroe Co., N. Y.

2 RAMS.

Marked with metallic labels in their ears, having thereon P. Martin and numbers 41, 189.

34 EWES.

Labeled same as rams and numbers 3, 4, 10, 11, 19, 26, 28, 29, 35, 38, 39, 40, 41, 42, 43, 58, 89, 115, 124, 125, 129, 130, 132, 133, 134, 155, 156, 160, 169, 172, 175, 181, 182, 187, 188.

PEDIGREE.

Descended from the importation from Spain by Col. David Humphreys through the flocks of S. Atwood, Conn.; Messrs. Hammond, W. R. Remele, Middlebury; R. P. Hall, Henry Lane, F. H. Dean, E. S. Stowell, Cornwall; N. Richards, Panton; C. B. Cook, Charlotte; N. A. Saxton, Waltham; P. Elitharp, C. N. Hayward, Bridport; Wm. Gage, Ferrisburgh; C. R. Jones, Hubbardton; W. R. Sanford, Orwell; V. Wright, A. J. Stowe, Weybridge; W. H. B. Rogers, Mendon, N. Y.; G. F. Martin, Rush, N. Y.; A. C. Bennet, Livonia, N. Y.

EWES PURCHASED.

In company with Geo. F. and P. M. Martin, six ewes were purchased of L. J. Wright, Weybridge, Vt.; these ewes were bred by C. N. Hayward and were sired by Chunk (95). They were bred together two years, when the interest of G. F. Martin was purchased by Peter, and the year following there was a division of the flock ; Peter Martin retaining four of the ewes first purchased and seven of the ewes that had been bred from them ; in 1867, five Atwood ewes were purchased of W. H. B. Rogers ; they were descended from Atwood ewes bred by Henry Lane and sired by Curley (116), In January, 1870, in company with G. F. Martin, forty 2-year-old ewes were purchased of F. H. Dean ; they were sired by Little Wrinkly (48) ; in March twenty more were purchased of the same breeder that were sired by the same ram ; both lots were equally divided ; in 1872 ten yearling ewes were purchased of G F. Martin of the same blood and breeding of the rest of the flock ; in 1875 six ewes were purchased of C. R. Jones, bred from ewes and rams bred by Messrs. Hammond and sired by David (118). (A very few ewes have been purchased that were not pure Atwoods, but the pedigrees of those herein recorded do not run to them and therefore they are omitted)

RAMS USED.

Torrent (97) was purchased in 1869 and used for a number of years ; Green Mountain, Jr. (96) was also purchased and used a number of years ; Maj. Hammond (114), Little Wrinkly (48), Old Wrinkly (102), Bull Dog (115), Chunk (149), Major (100), Keystone (101), Trump (105), Greasy (108), Triumph (106), Green Mountain, 3d (263), Little Phil (232), Rhoderick Dhu (512), Little Monitor (158), Robin Hood (511), Peter Martin's 204 (516) have all been used to breed up the present flock.

———o———

FLOCK 121.

Owned by GEORGE F. MARTIN, East Rush, Munroe Co.,N. Y.

3 RAMS.

Marked with metallic labels in their ears, having thereon G. F. Martin and numbers 52, 197, 199.

38 EWES.

Labeled same as rams with numbers 2, 5, 9, 21, 22, 25, 48, 96, 97, 100, 105 to 109, 114, 119 to 122, 127, 128, 132, 134, 135, 138 to 142, 144, 146 to 152.

PEDIGREE.

Descended from the importation from Spain by Col. David Humphreys through the flocks of S. Atwood, Conn.; Messrs. Hammond,

W. R. Remele, Middlebury; R. P. Hall, John Towle, Henry Lane, F. H. Dean, E S. Stowell, Cornwall; N. Richards, Panton; C. B. Cook, Charlotte; N. A. Saxton, Waltham; P. Elitharp, C. N. Hayward, Bridport; Wm. Gage, Ferrisburgh; C. R. Jones, Hubbardton; A. E. Fuller, Pomfret; V. Wright, A. J. Stowe, Weybridge; W. H. B. Rogers, Mendon, N. Y.; Peter Martin, Rush, N. Y.; A. C. Bennet, Livonia, N. Y.

EWES PURCHASED.

Besides the purchases made in company with Peter Martin of F. H. Dean, of forty two-year old ewes in Jan., 1870, and twenty of the same in March, 1872, there was a purchase of five ewes of W. H. B. Rogers (these ewes were purchased in company with Peter Martin at the time he made the purchase of Mr. Rogers; they were divided in 1870). G. F. Martin at a later date purchased of Peter Martin a choice of four of his share in the division of the first forty ewes purchased of Mr. Dean, and in June, 1872, his entire interest in the last purchase from the same breeder. In Jan., 1874, seven ewe tegs were purchased of Peter Martin; in Jan., 1875, four ewes were purchased of C. H. Jones; they were sired by David (118), their dams were Hammond and Stowell ewes, or bred direct from those flocks; in Feb., 1875, two ewes bred by E. S. Stowell of C. R. Jones; they were sired by Diamond Dust (122) and of the same, five that were bred by J. Towle; also nine that were bred by C. R. Jones; they were sired by David (118) and from ewes bred from Hammond and Stowell's flocks.

RAMS USED.

Torrent (97), purchased in company with Peter Martin, Green Mountain, Jr. (96), Keystone (101), Major (100), Trump (105), Torrent, Jr. (111), Green Mountain, 3d (263), Smuggler (423), Compact (113), Robin Hood (511), Rhoderick Dhu (512) and Infantado bred within the flock, sired by Triumph (106) have all been used to breed up the flock. Compact, Robin Hood and Infantado are now the stock rams of the flock.

———o———

FLOCK 122.

Owned by T. F. & C. D. McCONNELL, Ripon, Wis.

7 RAMS.

Marked with metallic labels in their ears, having thereon T. F. & C. D. McC. and numbers 79 to 85.

78 EWES.

Labeled same as rams with numbers 1 to 78.

PEDIGREE.

Descended from importations from Spain through the flocks of
D. Humphreys, S. Atwood, Conn.; A. Cock, E. Lawrence, Long Is-
land; D. Buffum, Wm. Bailey, J. I. Bailey, Rhode Island; Wm.
Jarvis, Weathersfield; Messrs. Rich, Leonard Beedle, E. R. Robin-
son, D. E. Robinson, Lucius Robinson, L. C. Remele, T. Stickney &
Son, Zebulon Frost, J. M. Ormsbee, D. & G. Cutting, H. W. Walker,
E. D. Bush, M. W. C. Wright, D. J. Wright, Geo. H. Hall, Shore-
ham; Messrs Hammond, S. W. Remele, D. Hooker, W. R. Remele,
Middlebury; Wm. Wright, Wm. Baldwin, S. T. Baker, A. M. Bald-
win, J. Q. Stickney, A. F. Ellsworth, A. H. Hubbard, Whiting;
Baker & Smith, P. Elitharp, Bridport; S. Root, Hubbardton; L.
Webster, Sudbury; C. B. Cook, Charlotte; W. R. Sanford, Orwell;
V. Wright, Weybridge; R. P. Hall, J. Towle, E. S. Stowell, B. S.
Fields, F. Hooker, Cornwall; N. A. Saxton, Waltham; A. A.
Farnsworth, Brooksville.

EWES PURCHASED.

Nov. 6, 1876, forty-eight ewes were purchased of H. W. Walker;
Nov. 7, 1876, nine ewes were purchased of D. J. Wright; Nov. 8,
1876, eight ewes were purchased of T. Stickney & Son; the same
day four ewes and four ewe tegs were purchased of W. P. Wright;
Nov. 9, 1876, ten ewes bred by A. M. Baldwin were purchased of
E. E. Stickney of the firm of T. Stickney & Son.

RAMS PURCHASED.

McConnell's Stickney Ram (435), McConnell's Hall Ram (426),
the E. D. Bush Ram (437) were purchased as given in their pedigrees; two other rams were also purchased of T. Stickney & Son,
one of which was retained in the flock at the time of registration;
three ram tegs were also purchased of W. P. Wright at the same
time of the purchase of ewes and ewe tegs from the same flock.
These seven rams, with the seventy-eight of the eighty-one ewes
purchased, make up the flock registered.

—o—

FLOCK 123.

Owned by WILLIAM McCAULEY, Middlebury, Addison Co., Vt.

6 RAMS.

Marked with metallic labels in their ears, having thereon W.
McCauley, and numbers 1 to 6.

37 EWES.

Labeled same as rams with numbers 7 to 43.

PEDIGREE.

Descended from importations from Spain through the flocks of D. Humphreys, S. Atwood, Conn.; A. Cock, Long Island; Wm. Jarvis, Weathersfield; Leonard Beedle, Messrs. Rich, E. R. Robinson, L. C. Remele, Shoreham; P. P. Hall, E. Sanford, E. S. Stowell, F. Hooker, S. S. Rockwell, Cornwall; Messrs. Hammond, W. R. Remele, D. Hooker, S. W. Remele, Middlebury; A. A. Farnsworth, Brooksville; W. R. Sanford, Orwell; N. A. Saxton, Waltham; N. Richards, Panton; C. B. Cook, Charlotte; P. Elitharp, C. N. Hayward, Bridport; J. Marsh, Hinesburgh.

EWES PURCHASED.

In 1863 twenty ewes were purchased of S. S. Rockwell that were bred by him from Atwood and Beedle stock.

RAMS USED.

Eureka (58) three years; Chunk (95) one year; Nevada (168) one year; Peerless (248) one year; Little Wrinkly (48) two years; Allright (169) one year; Wrinkly (292) one year, and rams bred within the flock sired by the above. Besides these, Fremont, Jr. (215) and Gold Dust (334) were used, but the produce has all been sold. In 1877 Mingle (440) was used.

——o——

FLOCK 124.

Owned by JOHN H. MEAD, West Rutland, Rutland Co., Vt.

48 RAMS.

Marked with metallic labels in their ears, having thereon 74 to 92, 93 to 119, 144, 145.

97 EWES.

Labeled same as rams, with numbers 1 to 73, 120 to 143.

PEDIGREE.

Descended from importations from Spain through the flocks of A. Cock, Long Island; D. Humphreys, S. Atwood, Conn.; Wm. Jarvis, Weathersfield; Messrs. Rich, L. Beedle, E. R. Robinson, L. C. Remele, Shoreham; Baker & Smith, P. Elitharp, Bridport; Messrs. Hammond, Middlebury; F. H. Dean, E. Hamilton, H. Peck, Cornwall; N. T. Sprague, Brandon; A. Hull, Wallingford; J. B. Proctor, Rutland.

EWES PURCHASED.

In Sept., 1865, A. J. Mead (father of the present owner of the flock) purchased fifteen ewes of J. B. Proctor; a part of these ewes were bred by J. T. & V. Rich and a part were bred from sheep from the same flock. In the spring of 1868, J. H. Mead purchased the flock.

RAMS USED.

Green Mountain (391) and Bull Dog (392); of those bred within the flock, Pony (394), Young Prince (393), Big Wrinkly (395) and Chunk (396) have been used; Pony and Chunk having been used most.

———o———

FLOCK 125.

Owned by MRS. MARY W. MEAD, West Cornwall, Addison Co., Vt.

3 RAMS.

Marked with metallic labels in their ears, having thereon Mrs. M. W. M. and numbers 44, 45, 46.

53 EWES.

Labeled same as rams with numbers 1 to 48, 51 to 60.

PEDIGREE.

Descended from importations from Spain through the flocks of D. Humphreys, S. Atwood, Conn.; A. Cock, Long Island; Messrs. Hammond, W. R. Remele, S. Piper, J. Piper, Middlebury; V. Wright, Weybridge; R. P. Hall, E. S. Stowell, Peet & Mead, R. J. Jones, Cornwall; W. R. Sanford, Orwell; C. B. Cook, Charlotte; Messrs. Rich, E. R. Robinson, L. C. Remele, E. G. Farnham, E. A. Birchard, Shoreham; Baker & Smith, P. Elitharp, Bridport.

EWES PURCHASED.

Horace Mead purchased in Feb., 1864, twenty Atwood sheep of Anson W. Frost, of Cornwall, that he had purchased four months previous of Peet & Mead; these ewes were bred from the flock of W. R. Sanford. After the death of Horace Mead these sheep descended to his widow, the present owner of the flock.

RAMS USED.

The Peet and Mead Ram (228), Keeler Ram (287) and Allright (169) and to some extent those bred within the flock.

FLOCK 126.

Owned by C. C. MINER, BRIDPORT, ADDISON CO., VT.

3 RAMS.

Marked with metallic labels in their ears, having thereon C. C. Miner and numbers 16, 17, 18.

15 EWES.

Labeled same as rams and numbers 1 to 15.

PEDIGREE.

Descended from importations from Spain through the flocks of D Humphreys, S. Atwood. Conn.; Messrs. Rich, L. Beedle, E. R. Robinson, E. A. Birchard, L. C. Remele, Shoreham ; Messrs. Hammond, W. R. Remele, Middlebury ; Baker & Smith, P. Elitharp, B. Myrick, J. J. Crane, C. N. Hayward, H. C. Burwell, D. F. Doty, C. P. Crane, C. E. Crane, Bridport ; C. B. Cook, Charlotte ; N. A. Saxton, Waltham ; N. Richards, Panton ; V. Wright, Weybridge ; R. P. Hall, S. S. Rockwell, Cornwall.

EWES PURCHASED.

In 1864 a pair of twin yearling ewes were purchased of J. J. Crane ; they were sired by Ironsides (80); their dam was bred by P. Elitharp. Ten ewes were leased of S. S. Rockwell, to be bred on shares to Eureka (58) ; one ewe and one ewe teg from these last are now in the flock. Two ewes were purchased of C. E. Crane in 1877, and two of L. Dodge, of Weybridge, that he purchased of the estate of V. Wright; one ewe was also purchased of C. P. Crane. Except the ewes leased of S. S. Rockwell, those purchased were Atwood ewes.

RAMS USED.

Eureka (58) and the stock rams of D. F. Doty, J. J. Crane and C. P. Crane.

———o———

FLOCK 127.

Owned by F. & L. E. MOORE, SHOREHAM, ADDISON CO., VT.

32 RAMS.

Marked with metallic labels in their ears, having thereon F. & L. E. Moore, and numbers 99 to 129, 150.

149 EWES.

Labeled same as rams, and numbers 1 to 98. 130 to 149, 151 to 181.

PEDIGREE.

Descended from importations from Spain through the flocks of A. Cock, Long Island; D. Humphreys, S. Atwood, Conn.; Wm. Jarvis, Weathersfield; Messrs Rich, L. C. Remele, L. Beedle, E. R. Robinson, E. A. Birchard, T. Stickney & Son, E. D. Bush, B. B. Tottingham, A. C. Harris, E. Fanning, D. E. Robinson, W. H. Cook, H. W. Jones, A. J. Towner, Geo. D. Bryant, Shoreham; Messrs. Hammond, W. R. Remele, D. Hooker, S. W. Remele, O. Severance, Middlebury; Baker & Smith, P. Elitharp, C. N. Hayward, W. W. Winchester, Bridport; C. B. Cook, Charlotte; N. A. Saxton, Waltham; N. Richards, Panton; A. M. Baldwin, J. Q. Stickney, Whiting; L. Webster, Sudbury; S. Root, Hubbardton; R. P. Hall, John Towle, F. Hooker, Cornwall; A. A. Farnsworth, Brooksville.

EWES PURCHASED.

In 1862 or '63 a few ewe tegs were purchased of D. E. Robinson; in 1868 thirty-four ewes and ewe tegs were purchased of H. W. Jones; these were descendants from the Robinson flock except a few that were sired by Little Wrinkly (48); in 1872, fifty Atwood and Rich ewes were purchased of A. C. Harris; they were all sired by the stock rams of Messrs. Hammond, Sweepstakes (32), Gold Drop (64), Silver Mine (61) and Green Mountain (70) having sired the most of them. Nov 1st, 1873, twenty-eight ewes were purchased of J. T. & V. Rich; nine ewes were purchased bred by E. Fanning, W. H. Cook, and A. M. Baldwin; these were all Atwood and Robinson, Rich or Stickney blood.

RAMS USED.

The old Tottingham (40) at first; Small Tom (88) for three years; Little Corporal (289) to a part of the flock one year; Birchard & Tottingham (202) was used two years to a part of the flock; Little Wrinkly (48) was used to a few ewes; Duke (275) and Dean (236) were used for a number of years; Don Pedro (276) was used a number of years; Addison Chief (278) was used in 1875, and Fortune (475) was used in 1876 and '77, Banker (471) a very little in the last year.

---o---

FLOCK 128.

Owned by C. P. MORRISON, Addison, Addison Co., Vt.

18 RAMS.

Marked with metallic labels in their ears, having thereon M. & N. and numbers 1 to 18.

53 EWES.

Labeled same as rams, with numbers 19 to 71.

PEDIGREE.

Descended from importations from Spain through the flocks of A. Cock, Long Island; D. Humphreys, S. Atwood, Conn.; Messrs. Rich, L. C. Remele, E. R. Robinson, E. A. Birchard, A. Smith, Shoreham; Messrs. Hammond, W. R. Remele, Middlebury; Baker & Smith, P. Elitharp, C. N. Hayward, H. C. Burwell, J. J. Crane, D. E. Hill, D. F. Doty, B. Myrick, Bridport; F. D. Barton, Waltham; W. R. Sanford, Orwell; N. A. Saxton, Waltham; N. Richards, Panton; C. B. Cook, Charlotte; R. P. Hall, E. S. Stowell, Cornwall; C. G. Seager, Addison.

EWES PURCHASED.

In Oct., 1851, C. G. Seager purchased of E. R. Robinson six Robinson ewes and two Atwood ewes of W. S. & E. Hammond; in August, 1852, twenty-five more Robinson ewes were purchased of E. R. Robinson; in 1860, C. P. Morrison became a partner in the flock, and took the sheep to his farm to breed. In 1877, P. B. Norton became an owner in the flock, purchasing the interest of C. G. Seager; during that year the sheep were labeled M. & N. C. P. Morrison is now sole owner of the flock.

RAMS USED.

The stock rams of Messrs. Hammond and Robinson until some were raised within the flock of sufficient excellence for stock rams. Sea Lion (83) was used two years; P. Elitharp's rams one year; one of F. D. Barton's rams two years; in 1876 and 1877, Bismarck (221), Stub (222) and Black Top (225) were used.

——o——

FLOCK 129.

Owned by E. B. MUSSEY, Middlebury, Addison Co., Vt.

7 RAMS.

Marked with metallic labels in their ears, having thereon E. B. Mussey, and numbers 1 to 7.

18 EWES.

Labeled same as rams, sixteen with numbers 8 to 28, and two with A. Chapman and numbers 1, 2.

PEDIGREE.

Descended from the importation from Spain by Col. David Humphreys through the flocks of S. Atwood, Conn.; Messrs. Hammond, W. R. Remele, O. Severance, Middlebury; A. A. Farnsworth,

Brooksville ; E. S. Rowley, E. Meach, Shelburn ; H. Thorp, C. B. Cook, Charlotte ; R. P. Hall, Cornwall ; R. Gleason, Benson ; W. R. Sanford, Orwell.

EWES PURCHASED.

In Dec., 1877, two ewes were purchased of A. Chapman ; they were bred by E. S. Rowley ; they were in lamb by Custer (551); also seven ewes were purchased of A. A. Farnsworth and six of S. W. Remele ; all were Atwood ewes.

RAMS.

Wanderer (439) is now the stock ram of the flock. E. S. Stowell's 274 (500) and Rarus (569), were used in part in 1877.

———o———

FLOCK 130.

Owned by the Estate of JOSEPH R. NASH, New Haven, Addison Co., Vt.

50 EWES.

Marked with metallic labels in their ears, having thereon J. R. Nash and numbers 1 to 50.

PEDIGREE.

Descended from importations from Spain through the flocks of D Humphreys, S. Atwood, Conn.; A. Cock, Long Island ; Wm. Jarvis, Weathersfield ; Messrs. Hammond, W. R. Remele, Middlebury ; R. P. Hall, E. S. Stowell, M. Bingham, Cornwall ; A. A. Farnsworth, Brooksville ; S. & S. D. Doud, New Haven ; S. G. Holyoke, St. Albans ; Messrs. Rich, E. R. Robinson, L. C. Remele, L. Wolcott, L. Beedle, Shoreham ; Baker & Smith, P. Elitharp, H. Cross, Moses Hamilton, Bridport.

EWES PURCHASED.

A lot of ewes were purchased of M. Peck, a part were Robinson ewes bred by Moses Hamilton, the rest Atwood ewes bred by M. Peck from Victor Wright stock.

RAMS USED.

Rams from the flocks of E. Hammond & Son, H. W. Hammond, A. A. Farnsworth, S. & S. D. Doud, and Merrill Bingham have been used to breed up the present flock.

FLOCK 131.

Owned by E. L. NEEDHAM, Whiting, Addison Co., Vt.

12 RAMS.

Marked with metallic labels in their ears, having thereon E. L. Needham, and numbers 28 to 39.

34 EWES.

Labeled same as rams, and numbers 1 to 27, 40 to 46.

PEDIGREE.

Descended from importations from Spain through the flocks of A. Cock, Long Island; D. Humphreys, S. Atwood, Conn.; Wm. Jarvis, Weathersfield; Messrs. Hammond, W. R. Remele, S. W. Remele, D. Hooker, Middlebury; R. P. Hall, J. Towle, F. Hooker, Cornwall; Messrs. Rich, E. R Robinson, T. Stickney, G. Rich, L. Beedle, Shoreham; W, R. Sanford, Orwell; Baker & Smith, P. Elitharp, Bridport; A. H. Hubbard, Whiting.

ORIGIN.

Twenty ewes were leased on shares of Gasca Rich; they were bred from the flocks of J. T. & V. Rich and W. R. Sanford. The increase from these ewes were afterwards purchased by E. L. Needham.

RAMS USED.

Besides those bred within the flock those from the flock of A. H. Hubbard.

———o———

FLOCK 132.

Owned by GEORGE N. PAYNE, Bridport, Addison Co., Vt.

68 EWES.

Marked with metallic labels in their ears, having thereon G. N. Payne, and numbers 1 to 68.

PEDIGREE.

Descended from importations from Spain through the flocks of A. Cock, Long Island; D. Humphreys, S. Atwood, Conn.; Wm. Jarvis, Weathersfield; Messrs. Rich, L. Beedle, E. R. Robinson, L. C. Remele, E. A. Birchard, Shoreham; Baker & Smith, P. Elitharp, C. N. Hayward, J. J. Crane, B. Myrick, H. C. Burwell, Bridport; N. A. Saxton, Waltham; N. Richards, Panton; Messrs. Hammond, W. R. Remele, Middlebury; W. R. Sanford, Orwell; V. Wright, Weybridge; R. P. Hall, Cornwall.

EWES PURCHASED.

In the fall of 1864 twenty-four Robinson ewe tegs were purchased of E. A. Birchard.

RAMS USED.

At the time of the purchase of ewes a ram of the same blood was purchased from the same flock, and was used six or seven years. Young Eureka (81), Old Dea. James Ram (52) were used three years. Eureka, 3d (223) has been used extensively the past year or two. Besides those named, rams bred within the flock have been used.

———o———

FLOCK 133.

Owned by LYMAN H. PAYNE, MIDDLEBURY, ADDISON CO., VT.

12 RAMS.

Marked with metallic labels in their ears, having thereon L. H. Payne and numbers 49 to 60.

58 EWES.

Labeled same as rams with numbers 1 to 48, 61 to 70.

PEDIGREE.

Descended from importations from Spain through the flocks of A. Cock, Long Island; D. Humphreys, S. Atwood, Conn.; Wm. Jarvis, Weathersfield; Messrs. Rich, Leonard Beedle, E. R. Robinson, L. C. Remele, E. A. Birchard, T. Stickney, Shoreham; R. P. Hall, E. S. Stowell, E. S. Dana, S. S. Gibbs, Cornwall; Messrs. Hammond, W. R. Remele, Middlebury; V. Wright, Weybridge; W. R. Sanford, Orwell; Baker & Smith, P. Elitharp, Bridport; C. B. Cook, Charlotte.

EWES PURCHASED.

March 24, 1877, forty ewes were purchased of E. S. Dana; twenty of these were Robinson ewes bred by S. S. Gibbs, and twenty were from these ewes and sired by Allright (169)

RAMS USED.

Allright (169) was used in 1877, and the ram and ewe tegs were sired by him.

FLOCK 134.

Owned by JAMES E. PARKER, Whiting, Addison Co., Vt.

16 EWES.

Marked with metallic labels in their ears, having thereon J. E. Parker, and numbers 1 to 16.

PEDIGREE.

Descended from importations from Spain through the flocks of A. Cock. Long Island; D. Humphrey, S. Atwood, Connecticut; Wm. Jarvis, Weathersfield; Messrs. Rich, L. Beedle, L. C. Remele T. Stickney & Son, E. R. Robinson, E. A. Birchard, Shoreham; A. F. Ellsworth, J. Q. Stickney, E. Rich, W. P. Wright, C. H. Ketchum, Whiting; R. P. Hall, F. H. Dean, J. Towle, A. J. Wooster, B. S. Field, Cornwall; Messrs. Hammond, W. R. Remele, Middlebury; W. R. Sanford, Orwell; L. Webster, Sudbury; S. Root, Hubbardton; Wm. Gage, Ferrisburgh; C. B. Cook, Charlotte; Baker & Smith, P. Elitharp, Bridport.

EWES PURCHASED.

In August, 1871, five Atwood and Robinson ewes were purchased of Wm. P. Wright.

RAMS USED.

C. H. Ketchum 51 (454) and Fremont, Jr. (215) were used, and the sixteen sheep recorded are all descendants of these rams and the five ewes purchased of W. P. Wright.

——o——

FLOCK 135.

Owned by JAMES W. PECK, West Cornwall, Addison Co., Vt.

16 RAMS.

Marked with metallic labels in their ears, having thereon J. W. Peck and numbers 42 to 57.

41 EWES.

Labeled same as rams and numbers 1 to 41.

PEDIGREE.

Descended from importations from Spain through the flocks of Andrew Cock, Long Island; David Humphreys, Stephen Atwood, Conn.; Messrs. Rich, L. Beedle, E. R. Robinson, L. C. Remele, E. A. Birchard, G. D. Bryant, Shoreham; Baker & Smith, P. Elitharp, C.

A. Landers, C. N. Hayward, D. C. Barber, Bridport; R. P. Hall, W. H
Delong, R. J. Jones, S. S. Gibbs, Cornwall; W. R. Sanford, Orwell;
F. D. Douglas, Whiting; Messrs. Hammond, W. R. Remele, Middlebury; N. A. Saxton, Waltham; C. B. Cook, Charlotte; N. Richards, Panton; J. Sheldon, Fairhaven.

EWES PURCHASED.

Dec. 13, 1864, six ewes were purchased of F. D. Douglas, certified by him to be "descended purely from the flocks of Erastus
Robinson and J. T. & V. Rich;" in 1877, fifteen ewes were purchased of F. G. Convers that were bred by D. C Barber from Hammond and Hayward ewes and by a ram bred by C. A. Landers of
Robinson blood.

RAMS USED.

Fortune (249), Little Wrinkly (48), Eureka (58), Green Mountain (407), and one of R. J. Jones's stock rams have all been used;
other rams have been used, the stock from which has all been sold.
The last ram used has been Delong's Remele Ram (361). At the
time of the purchase of the ewes of F. G. Convers, ten rams were
also purchased of the same, of the same blood and breeding by D.
C. Barber.

FLOCK 136.

Owned by H. M. PERRY & G. A. CUTTING, EAST SHOREHAM, VT.

22 RAMS.

Marked with metallic labels in their ears, having thereon H. M.
Perry and numbers 51, 87 to 100, 102 to 108.

96 EWES.

Labelled same as rams with numbers 1 to 50, 52 to 86, 101, 109
to 118.

The Pedigree and History of this flock are substantially the same
as Flock 61, belonging to the estate of German Cutting, at page
228, and to that flock we would refer, as it is thought unnecessary to
repeat them here.

FLOCK 137.

Owned by E. PECK & SONS, GENEVA, KANE CO., ILLINOIS.

3 RAMS.

Marked with metallic labels in their ears, having thereon
E. Peck & Sons and numbers 50, 51, 52.

79 EWES.

Labeled same as rams, with numbers 1 to 49, 53 to 82.

PEDIGREE.

Descended from importations from Spain through the flocks of
D. Humphreys, S. Atwood, J. N. Blakeslee, J. R. Nettleton, Conn.;
A. Cock, Long Island; Messrs. Hammond, W. R. Remele, Middle-
bury; N. A. Saxton, Waltham; Messrs. Rich L. Beedle, M. W. C.
Wright, D. J. Wright, L. C. Remele, E. R. Robinson, S. L. Bissell,
E. N. Bissell, T. Stickney & Son, Jasper Barnum, Lucius Robinson,
E. D. Bush, E. A. Birchard, H. S. Brookins, Q. C. Rich, Shoreham;
Baker & Smith, P. Elitharp, Bridport; W. R. Sanford, Orwell;
R. P. Hall, E. S. Stowell, J. Towle, Cornwall; A. H. Hubbard,
Whiting.

EWES PURCHASED.

Sept. 21, 1875, forty-nine ewes were purchased of E. N. Bissell;
they were bred in the flocks of E. N. Bissell, D. J. Wright, H. S.
Brookins and Q. C. Rich.

RAMS USED.

At the time the ewes were purchased, twenty-four rams, bred in
the same flocks, were purchased of the same party; three of these
were reserved in the flock; one of them, Little Wrinkly (413), with
Peck & Sons' Stickney Ram (412), have since been the stock rams
of the flock; they sired ewes 53 to 82.

——o——

FLOCK 138.

Owned by L. S & L. W. PEET, Cornwall, Addison Co., Vt.

20 RAMS.

Marked with metallic labels in their ears, having thereon L.
S. & L. W. Peet, and numbers 81 to 100.

80 EWES.

Labeled same as rams with numbers 1 to 80.

PEDIGREE.

Descended from importations from Spain through the flocks of
D. Humphreys, S. Atwood, Conn.; A. Cock, Long Island; Wm. Jar-
vis, Weathersfield; Messrs. Hammond, W. R. Remele, S. W. Rem-
ele, D. Hooker, A. C. Hooker, O. Severance, Middlebury; L. D.
Gregory, V. Wright, Weybridge; A. A. Farnsworth, Brooksville;
N. A Saxton, Waltham; R. P. Hall, E. S. Stowell, J. Towle, R. J.

Jones, J. B. Hamblin, M. J. Ellsworth, J. Ellsworth, S. S. Gibbs, B. S. Fields, E. Sanford, R. R. Wright, Cornwall ; W. R. Sanford, R. D. Hall, Orwell ; R. Gleason, Benson ; J. Hinds, W. M. Lincoln, Brandon ; A. Hull, Wallingford ; O. Smith, Pittsford ; Messrs. Rich, E. R. Robinson, L. C. Remele, E. A. Birchard, A. Clark, Shoreham ; Baker & Smith, P. Elitharp, Bridport.

EWES PURCHASED.

This flock was founded by L. S. Peet, the senior member of this firm, and his son-in-law, Lorin C. Mead. In January, 1861, they purchased of W. R. Sanford & Son ten ewe tegs ; in December following ten ewes, and in Oct., 1862, ten more ewes ; all these were Atwood blood. Atwood ewes have also been purchased of R. D. Hall bred from the flock of W. R. Sanford ; from A. C. Hooker bred from the flock of D. Hooker, he from Messrs. Hammond, S. W. Remele, and A. A. Farnsworth ; also from J. Towle from the old Hall flock. A few ewes of Atwood and Robinson blood were purchased of J. B. Hamblin.

RAMS USED.

Two ram tegs were purchased of W. R. Sanford & Son, one Jan. 5, 1861, and one Dec. 20, 1861 ; the Peet & Mead (228) was purchased in August, 1865 ; these rams were used a number of years or until there was a division of the flock in 1866 ; soon after this a ram was purchased of E. S. Stowell ; one has since been purchased of A. A. Farnsworth, Peet's Farnsworth Ram (411), and one since of R. J. Jones ; these were all Atwood rams. Other rams bred within the flock have been used. L. W. Peet, son of L. S. Peet, is now a partner in the ownership of the flock.

——o——

FLOCK 139.

Owned by MARTIN M. PECK, Cornwall, Addison Co., Vt.

4 RAMS.

Marked with metallic labels in their ears, having thereon M. M. Peck and numbers 14, 15, 18, 19.

18 EWES.

Labeled same as rams with numbers 1 to 13, 16, 20 to 23.

PEDIGREE.

Descended from importations from Spain through the flocks of A. Cock, Long Island ; D. Humphreys, S. Atwood, Conn.; D. Buffum, Wm. Bailey, J. I. Bailey, Rhode Island ; Wm. Jarvis, Weathersfield ; Messrs. Rich, L. Beedle, E. R. Robinson, L. C. Remele,

E. A. Birchard, J. M. Ormsbee, D. & G. Cutting, T. Stickney & Son, J. T. Stickney, A. Clark, Shoreham ; Messrs. Hammond, W. R. Remele, Middlebury ; R P. Hall, S. S. Rockwell, W. McCauley, M. J. Ellsworth, C. Benedict, E. S. Stowell, E Sanford, Cornwall ; S T. Baker, Whiting ; Baker & Smith, P. Elitharp, C. N. Hayward, J. O. Hamilton, J. J. Crane, Bridport ; C. B. Cook, Charlotte ; Henry Palmer, C. G. Seager, P. Cassidy, C. P. Morrison, Addison ; F. D. Barton, Waltham ; W. R. Sanford, Orwell.

EWES PURCHASED.

In 1874, ten ewes were purchased of H. T. Peck, purchased but a short time before of D. E. Taylor, then of Cornwall ; these ewes Taylor purchased of Henry Palmer, who bred them from ewes purchased of Wm. McCauley and the stock rams of the Seager flock of Addison. The ewes sold by McCauley were bred from ewes he purchased of S. S. Rockwell and were sired by Little Wrinkly (48).

RAMS USED.

The stock rams of C H. James ; his Stickney ram, those he purchased of M. J. Ellsworth, and Greasy (297).

———o———

FLOCK 140.

Owned by ALBRO E. PERKINS, POMFRET, WINDSOR CO., VT.

11 RAMS.

Marked with metallic labels in their ears, having thereon A. E. Perkins and numbers 1 to 11.

39 EWES.

Labeled same as rams, and numbers 12 to 50.

PEDIGREE.

Descended from the importation from Spain by Col. David Humphreys through the flocks of S. Atwood, Conn.; Messrs. Hammond, W. R. Remele, Middlebury ; O. C. Bacon, Waltham ; R. P. Hall, E. S. Stowell, H. F. Dean, Cornwall ; V. Wright, Weybridge ; W. R. Sanford, Orwell ; A E. Fuller, Pomfret.

EWES PURCHASED.

In Sept., 1857, thirty-four ewes were purchased of W. R. Sanford ; and in the following year, in October, twelve more ewes were purchased of the same ; in 1859 twelve ewes were purchased of E. Hammond ; in August, 1860, seven ewes were purchased of W. R. Sanford.

RAMS USED.

At the time of the last purchase of the ewes of Mr. Sanford four ram tegs were purchased, and in December following three more ram tegs ; some of these were used as stock rams. Perkin's Bacon Ram (184) was used as a stock ram ; a ram was purchased of H. F. Dean that was sired by Sweepstakes (32) and out of an Atwood ewe bred by E. Hammond ; Stowell Ram (256) was used a number of years ; the stock rams of A E. Fuller have been used to some extent ; of those bred within the flock and used as stock rams have been Old Deacon (405), Perkins' Sweepstakes (403), Constitution (404) and A. E. Perkins' No 4 (406.)

———o———

FLOCK 141.

Owned by J. H. & A. W. PETERS, Bradford, Orange Co., Vt.

8 RAMS.

Marked with metallic labels in their ears, six having thereon J. H. & A. W. P. and numbers 1 to 6, and two having on their labels O. & E. S. Hall and numbers 42, 90.

30 EWES.

Labeled same as rams, twenty-eight having on their labels J. H. & A. W. P. and numbers 7 to 34, and two with O. & E. S. Hall and numbers 7, 75.

PEDIGREE.

Descended from the importation from Spain by Col. David Humphreys through the flocks of S. Atwood, Conn.; Messrs. Hammond, Middlebury ; O. & E. S. Hall, East Randolph.

EWES PURCHASED.

In 1871, a few ewes were purchased of O. & E. S. Hall ; in 1872 a few more, and later still a few more, the last being two ewes in 1878.

RAMS USED.

Dexter (408), Black Buck (410) and Young America (409) have all been used to breed up the present flock ; two ram tegs were purchased of O. & E. S. Hall in the winter of 1878.

———o———

FLOCK 142.

Owned by W. B. PORTER, North Tunbridge, Orange Co., Vt

7 RAMS.

Marked with metallic labels in their ears, having thereon W. B. Porter and numbers 39, 46, 47, 48, 50 to 52.

42 EWES.

Labeled same as rams with numbers 1 to 16, 18 to 23, 25 to 38, 40 to 45.

PEDIGREE.

Descended from the importation from Spain by Col. David Humphreys through the flocks of S. Atwood, Conn.; Messrs. Hammond, Middlebury ; O. & E. S. Hall, East Randolph.

EWES PURCHASED.

In 1872, five ewes were purchased of O. & E. S. Hall; in 1873, eight more from the same flock, and in 1874 five more, all Atwood ewes bred direct from the flock of Messrs. Hammond.

RAMS USED.

Bull Dog (104), Queen Ram (415), Gold Drop (414) and other stock rams belonging to O. & E. S. Hall ; rams bred within the flock have been used, among which are Defiance (417), Stub (418) and Hayes (419), the present stock rams of the flock.

———o———

FLOCK 143.

Owned by J. F. RANDALL, West Cornwall, Addison Co., Vt.

7 RAMS.

Marked with metallic labels in their ears, having thereon J. F. Randall, and numbers 1, 2, 36 to 40.

42 EWES.

Labeled same as rams and numbers 3 to 35, 41 to 50.

PEDIGREE.

Descended from importations from Spain through the flocks of A. Cock, Long Island ; D. Humphreys, S. Atwood, Conn.; Wm. Jarvis, Weathersfield ; Messrs. Rich, L. Beedle, L. C. Remele, E. R. Robinson, F. & L. E. Moore, B. B. Tottingham, G. D. Bryant, J. Forbes, Jr., T. Stickney & Son, E. A. Birchard, Levi Wolcott, E. D. Bush, A. C. Harris, Shoreham ; Messrs. Hammond, W. R. Remele, Middlebury ; R P. Hall, E. S. Stowell, F. H. Dean, W. McCauley, S. S. Rockwell, Merrill Bingham, Cornwall ; J. Q. Stickney, Whiting ; S. Root, Hubbardton ; L. Webster, Sudbury ; W. R. Sanford, Orwell ; Baker & Smith, P. Elitharp, Bridport.

EWES PURCHASED.

In 1876, eight ewes were purchased of E. G. Farnham ; four of these ewes were bred by G. D. Bryant and four by J. Forbes, Jr.;

two of these are retained in the flock, one of each lot, also two ewes bred from them sired by M. Bingham's McCauley Ram that was sired by Fremont, Jr. (215); dam by Little Wrinkly (48), G. dam bred by S. S. Rockwell; four ewe tegs sired by the same ram were purchased of M. Bingham; their dams were Atwood and Robinson ewes bred by M. Bingham; in 1878, twenty-four ewes were purchased of J. W. Peck, that were bred by F. & L. E. Moore, F. H. Dean, B. B. Tottingham and M. Bingham.

RAMS USED.

In 1877, a ram bred by L. P. Clark, M. Bingham's McCauley Ram and O. A. Field's 1 (553) were used and the ram tegs 36 to 40 and ewe tegs 41 to 50 were sired by them; two ram tegs were purchased of M. Bingham in January, 1878.

——o——

FLOCK 144.

Owned by S. H. RAWSON, WHITING, ADDISON Co., VT.

9 RAMS.

Marked with metallic labels in their ears, having thereon S. H. Rawson, and numbers 51 to 59.

27 EWES.

Labeled same as rams, seventeen marked on labels S. H. Rawson and numbers 1 to 17, and ten marked on labels T. S. & Son and numbers 9, 14, 17, 27, 28, 72, 74, 75, 84, 95.

PEDIGREE.

Descended from importations from Spain through the flocks of A. Cock, Long Island; D. Humphreys, S. Atwood, Conn.; Wm. Jarvis, Weathersfield; Messrs. Rich, L. Beedle, T. Stickney & Son, L. C. Remele, E R. Robinson, E. A. Birchard, C. J. Rich, Shoreham; Baker & Smith, P. Elitharp, Bridport; Messrs. Hammond, Middlebury; C. B. Cook, Charlotte.

EWES PURCHASED.

Sixteen ewes were purchased of C. J. Rich that were bred from the old Robinson flock, being a part of that flock received by C. J. Rich and wife in the division of the A. Russell estate; eleven ewes were also purchased of T. Stickney & Son in the fall of 1877.

RAMS USED.

At the same time of the purchase of the ewes of C. J. Rich two rams of the same flock and blood were also purchased; in 1874 a stock ram was purchased of T. Stickney & Son.

FLOCK 145.

Owned by L. C. REMELE, RICHVILLE (in Shoreham), ADDISON Co., VT.

8 RAMS.

Marked with metallic labels in their ears, having thereon L. C. Remele and numbers 1 to 8.

40 EWES.

Labeled same as rams with numbers 9 to 48.

PEDIGREE.

Descended from importations from Spain through the flocks of D. Humphreys, S. Atwood, Conn.; A. Cock, Long Island; Wm. Jarvis, Weathersfield; Messrs. Rich, L Beedle, T. Stickney & Son, J. T. Stickney, E. R. Robinson, Shoreham; Messrs. Hammond, W. R. Remele, Middlebury; R. P. Hall, J. Towle, Cornwall; V. Wright, Weybridge; C. B. Cook, Charlotte; W. R. Sanford, Orwell; J. Q. Stickney, Whiting; S. Root, Hubbardton; L. Webster, Sudbury; Baker & Smith, P. Elitharp, Bridport.

EWES PURCHASED.

In 1841, one ewe and one ewe teg were purchased of Wm. Jarvis; in 1845 one ewe teg was purchased from the same flock; in 1844 five ewes were purchased of J. Thurman Rich; soon after eight Jarvis ewes were purchased of W. R. Sanford; at a later date one Atwood ewe was purchased of C. B. Cook and five of R. P. Hall; these six ewes were all bred by S. Atwood.

RAMS USED.

In 1845, one ram teg was purchased of Wm. Jarvis; the stock rams of J. Thurman Rich, T. Stickney & Son, J. T Stickney, J. Q. Stickney, J. T. & V. Rich, Messrs. Hammond, V. Wright, R P. Hall, J. Towle, have been used; also those bred within the flock extensively.

———o———

FLOCK 146.

Owned by STEPHEN W. REMELE, MIDDLEBURY, ADDISON CO., VT.

8 RAMS.

Marked with metallic labels in their ears, having thereon S. W. Remele and numbers 35 to 42.

34 EWES.

Labeled same as rams with numbers 1 to 34.

87

PEDIGREE.

Descended from importations from Spain through the flocks of
D. Humphreys, S. Atwood, Conn.; A. Cock, Long Island; Wm.
Jarvis, Weathersfield; Messrs. Hammond, W. R. Remele, Cherbino
& Williamson, Middlebury; R. P. Hall, F. H. Dean, H. Lane, Corn-
wall; Messrs. Rich. L. Beedle, E. R. Robinson, E. A. Birchard,
Geo. H. Hall, Shoreham; Baker & Smith, P. Elitharp, C. N. Hay-
ward, J. J. Crane, D. F. Doty, H. C. Burwell, B. Myrick, Bridport;
C. B. Cook, H. Thorp, Charlotte; V. Wright, Weybridge; E. S.
Rowley, E. Meach, Shelburn; N. Richards, Panton; W. R. Sanford,
Orwell; A. A. Farnsworth, Brooksville; N. A. Saxton, Waltham.

EWES PURCHASED.

In December, 1850, ten Atwood ewes were purchased of R. P.
Hall and five of W. R. Remele.

RAMS USED.

The stock rams of R. P. Hall, W. R. Remele and Messrs. Hammond were used for the first fifteen or twenty years, besides those
bred within flock; Sweepstakes (32) and Gold Drop (64) more than
any others of Messrs. Hammond's stock rams; the former sired
Washoe (136) which was used as a stock ram in the flock. Two
rams from the flock of F. H. Dean were used one year; Sherman
(848) was used two years; in 1877, Rarus (569) was used. The
first introduction of Robinson blood was in 1877, when Sam (368)
was used to a few ewes.

———o———

FLOCK 147.

Owned by WILLIAM R. REMELE, Middlebury, Addison Co.,
Vt.

10 RAMS.

Marked with metallic labels in their ears, having thereon W.
R. Remele and numbers 5 to 12, 14, 15.

60 EWES.

Labeled same as rams, with numbers 1 to 4, 13, 16 to 70.

PEDIGREE.

Descended from importations from Spain through the flocks of
D. Humphreys, S. Atwood, Conn.; A. Cock, Long Island; Wm. Jarvis, Weathersfield; Messrs. Hammond, Cherbino & Williamson,
Middlebury; A. A. Farnsworth, Brooksville; N. A. Saxton, Waltham; R. P. Hall, H. Lane, C. D. Lane, Cornwall; C. B. Cook,

Charlotte ; Messrs. Rich, L. Beedle, E. R. Robinson, E. A. Birchard,
Shoreham ; Baker & Smith, P. Elitharp, C. N. Hayward, D. F. Doty,
J. J. Crane, H. C. Burwell, B. Myrick, Bridport ; N. Richards, Pan-
ton ; W. R. Sanford, Orwell.

ORIGIN.

The foundation of this flock was in 1845, by a partnership with
R. P. Hall in the sheep he purchased of Stephen Atwood. This
partnership was dissolved in 1849. In 1851 Mr. Remele was a partner with Messrs. Hammond, W. R. Sanford and R. P. Hall in the
purchase of French and Silesian Merinos, but they were soon after
sold from the flock, none of the blood being retained.

RAMS USED.

The stock rams of Messrs. Hammond and Sanford, including
Old Black (9) (in which an interest was owned), Young Matchless
(17), Old Greasy (18), Long Wool (21), Sweepstakes (32), Kilpatrick (71), Blucher (77), Green Mountain (70), Gold Drop (64),
Bull Dog (104). The stock rams of R. P. Hall were also used ;
besides these, rams bred within the flock have been used along with
those named. The Robinson blood was first introduced in 1875 by the
use of Sam (368); this ram sired all the lambs of 1876 ; a ram sired
by Sam (368) was used in 1877 ; Allright (169) was used in 1876,
but the produce has been sold ; none are among the flock registered.

——o——

FLOCK 148.

Owned by ELISHA RICH, Whiting, Addison Co., Vt.

16 RAMS.

Marked with metallic labels in their ears, having thereon E.
Rich and numbers 1 to 16.

84 EWES.

Labeled same as rams and numbers 17 to 100.

PEDIGREE.

Descended from importations from Spain through the flocks of
A. Cock, Long Island; D. Humphreys, S. Atwood. Conn.; Wm.
Jarvis, Weathersfield ; Messrs. Rich, L. Beedle, E. R. Robinson, L.
C. Remele, T. Stickney & Son, W. H. Cook, Shoreham ; A. Bald-
win, J. Q. Stickney, Whiting ; Messrs. Hammond, W. R. Remele,
Middlebury ; W. R. Sanford, Orwell ; S. Root, Hubbardton ; L.
Webster, Sudbury ; V. Wright, Weybridge ; C. B. Cook, Charlotte ;
Baker & Smith, P. Elitharp, Bridport.

EWES PURCHASED.

Jan. 23, 1855, forty ewes were purchased of Mrs. E. R. Robinson, being a part of the flock bred by her husband; these ewes were in lamb by the Lute Robinson Ram (39).

RAMS USED.

Old Robinson (38) to a part of the flock in 1855 and a ram teg sired by Lute Robinson to the remainder; this last was used two or three years; a few ewes were then taken to Sweepstakes (32) and from him was raised E. Rich's Sweepstakes (53); this was used as the stock ram of the flock eight years; a ram sired by him was used six or seven years; in 1873, Fremont, Jr. (215) was used to twenty ewes, and the remainder to a ram owned by T. Stickney & Son; a ram sired by Fremont (126), dam a Robinson ewe, was used in 1874; one bred by A Baldwin by a Stickney ram, dam a Robinson ewe, in 1876; besides the rams named others bred within the flock have been used.

——o——

FLOCK 149.

Owned by Q. C. RICH, Shoreham, Addison Co., Vt.

31 EWES.

Marked with metallic labels in their ears, having thereon Q. C. Rich and numbers 1 to 31.

PEDIGREE.

Descended from importations from Spain through the flocks of A. Cock, Long Island; D. Humphreys, S. Atwood, Conn.; Wm. Jarvis, Weathersfield; Messrs. Rich, E. R. Robinson, L. C. Remele, A. C. Harris, E. D. Bush, T. Stickney & Son, L. Beedle, E. A. Birchard, B. B. Tottingham, Shoreham; Messrs. Hammond, Middlebury; C. B. Cook, Charlotte; Baker & Smith, P. Elitharp, Bridport; J. Q. Stickney, Whiting; S. Root, Hubbardton; L. Webster, Sudbury; J. H. Thomas, Orwell; O. Merrill, Addison.

EWES PURCHASED.

In the fall of 1860, twenty Rich ewes were purchased of the estate of Hiram Rich; they were descended from ewes purchased of J. T. & V. Rich in 1855.

RAMS USED.

The stock rams of B. B. Tottingham, E. D. Bush and G. H. Hall have been used, besides those bred within the flock; those from the rams of G. H. Hall have all been sold.

FLOCK 150.

Owned by NELSON RICHARDS, Vergennes, Addison Co., Vt.

9 RAMS.

Marked with metallic labels in their ears, having thereon N. Richards and numbers 1 to 9.

41 EWES.

Labeled same as rams, with numbers 10 to 50.

PEDIGREE.

Descended from importations from Spain through the flocks of. D. Humphreys, S. Atwood, Conn.; A. Cock, Long Island; Messrs. Hammond, W. R. Remele, Middlebury; F. D. Barton, N. A. Saxton, Waltham; C. B. Cook, Charlotte; W. R. Sanford, Orwell; V. Wright, Weybridge; Messrs. Rich, E. R. Robinson, L. Beedle, L. C. Remele, Shoreham; Charles Merrill, Addison.

EWES PURCHASED.

About the year 1850, one Atwood ewe and a pair of ewe tegs were purchased of N. A. Saxton; the ewe was bred by W. S. & E. Hammond; in Sept., 1853, one ewe was purchased of N. A. Saxton; three ewes that were bred by Messrs. Hammond were purchased of Almon Lawrence, of Hinesburgh; two Atwood ewes were purchased of C. B. Cook; a few ewes of Rich blood were purchased of Charles Merrill, two of which were retained in the flock.

RAMS USED.

At first the stock rams of Messrs. Hammond, and those bred by them; America (35) was used most of any until Ironsides (80) was bred; Eureka (58) and Golden Fleece (91) were used; two rams sired by Eureka were used some years; a ram bred by F. D. Barton is the last ram used in the flock.

---o---

FLOCK 151.

Owned by VIRTULAN RICH, Richville (in Shoreham), Addison Co., Vt.

58 RAMS.

Marked with metallic labels in their ears, having thereon J. T. & V. R., and numbers 301, 310, 312, 328, 331, 338, 339, 343, 344, 347, 349, 350, 355, 356 to 374, 376; 379 to 382, 384 to 395, 397 to 399, 401 to 404.

103 EWES.

Labeled same as rams, with numbers 1, 2, 4 to 10, 13, 14, 16, 17, 20, 21, 23 to 29, 31, 33, 36 to 39, 41, 42, 44, 48, 51, 52, 54, 55, 57, 59 to 61, 64, 65, 68, 69, 73, 74, 75, 77 to 79, 94, 96, 139 to 179.

PEDIGREE.

Descended from importations from Spain through the flocks of A. Cock, Long Island; D. Humphreys, S. Atwood, Conn.; Wm. Jarvis, Weathersfield; Messrs. Hammond, Middlebury; R. P. Hall, J. Towle, Cornwall; A. H. Hubbard, Whiting; W. M. Lincoln, Brandon; E. Porter, Rutland; H G. Hibbard, Orwell; T. Stickney & Son, E. R. Robinson, L. Beedle, L. C. Remele, Shoreham; Baker & Smith, P. Elitharp, Bridport; C. B. Cook, Charlotte.

HISTORY.

This is the flock founded in 1823 by Hon. Charles Rich with ewes and rams brought from Long Island, where they were purchased of A. Cock. (For the history of that flock see pages 56 and 164 of this work.) This is the oldest flock of pure bred Merinos now existing in the United States without changes except by regular family descent. It has been kept on the same farm, and owned in the same family for three generations, or a period of 56 years. No other ewes have been added to the flock. Hon. Charles Rich died Oct. 16, 1824; after his death the flock descended to his two sons, Charles Rich and J. Thurman Rich. That portion of the flock inherited by Charles Rich has descended to us through Tyler, Stickney and Erastus R. Robinson; that inherited by J. Thurman Rich at his death Oct. 12, 1846, descended to his sons, John T. and Virtulan Rich. John T. Rich died Sept. 27, 1876; after his death the flock became the property of Virtulan Rich, the present owner.

RAMS USED.

For the largest part of the 56 years rams bred within the flock; the first introduction of Jarvis blood was in 1841 by a ram bred by Consul Jarvis and selected from his flock; afterwards Consul (2) was used to a limited extent; soon after the Atwood blood was introduced by using Atwood (12) to a few ewes in 1845, and further by about twenty ewes being served by Old Black (9) soon after; the old Robinson Ram (38) was used to a part of the flock, and also the Tottingham (40) to a limited extent; Hall Ram (98) was used to a very few ewes; Gold Drop (64) also to a few; Gen. Fremont (126) was used several times to a few ewes; in the later history of the flock Hibbard (214) was used two years, 1874 and 1875; of those bred within the flock more care has been taken in selection and coupling than to hand down names; very few of the stock rams of the flock have received names that can now be given; the exceptions are Mountaineer (51), Rich's Stock Ram (462), J. T. & V. R.'s

301 (470), Banker (471). In crossing with other families the rule with J. Thurman Rich was, to breed back, using rams that should pedigree through the sires direct to the Cock flock.

——o——

FLOCK 152.

Owned by JOHN T. RICH, Elba, Lapeer Co., Mich.

32 RAMS.

Marked with metallic labels in their ears, having thereon J. T. Rich, and numbers 1 to 32.

91 EWES.

Labeled same as rams, with numbers 1 to 91.

PEDIGREE.

Descended from importations from Spain through the flocks of A. Cock, Long Island; D. Buffum, Wm. Bailey, J. I. Bailey, Rhode Island; D. Humphreys, S. Atwood, Conn.; Wm. Jarvis, Weathersfield; Messrs. Rich, L. Beedle, E. R. Robinson, L C. Remele, T. Stickney & Son, J. M Ormsbee, D. & G. Cutting, Shoreham; S. T. Baker, Whiting; Baker & Smith, P. Elitharp, Bridport; N. A. Saxton, Waltham; C B. Cook, Charlotte; Charles Rich, John W. Rich, Thomas Slayton, Lapeer, Mich.

HISTORY.

In 1848, John W. Rich, father of the present owner, purchased a few ewes and two or three rams from the flocks of J. T. & V. Rich. T. Stickney and D. & G. Cutting and took them to Lapeer, Mich.; these purchases were made partly on account of Hon. Charles Rich, formerly the owner of the part of the Rich flock sold E. R. Robinson and Tyler Stickney. Hon. Charles Rich and John W. Rich soon after went into partnership and the sheep became common property; the ewes were bred to the rams taken to Lapeer with them until one was purchased of J. T. & V. Rich in 1851, and used a number of years; this ram was called Bonaparte (383). In 1853, the firm purchased ten ewes of T. Stickney, and added them to the flock. About the same time Thomas Slayton, of Lapeer, purchased a few ewes and a ram of T. Stickney and bred them pure within their own blood until 1859, when the firm purchased his entire flock thus bred. The ram Mr. Slayton purchased of T. Stickney was called Captain (384); he was used by Messrs. Rich to some extent. After the purchase of Mr. Slayton, rams bred within the joint flocks were used until 1863, when J W. Rich purchased a ram teg of T. Stickney, and another of David Cutting; these rams, Stickney (385) and Cutting (386) were used a number of years or until they died, when rams sired by them were used. In 1872, John

W. Rich died, but previous to his death he purchased the interest of Charles Rich, except ten ewes, a division of the best twenty ; the flock at this time with the exception of the ten ewes named came into the possession of the present owner, John T. Rich.

RAMS USED BY J. T. RICH.

From 1872 to 1874 those bred within the flock ; in 1874, a ram was purchased of T. Stickney & Son; this ram, Stickney No. 2 (387), and one of his get are all that have been used since 1874.

——o——

FLOCK 153.

Owned by S. S. ROCKWELL & H. E. SANFORD, West Cornwall, Addison Co., Vt.

100 RAMS.

Marked with metallic labels in their ears, having thereon H. E. Sanford, and numbers 51 to 90, 234 to 250, 252 to 262, 296 to 327.

228 EWES.

Labeled same as rams, with numbers 1 to 50. 91 to 233, 263 to 295, 328, 329.

PEDIGREE.

Descended from importations from Spain through the flocks of A. Cock, Long Island ; D. Humphreys, S. Atwood, Conn.; Wm Jarvis, Weathersfield ; Messrs. Rich, L. Beedle, T. Stickney & Son, E. R. Robinson, L. C. Remele, E. A. Birchard, Rollin Birchard, E. G. Farnham, Shoreham ; Baker & Smith. P. Elitharp, C. N. Hayward, D. F. Doty, J. J. Crane, H. C. Burwell, B. Myrick, Bridport ; Messrs. Hammond, W. R. Remele, D. Hooker, S. W. Remele, Middlebury ; A. A. Farnsworth, Brooksville ; C. B. Cook, Charlotte; N. A. Saxton, Waltham ; C. I. Benedict, Arlington ; J. Marsh, Hinesburgh; W. R. Sanford, Orwell ; V. Wright, Weybridge ; R. P. Hall, E. S. Stowell, F. Hooker, A. H. Sperry, R. J. Jones, W. H. Delong, Cornwall ; J. G. Barker, Leicester ; H. M. Graves, Salisbury ; L. P. Clark, Addison ; J. Q. Stickney, Whiting ; S. Root, Hubbardton ; L. Webster, Sudbury.

HISTORY.

The foundation of this flock was about the year 1830 by John Rockwell, father of the senior member of this firm ; he purchased a few sheep of Leonard Beedle ; evidence has been furnished the committee that he bred these sheep pure until they came into possesion of Simeon S. Rockwell, who also bred them pure. Oct. 15, 1860, fifteen Atwood ewes were purchased of E. C. Everest, of Vergennes, that he purchased of Joseph Marsh, who bred them from the flock he purchased of S. Atwood and W. S. & E. Hammond. After H. E. Sanford became a partner in the flock, he purchased in Dec., 1874, five ewes and five ewe tegs of Atwood blood of F. Hooker; in April, 1876, four Atwood and Robinson ewes were purchased of A. H. Sperry. About the same time a party of a few ewes, Robinson blood, were purchased of E. G. Farnham that he purchased of their breeder, J. G. Barker.

RAMS USED.

S. S. Rockwell used rams from the flocks of R. P. Hall, Messrs. Hammond, V. Wright, C. I. Benedict; Eureka (58) was purchased in 1862 and was used as the stock ram of the flock until his death in 1868; since that time rams bred within the flock have been used; an interest was at one time owned in Bonaparte (176) and he was used extensively in the flock; Fortune (249) has been used to some extent; Delong's 100 (360) also to some extent; Plato (381) has been the stock ram of the flock for some years; Fremont, Jr. (215) has been used to one or two ewes; Eureka, 3d (223) was used in 1876 to a number of ewes; H. E. Sanford's 52 (382) was used to a few ewes before being sold; General (210) was used to a few ewes in 1877.

—o—

FLOCK 154.

Owned by JUAN ROBINSON, Keeler's Bay, Grand Isle Co., Vt.

1 RAM.

Marked with metallic label in his ear, having thereon H. H. and number 54.

10 EWES.

Labeled same as ram with numbers 11, 12, 13, 15, 19, 22, 24, 25, 50, 53.

PEDIGREE.

Descended from the importation from Spain by Col. David Humphreys through the flocks of S. Atwood, Conn.; Messrs. Hammond, Middlebury; V. Wright, Weybridge; H. Thorp, C. B. Cook, Charlotte; H. Newell, Shelburn; H. Harrington, Keeler's Bay.

ORIGIN.

The ewes and ram recorded above were all purchased of H. Harrington in 1877.

FLOCK 155.

Owned by MARCELLUS ROYCE, ORWELL, ADDISON CO., VT.

3 RAMS

Marked with metallic labels in their ears, having thereon M. Royce, and numbers 33, 34, 35.

32 EWES.

Labeled same as rams and numbers 1 to 32.

PEDIGREE.

Descended from importations from Spain through the flocks of A. Cock, Long Island; D. Humphreys, S. Atwood, Conn.; D. Buffum, Wm. Bailey, J. I. Bailey, Rhode Island; Wm. Jarvis, Weathersfield; Messrs. Rich, L. C. Remele, E. R. Robinson, D. & G. Cutting, J. M. Ormsbee, E. A. Birchard, Shoreham; N. A. Saxton, Waltham; C. B. Cook, Charlotte; Baker & Smith, P. Elitharp, Bridport; Messrs. Hammond, Middlebury; E. L. Needham, S. T. Baker, Whiting.

EWES PURCHASED.

The foundation of this flock was by a purchase of ten or twelve Rich ewes of E. L. Needham.

RAMS USED.

Those raised within the flock; the exceptions are, the Tottingham Ram (40) to ten ewes in 1861, Gold Drop (64) to four ewes in 1862; a ram teg was purchased of E. Hammond and used a few years; in 1863, D. Cutting's stock ram was used.

——o——

FLOCK 156.

Owned by HENRY ROBBINS, CORNWALL, ADDISON CO., VT.

3 RAMS.

Marked with metallic labels in their ears, having thereon H. Robbins, and numbers 23, 24, 25.

22 EWES.

Labeled same as rams, with numbers 1 to 22.

PEDIGREE.

Descended from the importation from Spain by Col. David Humphreys through the flocks of S. Atwood, Connecticut; Messrs. Hammond, W. R. Remele, Middlebury; R. P. Hall, E. S. Stowell,

B. S. Field, F. H. Dean, H. Lane, J. Towle, Cornwall; W. R. Sanford, Orwell; N. A. Saxton, C. & O. C. Bacon, Waltham; N. Cushing, Woodstock.

EWES PURCHASED.

April 21, 1859, ten Atwood ewes were purchased of C. & O. C. Bacon; in the fall of the same year ten more ewes of the same blood were purchased from the same flock; in December, 1860, two Atwood ewes were purchased of H. W. Hammond.

RAMS USED.

Sweepstakes (32), Cross Tom (36), Gold Drop (64), Sanford Lane (60), Bonaparte (73), Tempest (230), Robbins' Ram (373), and an Atwood ram bred by Nathan Cushing have all been used, besides some that were bred within the flock.

——o——

FLOCK 157.

Owned by C. B. RUSSELL, North Hero, Vt.

23 RAMS.

Marked with metallic labels in their ears, having thereon C. B. Russell and numbers 1 to 14, 51 to 59.

44 EWES.

Labeled same as rams with numbers 15 to 50, 60 to 67.

PEDIGREE.

Descended from the importation from Spain by Col. David Humphreys through the flocks of Stephen Atwood, Conn.; Messrs. Hammond, W. R. Remele, Middlebury; S. G. Holyoke, St. Albans; W. R. Sanford, Orwell; F. D. Barton, Waltham; Henry Thorp, Charlotte; Chapman & Henry Giddings, Lyman Hunt, Fairfax; R. P. Hall, Cornwall; V. Wright, Weybridge.

EWES PURCHASED.

Flock commenced in 1857 by a purchase of one Atwood ewe of S. G. Holyoke and two years after two more ewes same blood were purchased of Mr. Holyoke; Feb. 21, 1866, two Atwood ewes were purchased of Lyman Hunt; Sept. 6, 1872, Mr. Hunt's entire flock, consisting at that time of fourteen ewes, twelve ram tegs and three ewe tegs were purchased, all of Atwood blood.

RAMS USED.

A ram owned by Orange Phelps bred by E. Hammond several years; a ram bred within the flock, sired by the Phelps ram, was used four years ; an Atwood ram purchased of S. G. Holyoke, sired by his stock ram by Hammond's Onward was used three years ; Blucher (77) to one or two ewes ; North Hero (376) was purchased in 1876, and has been the stock ram of the flock since.

———o———

FLOCK 158.

Owned by EDGAR SANFORD, West Cornwall, Addison Co., Vt.

28 RAMS.

Marked with metallic labels in their ears, having thereon E. Sanford, and numbers 48 to 55, 75 to 84, 102 to 111.

83 EWES.

Labeled same as rams, with numbers 1 to 47, 56 to 74, 85 to 101.

PEDIGREE.

Descended from importations from Spain through the flocks of A. Cock, Long Island ; D. Humphreys, S. Atwood, Conn.; Wm. Jarvis, Weathersfield ; Messrs. Rich, L. C. Remele, E. R. Robinson, Geo. & E. G. Farnham, E. A. Birchard, L. Beedle, Shoreham ; Baker & Smith, P. Elitharp, C. N. Hayward, D. F. Doty, B. Myrick, H. C. Burwell, J. J. Crane, Bridport ; N. A. Saxton, Waltham ; C. B. Cook, Charlotte ; N. Richards, Panton ; R. P. Hall, E. S. Stowell, C. R. Ford, Henry Lane, M. J. Ellsworth, Rockwell & Sanford, Charles Benedict, Truman Eells, C. D. Lane, T. B. Holley, E. Hamilton, R. J. Jones, W. H. Delong, F. H. Dean, Cornwall.

EWES PURCHASED.

In 1852 or 1853, twenty-three ewe tegs and ten yearling ewes (all Atwood blood) were purchased of P. Elitharp ; a part were sold and a part retained as breeding stock ; two Atwood ewes were purchased of C. D. Lane ; they were sired by an Elitharp ram ; their dams were bred by C. B. Cook ; four Atwood ewes were puchased of T. B. Holley ; two were sold and two retained ; they were sired by an Elitharp ram, their dams bred by R. R. Hall ; two ewes were purchased of Truman Eells ; they were sired by an Atwood ram bred by R. P. Hall ; their dams were Rich ewes ; nine Robinson ewes were purchased of E G. Farnham in 1856 ; they were sired by Sanford Gibbs (56) ; two Robinson ewes were purchased in 1861 or '62 of E. Hamilton ; in 1865 or '66 nine Atwood and Robinson ewes were purchased of M. J. Ellsworth.

RAMS USED.

The first ram used was the Old Robinson Ram (38); the next the Sanford & Treadway (54); the next the Sanford & Gibbs Ram (56); he was used four or five years; the Sanford Lane (60) was next used; a ram sired by the Sanford & Gibbs (56), dam one of the ewes of the flock, was used; Prince (86) was bred within the flock and used one year; a ram bred by J. T. & V. Rich was owned in company with A. J. Benedict and used one year; Ford Ram (233) was used one year; Sanford's Ford Ram (446) one year; a ram bred within the flock sired by Little Wrinkly (48) was used three years; Plato (381) was used one year; Sanford's Bonaparte Ram (447) was used in 1875; in 1876, Eureka, 3d (223) was used to five ewes and a ram bred within the flock sired by the Little Wrinkly Ram was used to the rest of the flock; in 1877, the last named ram to two ewes, Delong's 100 (360) to two ewes, and Delong's Remele Ram (361) to the remainder. The tegs in rams from 102 to 111 and and in ewes from 85 to 101 were sired by these rams.

————o————

FLOCK 159.

Owned by W. R. & CHARLES SANFORD, Orwell, Addison Co., Vt.

11 RAMS.

Marked with metallic labels in their ears, having thereon C. Sanford and numbers 280 to 290.

36 EWES.

Labeled same as rams with numbers 234 to 252, 373 to 279, 291 to 299, 391.

PEDIGREE.

Descended from the importation from Spain by Col. David Humphreys, through the flocks of S. Atwood, Conn.; Messrs. Hammond, W. R. Remele, Middlebury; Victor Wright, Weybridge; E. S. Stowell, R. J. Jones, Cornwall; O. H. & W. O. Bascom, Orwell; Geo. H. Hall, Shoreham.

EWES PURCHASED.

In 1875 eighteen ewes were purchased of O. H. & W. O. Bascom. These ewes were unmixed descendants of Atwood ewes, and rams bred by W. R. Sanford.

RAMS USED.

In 1876, Allright (169) and in 1877, Ben (339) were used.

FLOCK 160.

Owned by E. D. SEARL, Cornwall, Addison Co., Vt.

29 RAMS.

Marked with metallic labels in their ears, twenty-six having thereon E. D. Searl and numbers 11 to 32, 36, 38, 39, 41, and three having on their labels F. D. Douglas and numbers 4, 5, 6.

12 EWES.

Labeled same as rams with numbers 1 to 8, 10, 33, 34, 35.

PEDIGREE.

Descended from importations from Spain through the flocks of D. Humphreys, S. Atwood, Conn.; A. Cock, Long Island; D. Buffum, Wm. Bailey, J. I. Bailey, Rhode Island; Wm. Jarvis, Weathersfield; Messrs. Hammond, W. R. Remele, D. Hooker, S. W. Remele, A. Chapman, Middlebury; Wm. Gage, Ferrisburgh; V. Wright, A. J. Stowe, Weybridge; N. A. Saxton, Waltham; F. H. Dean, E. S. Stowell, F. Hooker, A. J. Wooster, Cornwall; Messrs. Rich, T. Stickney & Son, L. C. Remele, E. R. Robinson, H. W. Walker, E. A. Birchard, Z. Frost, A. Frost, L. Beedle, J. M. Ormsbee, D. & G. Cutting, E. A. Jennings, D. Rich, Shoreham; S. T. Baker, F. D. Douglas, G. F. Casey, Nelson Remele, C. F. Church, J. Q. Stickney, Whiting; S. Root, Hubbardton; L. Webster, Sudbury; J. Sheldon, Fairhaven; Baker & Smith, P. Elitharp, C. N. Hayward, D. F. Doty, J. J. Crane, B. Myrick, H. C. Burwell, Bridport; C. B. Cook, H. Thorp, Charlotte; E. S. Rowley, E. Meach, Shelburn.

EWES PURCHASED.

Ewes 1 to 8, 10 were purchased of F. H. Dean, in April, 1877, when tegs; ewes 33, 34, 35 were purchased of E. A. Jennings.

RAMS PURCHASED.

The rams were purchased of F. H. Dean (Atwood and Robinson blood), E. A. Jennings (Robinson blood); C. F. Church (Stickney blood); G. F. Casey (Atwood, Stickney, Cutting and Rich blood); Nelson Remele (Atwood and Stickney blood); and F. D. Douglas, (Atwood, Rich and Cutting blood).

FLOCK 161.

Owned by MOSES S. SHELDON, East Middlebury, Addison Co., Vt.

5 RAMS.

Marked with metallic labels in their ears, having thereon M. Sheldon and numbers 30 to 34.

29 EWES.

Labeled same as rams, and numbers 1 to 29.

PEDIGREE.

Descended from importations from Spain through the flocks of D. Humphreys, S. Atwood, Conn.; A. Cock, Long Island; D. Buffum, Wm. Bailey, J. I. Bailey, Rhode Island; Wm. Jarvis, Weathersfield; N. Cushing, Woodstock; D. Davis, Windsor; Messrs. Hammond, W. R. Remele, S. W. Remele, D. Hooker, E. R. Clay, J. G. Wellington, Middlebury; Messrs. Rich, L. Beedle, E. R. Robinson, L. C. Remele, J. M. Ormsbee, T. Stickney, D. & G. Cutting, E. A. Birchard, Shoreham; E. S. Casey, S. T. Baker, Whiting; A. W. Dana, V. Wright, Weybridge; R. P. Hall, Henry Lane, E. Sanford, Cornwall; C. B. Cook, Charlotte; N. A. Saxton, Waltham; Baker & Smith, P. Elitharp, J. O. Hamilton, Bridport; A. A. Farnsworth, Brooksville.

EWES PURCHASED.

December 27, 1869, fourteen ewes and seven ewe tegs were purchased of J. G. Wellington; they were of Atwood, Robinson and Stickney blood; three were bred by E. S. Casey from Stickney ewes and sired by Sanford Gibbs (56); one an Atwood ewe sired by Old Greasy (74), bred by A. W. Dana; two Atwood ewes bred by D. Davis from Hammond and Cushing stock; eight were ewes bred by J. G. Wellington from the above ewes and rams bred by E. Sanford, E. Hammond and E. R. Clay. The ewe tegs were from a part of these ewes, and sired by rams bred by the same three breeders. In 1873, ten ewes were purchased of Mrs. Victor Wright.

RAMS USED.

With the ewes mentioned above was purchased of J. G. Wellington two rams, one of which, Wellington (345), was used a number of year as the stock ram of the flock; afterwards a ram sired by Wellington was bred within the flock and used for three years; Allright (169) was used to a few ewes; for two years past Dunmore (346) and Brooksville (354) have been used as the stock rams of the flock.

FLOCK 162.

Owned by GEORGE H. SMITH, Bridport, Addison Co., Vt.

11 RAMS.

Marked with metallic labels in their ears, having thereon G. H. Smith and numbers 1 to 11.

31 EWES.

Labeled same as rams with numbers 12 to 42.

PEDIGREE.

Descended from importations from Spain through the flocks of D. Humphreys. S. Atwood. Conn ; A. Cock, Long Island ; D. Buffum, Wm. Bailey, J. I. Bailey, Rhode Island ; Wm. Jarvis, Weathersfield ; Messrs. Hammond, W. R Remele, Middlebury ; Messrs. Rich, L. C. Remele, E. R. Robinson, E. A. Birchard, J. M. Ormsbee, A. Clark, D. & G. Cutting, E. D. Bush, A. C. Harris, O. S. Jones, Shoreham ; S. T. Baker, Whiting ; N. A. Saxton, Waltham; C. B. Cook, Charlotte ; Baker & Smith, P. Elitharp, C. N. Hayward, D. F. Doty, J. J. Crane, B. Myrick, H. C. Burwell, Bridport ; N. Richards, Panton ; R. P. Hall, E. S. Stowell, Cornwall ; V. Wright, L. D. Gregory, Weybridge ; W. R. Sanford, Orwell.

EWES PURCHASED.

A half interest was purchased of S. Benton, of Cornwall, in ten old Atwood ewes he purchased of L. D. Gregory. In 1874, fourteen Robinson and four Cutting ewes were purchased of Orson S. Jones, that he purchased of H. B. Wright, bred from the Harris flock.

RAMS USED.

A ram bred by S. Benton, of the same breeding of the ewes ; Sea Lion (83) to some extent and a ram sired by him bred within the flock. Chunk (95), Elitharp and Burwell (79), Bismarck (221), Stub (222) and Eureka, 3d (223), have all been used to breed up the present flock, besides some raised within it.

——o——

FLOCK 163.

Owned by ROLLIN J. SMITH, Sudbury, Addison Co., Vt.

13 RAMS.

Marked with metallic labels in their ears, having thereon R. J. Smith and numbers 1 to 13.

55 EWES.

Labeled same as rams, with numbers 14 to 68.

PEDIGREE.

Descended from the importation from Spain by Col. David Humphreys through the flocks of S. Atwood, Conn.; Messrs. Hammond, W. R. Remele, Middlebury; W. R. Sanford, Orwell; V. Wright, Weybridge; C. R. Jones, Hubbardton.

EWES PURCHASED.

October 22, 1862, five ewes and one ewe teg were purchased of V. Wright; in 1863, four more ewes were purchased from the same flock; all were bred by V. Wright, and all were Atwood blood.

RAMS USED.

At first, V. Wright's and E. Hammond's stock rams: Old Greasy (74), V. Wright's Wrinkly (173), Sweepstakes (32), and Gold Drop (64). Later, Young Black Top (283), Little Wrinkly (341), Cromwell (350), Old Black (445), Hugo (453), Long Wool (509), have all been used in the flock as stock rams. Hassan (567) was purchased of C. R. Jones.

---o---

FLOCK 164.

Owned by N. T. SPRAGUE, Brandon, Rutland Co., Vt.

8 RAMS.

Marked with metallic labels in their ears, having thereon N. T. Sprague, and numbers 44 to 51.

51 EWES.

Labeled same as rams, with numbers 1 to 43. 501 to 508.

PEDIGREE.

Descended from importations from Spain through the flocks of D. Humphreys, S. Atwood, Conn.; A. Cock, Long Island; Wm Jarvis, Weathersfield; Alfred Hull, Wallingford; Messrs. Hammond, W. R. Remele, E. R. Clay, J. G. Wellington, Middlebury; Messrs. Rich, L. Beedle, E. R. Robinson, T. Stickney, L. C. Remele, B. B. Tottingham, E. A. Birchard, Shoreham; N. Cushing, Woodstock; D. Davis, Windsor; E. S. Casey, Whiting; A. W. Dana, V. Wright, Weybridge; R. P. Hall, H. Lane, E. Sanford, H. Peck, E. Hamilton, Cornwall; Baker & Smith, P. Elitharp, C. N. Hayward, D. F. Doty, J. J. Crane, B. Myrick, J. Hill, E. Fitch, H. C. Burwell, Bridport; C. B. Cook, Charlotte; N Richards, Panton; Moses S. Sheldon, Salisbury.

ORIGIN.

Alfred Hull purchased of S. Atwood thirteen ewes and one or two rams; from these he bred a flock of seventy-six; these were divided in 1857, N. T. Sprague receiving 36 as his share; these formed the basis of the flock. In 1877, eight yearling ewes were purchased of U. D. Twitchell, that were bred by Moses S. Sheldon, from stock that he purchased of J. G. Wellington and Mrs. V. Wright.

RAMS USED.

Rams bred within the flock at first; in 1863, the Peck ram (144) was purchased and was used for a number of years. Soon after, a Robinson ram was purchased of B. B. Tottingham; he was used one or two seasons; Silver Horns, Jr. (401), was purchased in 1876, and Sprague's Burwell (402) in 1877, and these are now the stock rams of the flock. Of those bred within the flock and used as stock rams, the most noted were Green Mountain (391), Gen. Grant (398), Tommy Sayres (399), and Heenan (400).

The flock descended from the ewes bred by Mr. Hull are kept at Mr. Sprague's farm, in Brandon, managed by Col. Merrill; those bred by Mr. Sheldon are kept at his farm in Middlebury, managed by John Spencer.

————o————

FLOCK 165.

Owned by ALBERT H. SPERRY, Cornwall, Addison Co., Vt.

11 RAMS.

Marked with metallic labels in their ears, having thereon A. H. Sperry and numbers 40 to 50.

39 EWES.

Labeled same as rams, with numbers 1 to 39.

PEDIGREE.

Descended from importations from Spain through the flocks of D. Humphreys, S. Atwood, Conn.; A. Cock, Long Island; Wm. Jarvis, Weathersfield; Messrs. Rich, L. Beedle, E. R. Robinson, L. C. Remele, E. A. Birchard, E. G. Farnham, Rollin Birchard, Shoreham; Baker & Smith, P. Elitharp, Bridport; C. B. Cook, Charlotte; R. P. Hall, E. S. Stowell, Cornwall; Messrs. Hammond, W. R. Remele, Middlebury; W. R. Sanford, Orwell.

EWES PURCHASED.

In 1868 or '9, fifty ewes were purchased of Rollin Birchard, bred from Robinson and Rich stock that he purchased of E. G. Farnham, and rams of Robinson blood bred by E. A. Birchard.

RAMS USED.

Rams bred by E. A. Birchard, W. R. Sanford and E. S. Stowell, have been used to breed up the present flock.

———o———

FLOCK 166.

Owned by JOSEPH T. STICKNEY, SHOREHAM, ADDISON CO., VT.

62 RAMS.

Marked with metallic labels in their ears, having thereon J. T. S., and numbers 104 to 112, 114 to 123, 125 to 136, 138 to 141, 143 to 146, 172, 178, 181, 184, 188, 192, 193, 195, 197, 199, 200, 203, 206 to 208, 210, 211, 221, 226 to 228, 234, 235.

108 EWES

Labeled same as rams with numbers 1, 3, 4, 5, 7 to 13, 16 to 22, 24 to 27, 29 to 35, 38 to 42, 44, 49, 53, 54, 59, 62, 84 to 93, 95 to 99, 147 to 161, 163 to 171, 173, 186, 187, 189 to 191, 196, 201, 202, 204, 205, 209, 212, 214, 220, 222 to 225, 229 to 233.

PEDIGREE.

Descended from importations from Spain through the flocks of A. Cock, Long Island ; D. Humphreys, S Atwood, Conn.; Wm. Jarvis, Weathersfield ; Messrs. Rich, T. Stickney, [E. R. Robinson, L C. Remele, A. Frost, Z. Frost, H. W. Walker, Miner Jones, E. R. Cudworth, E. Fanning, G. H. Hall, Shoreham ; Messrs. Hammond, W. R. Remele, Middlebury ; W. R. Sanford, Orwell ; John Preston, Leicester ; J. Q. Stickney, Whiting ; S. Root, Hubbardton ; L. Webster, Sudbury ; R. P. Hall, R. J. Jones, Cornwall ; C. B. Cook, Charlotte ; Baker & Smith, P. Elitharp, Bridport.

EWES PURCHASED.

May 11, 1857, ten two-year old ewes were purchased of Tyler Stickney ; in 1876, twenty-four breeding ewes and six ewe tegs were purchased of John Preston, of Leicester, that were bred direct from the flock of Tyler Stickney ; in 1876, nine ewe tegs were purchased of E. Fanning, bred from ewes from L. C. Remele, ten ewe tegs of Miner Jones from ewes from H. W. Walker, twenty-two tegs and five yearling ewes of E. R Cudworth from ewes bred from J. T. & V. Rich and W. R. Sanford ; all these forty-six were sired by the stock rams of this flock.

RAMS USED.

The stock rams of T. Stickney & Son, J. Q. Stickney's Fremont, Jr. (215), and those bred within the flock ; the exceptions being ten ewes to Towle Ram (164), twenty to Charlie (253) and eight to Major (181), but with the exception of one ewe sired by Major all the produce is sold from the flock ; J. T. Stickney's 146 (441), Centennial (442) and J. T Stickney's 142 (528) are now the stock rams of the flock and all the young stock in the flock have been sired by them.

——o——

FLOCK 167.

Owned by TYLER STICKNEY & SON (E. E. Stickney), East Shoreham, Addison Co., Vt.

51 RAMS.

Marked with metallic labels in their ears, having thereon T. S. & Son, and numbers 140, 143, 144, 149 to 151, 202, 206, 208 to 210, 212 to 214, 216, 218, 222 to 224, 227, 229, 231, 232, 234, 280 to 283, 287, 288, 290, 291, 293, 295, 301, 303, 307, 308, 311 to 313, 317 to 320, 322, 326 to 329, 332.

143 EWES.

Labeled same as rams, with numbers 1 to 4, 8, 11, 13, 16, 19, 23, 26, 29 to 33, 35 to 41, 43 to 47, 49 to 61, 64 to 68, 70, 73, 76 to 80, 82, 83, 87, 88, 90 to 93, 97, 98, 100 to 102, 104, 107, 111, 116, 119, 121, 125, 126, 129, 131, 134, 139, 235, 238, 240 to 242, 244 to 250, 253 to 256, 258 to 279, 284 to 286, 289, 292, 294, 296 to 300, 302, 304 to 306, 309, 311, 314 to 316, 321, 323 to 325, 330, 331, 333.

PEDIGREE.

Descended from importations from Spain through the flocks of A. Cock, Long Island ; D. Humphreys, S. Atwood, Conn.; Wm. Jarvis, Weathersfield ; Messrs. Rich, L. Beedle, E. R. Robinson, L. C. Remele, C. Delong, G. H. Hall, W. Perry, J. T. Stickney, Shoreham ; Baker & Smith, P. Elitharp, C. N. Hayward, D. F. Doty, J. J. Crane, H. C. Burwell, B. Myrick, Bridport ; W. R. Sanford, Orwell ; Messrs. Hammond, W. R. Remele, Middlebury ; R. P. Hall, R. J. Jones, Cornwall ; J. Q. Stickney, Whiting ; S. Root, Hubbardton ; L. Webster, Sudbury.

HISTORY.

The flock was founded in 1834 by Tyler Stickney, the senior member of the firm ; previous to this, he had been in co-partnership in breeding sheep with Charles Rich, and when they divided one ewe that afterwards became celebrated as the dam of Hero (4), and Fortune (5) fell to the share of Tyler Stickney ; this with 12 ewe tegs reserved by Charles Rich, when he sold his flock to E. R. Robinson, and which he afterwards sold to Tyler Stickney, formed the basis of this flock, which has been kept together, and the breeding directed by one person, for a period of over forty years; a longer period than any flock has been bred by any one person, and with a longer continuous existence as a flock than any, save one.

RAMS USED.

The same year Mr. Stickney purchased the ewes named above, he selected from the flock of Consul Wm. Jarvis, a ram teg which Mr. Jarvis stated to be the first selection he had ever permitted from his rams of any year before selecting for his own use; this ram, Consul (2), was used for eight years; Hero (5) was used until he was three years old; Atwood (12) was used to a few ewes; Old Robinson (38) was purchased in 1853, and used four or five years; a ram of Robinson blood bred by Charles Delong was purchased and used a few years; ten ewes were taken to Sweepstakes (32), Silver Mine (61) and Gold Drop (64); Sanford & Gibbs (56) was used to a few ewes; he sired Vermont (123) and Rough-and-Ready (178), both used in the flock as stock rams; Rich's Sweepstakes (53) was used to some extent; Fremont (126) was used nine years; twelve ewes were served by Bonaparte (176), and Silver Horns (177) in 1873; the same year, ten ewes were bred to Major (181), and ten to Birchard & Tottingham (202), and to the latter ten ewes in 1874; T. S. & Sons' 140 (216), sired by Birchard and Tottingham (202), (not Tottingham (40), as is erroneously stated in our list of stock rams) has been used to some extent; Old Cap (131) was used a few years; Fremont, Jr. (215), J.T.Stickney's 146 (441), and Centennial (442) have been used to some extent in the flock; T. S. & Sons' 150 (217) and T. S. & Sons' 217 (219) with (216) are now the stock rams of the flock. Other rams may have been used to a limited extent, but great care has been taken to use none but of pure Spanish Merino blood. It is but justice to the aged senior proprietor of this flock, to call attention to the immense improvement from Consul (2), bred in 1835, shearing about 14 pounds unwashed wool, at his best, to Fremont (126) shearing 34 ⅞ pounds in 1868, the latter, the heaviest shearing ram of his day, and for years not equalled in weight of fleece, also from an ewe shearing only 8 or 9 pounds unwashed wool to one shearing 20 pounds in 1878.

——o——

FLOCK 168.

Owned by JOHN Q. STICKNEY, WHITING, ADDISON CO., VT.

39 RAMS.

Marked with metallic labels in their ears, having thereon J. Q. S., and numbers 1 to 39.

72 EWES.

Labeled same as rams and numbers 40 to 111.

PEDIGREE.

Descended from importations from Spain through the flocks of
A. Cock, Long Island; D. Humphreys, S. Atwood. Conn.; Wm.
Jarvis, Weathersfield; Messrs. Rich, L. Beedle, E. R. Robinson, T.
Stickney, L. C. Remele, Shoreham; Messrs. Hammond, S. W. Rem-
ele, D. Hooker, Middlebury; A. A. Farnsworth, Brooksville; N. A.
Saxton. Waltham; C. B. Cook, Charlotte; R. P. Hall, J. Towle, F.
Hooker, Cornwall; S. Root, Hubbardton; L. Webster, Sudbury;
Wm. Baldwin, Whiting.

ORIGIN.

The flock was commenced in the fall of 1870, by twenty ewes
from the flock of Tyler Stickney; in 1871, John Q. Stickney came
into possession of one half of the flock of William Baldwin, for
history of which see flock 13.

RAMS USED.

From the flock of Tyler Stickney & Son; Fremont, Jr. (215),
and rams sired by him.

------o------

FLOCK 169.

Owned by S. L. STEVENS, Orwell, Addison Co., Vt.

40 RAMS.

Marked with metallic labels in their ears, having thereon C.
L. Stevens and numbers 1 to 40.

100 EWES.

Labeled same as rams, thirty with C. L. Stevens and numbers
41 to 45, 47 to 71; seventy with their labels marked W. M. Holmes
and numbers 1, 3 to 6, 8, 10, 14, 15, 17, 18, 20, 36, 37, 46, 53, 55,
58, 60 to 62, 69, 71, 73, 92, 101, 108, 111, 112, 114 to 117, 120, 121,
123, 125, 126, 129, 131, 135, 140 to 146, 148, 150, 151, 152, 154,
155, 156, 159, 161, 165 to 174, 177, 178, 179.

PEDIGREE.

Descended from importations from Spain through the flocks of
D. Humphreys, S. Atwood, Conn.; A. Cock, Long Island; Wm.
Jarvis, Weathersfield; Messrs. Hammond, W. R. Remele, Middle-
bury; W. R. Sanford, O. H. Bascom, Orwell; Messrs. Rich, L.
Beedle, L. C. Remele, E. R. Robinson, E. G. Farnham, Shoreham;
Baker & Smith, P. Elitharp, Bridport; R. P Hall, E. S. Stowell, S.
S. Rockwell, E. Sanford, Cornwall; C. I. Benedict, Arlington; N. &

N. Bottum, Shaftsbury; S. Steel, Shushan, N. Y.; H. Bramer, Hebron, N. Y; J. Harwood, S. Harwood, J. M. Harwood, West Rupert; Wm. M. Holmes, Greenwich, N. Y.

EWES PURCHASED.

In March, 1865, ten ewes were purchased of W. M. Holmes; in the fall of 1868, twelve ewe tegs were purchased of Mr. Holmes; in April, 1869, sixteen ewes and one ewe teg were purchased from the same flock; in July, 1874, the entire flock of Mr. Holmes was purchased, consisting of 94 ewes and 33 ram and ewe tegs; the. flock of Mr. Holmes was bred from ewes he purchased of E. G. Farnham, S. S. Rockwell, Edgar Sanford, A. H. Bascom, Henry Bramer, N. & N. Bottum, S. Harwood, J. Harwood and J. M. Harwood. Mr. Bramer and the Harwoods bred their stock from Messrs. Hammond's stock.

RAMS USED BY W. M. HOLMES.

Capt. Joe (388), Gruffy (389), Gold Mine (63), Madoc (204), Prince of Gold Drops (153), Rocket (203), and a few other pure bred Hammond rams; also Messrs. Hammond's stock rams during the time Mr Holmes was breeding.

RAMS USED BY S. L. STEVENS.

Steele Ram (390) and rams bred by O. H. Bascom, C. I. Benedict and E. Hammond & Son.

---o---

FLOCK 170.

Owned by WATSON C. STURDEVANT, Weybridge, Addison Co., Vt.

10 RAMS.

Marked with metallic labels in their ears, having thereon W. C. W. and numbers 1 to 10.

32 EWES.

Labeled same as rams with numbers 1 to 16, 98 to 113.

PEDIGREE.

Descended from the importation from Spain by Col. David Humphreys through the flocks of S. Atwood, Conn.; E. Hammond, E. D. Munger, W. R. Remele, Middlebury; R. P. Hall, E. S. Stowell, Cornwall; W. R Sanford, Orwell; N. A. Saxton, F. D. Barton, S. M. Stowe, Waltham; V. Wright, A. J. Stowe, Weybridge; L. P. Clark, Addison.

EWES PURCHASED.

In 1870, seven ewe tegs were purchased of E. D. Munger that were bred from the flocks of Messrs. Hammond and V. Wright and E. S. Stowell; they were sired by a son of Golden Fleece (91); in 1873, six ewes were purchased of S. M. Stowe, bred from stock purchased of E. Hammond by A. J. & S. M. Sowe; in September, 1876, seventeen ewes were purchased of F. D. Barton; the blood of all these ewes purchased was Atwood.

RAMS USED.

All the increase of the Munger and Stowe ewes bred previous to 1876 has been sold. In 1875, two ewes were served by a ram bred by L. P. Clark; in 1876, a ram bred by and belonging to F. D. Barton was used; in 1877, Victor (476) was purchased and used.

——o——

FLOCK 171.

Owned by EDWIN S. STOWELL, Cornwall, Addison Co., Vt.

25 RAMS.

Marked with metallic labels in their ears, having thereon E. S. Stowell and numbers 252, 260, 261, 266 to 269, 275, 276, 284 to 299.

82 EWES.

Labeled same as rams with numbers 201, 202, 204, 207, 209 to 211, 215 to 235, 237, 238, 240 to 243, 247 to 249, 252 to 254, 258, 260, 261, 263, 265 to 268, 270 to 275, 278 to 283, 300 to 321.

PEDIGREE.

Descended from the importation from Spain by Col. David Humphreys through the flocks of S. Atwood, Conn.; Messrs. Hammond, W. R. Remele, Middlebury; W. R. Sanford, Orwell; R. P. Hall, R. J. Jones, Cornwall; N. A. Saxton, Waltham.

EWES PURCHASED.

In 1853, ten ewes were purchased of W. S. & E. Hammond; in 1854, twenty-one ewe tegs were purchased of the same firm; in January, 1855, ten more were purchased from the same flock; in 1858, ten ewes of E. D. Munger, Middlebury, that were bred by W. R Remele; in 1859, two ewes were purchased of N. A. Saxton; in 1860, six ewes were purchased of George Hammond; the same year, eight ewes from the flock of W. R. Sanford were purchased of E. D. Griswold, of Orwell.

RAMS USED.

At first, the stock rams of Messrs. Hammond, until 1861, when those bred within the flock were used: Stowell's Sweepstakes (47),

Golden Fleece (91), Dew Drop (59), Nugget (187), King Solomon
(93), Red Leg (109), Panic (110), Col. Stowell (190), Golden Horn
(191), David (118), Iron Clad (194), Diamond Dust (122), and Goliath (197) ; these have all been used as stock rams, the last being
the present stock ram of the flock. In 1876, Allright (169) was
used to a few ewes ; four ewes and four rams were sired by him.

———o———

FLOCK 172.

Owned by C. C. STICKNEY, WHEELER, STEUBEN Co., N. Y.

25 RAMS.

Marked with metallic labels in their ears, having thereon C. C.
S. and numbers 1 to 25.

75 EWES.

Labelled same as rams with numbers 26 to 100.

PEDIGREE.

Descended from importations from Spain through the flocks of
A. Cock, Long Island ; D. Humphreys, S. Atwood, Conn.; Wm.
Jarvis, Weathersfield ; Messrs. Rich, L. Beedle. T. Stickney & Son,
E. R. Robinson, L C. Remele, J. T. Stickney, C. Delong, Shoreham ;
Baker & Smith, P. Elitharp, Bridport ; Messrs. Hammond, W. R.
Remele, Middlebury ; E. Rich, Whiting.

ORIGIN.

During the years 1864, '65, '66, '67, C C. Stickney in company
with J. R. Stickney took the farm of Tyler Stickney, in Shoreham,
Vt., on shares ; at the end of that time the surplus on hand was
divided, and C. C. Stickney took his share to Wheeler, N. Y.; this
formed the foundation of the flock ; a few were purchased from the
same flock and added to the others.

RAMS USED.

The ewes that were of breeding age were in lamb by Gen. Fremont (126) ; the first year a ram bred by T. Stickney & Son, sired
by Gen. Fremont (126), owned by Julius Stickney, was used ; after
that pure Stickney rams bred within the flock.

———o———

FLOCK 173.

Owned by HENRY THORP, CHARLOTTE, CHITTENDEN Co., VT.

12 RAMS.

Marked with metallic labels in their ears, having thereon H.
Thorp, and numbers 1, 18, 22, 23, 35 to 40, 71, 72.   40

70 EWES.

Forty-five labeled same as rams and numbers 1 to 32, 41 to 50, 52, 56, 57; twenty-five marked with labels having thereon L. J. O. and numbers 2, 3, 5, 6, 7, 21, 24, 25, 34, 39, 44, 47, 50, 63, 67, 72, 73, 77, 85, 130, 132, 141, 167, 177, 178.

PEDIGREE.

Descended from the importation from Spain by Col. David Humphreys through the flocks of S. Atwood, Conn.; Messrs. Hammond, W. R. Remele, Middlebury; Victor Wright, Weybridge; W. R. Sanford, Orwell; F. D. Barton, Waltham.

EWES PURCHASED.

The flock was commenced Sept., 1862, by a purchase of five Atwood ewes of Victor Wright; April 23, 1863, five Atwood ewes were purchased of the same; Sept. 14, 1863, seven Atwood ewes were purchased of H. W. Hammond; Sept. 15, 1863, two Atwood ewes were purchased of V. Wright, and Sept. 5, 1865, ten Atwood ewes were bought of the same; Sept. 26, 1874, twenty Atwood ewes were purchased of F. D. Barton that he had bred from the flocks of Messrs. Hammond and W. R. Sanford; August 29, 1877, twenty-two Atwood ewes were purchased of G. Hammond, agent for L. J. Orcutt; twenty-one of them were bred by W. R Sanford and one of them from the old Hammond flock; Oct 3, 1877, five Atwood ewes were purchased of G. Hammond, agent; two of them were bred by W. R. Sanford and three of them were of the old Hammond flock.

RAMS USED.

The several stock rams of Messrs. Hammond and V. Wright, to a limited extent; Sept., 1862, an Atwood ram teg was purchased of V. Wright; Sept., 1864, two Atwood ram tegs were purchased of the same, and in Oct., 1868, Vanderbilt (261) was purchased of his estate; Black Diamond (347) was also purchased of the same; all these were used more or less in the flock; Sept., 1872, two Atwood rams were purchased of F. D. Barton; one of them, Thorp's Barton Ram (375) was used as a stock ram; H. Thorp's No. 1 (378) has been used as the stock ram of the flock for two years past. These are all the rams that have been used in the flock, except one year when Sam (363) was used to a very few ewes, but the produce has been disposed of, and none remains in the flock.

———o———

FLOCK 174.

Owned by JOHN TOWLE, CORNWALL, ADDISON Co., VT.

5 RAMS.

Marked with metallic labels in their ears, having thereon John Towle and numbers 69, 70, 71, 80, 81.

47 EWES.

Labeled same as rams, and numbers 23 to 69.

PEDIGREE.

Descended from the importation from Spain by Col. David Humphreys through the flocks of S. Atwood, Conn.; Messrs. Hammond, W. R. Remele, Middlebury; V. Wright, Weybridge; R. Gleason, Benson; W. R. Sanford, Orwell.

ORIGIN.

John Towle purchased the flock in 1869, as stated in the history of the flock of R. P. Hall on page 161 of this work, and has bred it since without the introduction of other ewes into the flock.

RAMS USED.

Those bred within the flock except a part of the ewes have been served in three years to rams bred by L. C Remele. but the stock from these has been sold and none is included in the flock registered; J. Towle's 69 (477), J. Towle's 70 (478) and J. Towle's 71 (479) are now the stock rams of the flock.

———o———

FLOCK 175.

Owned by E. TOWNSEND, Pavilion Centre, Genesee Co., N.Y.

3 RAMS.

Marked with metallic labels in their ears, two having thereon E. Townsend and numbers 122, 172 and one L. P. Clark and number 114.

143 EWES.

Labeled same as first two rams with numbers 3 to 8, 10 to 12, 14 to 23, 26 to 98, 101 to 114, 116 to 142, 144 to 151, 153, 154.

PEDIGREE.

Descended from the importation from Spain by Col. David Humphreys, through the flocks of S. Atwood, Conn.; Messrs. Hammond, C. B. Currier, Middlebury; F. H. Dean. E. S Stowell, R. J. Jones, Cornwall; G. Miner, P. Elitharp, C. N. Hayward, J. J. Crane, Bridport; Victor Wright, Weybridge; F. D. Barton, Waltham; C. B. Cook, Henry Thorp, Charlotte; L. S. Drew, Burlington; W. R San. ford, O. H. & W. O. Bascom, Orwell; A. E. Fuller, Pomfret; J. E. Parker, Whiting; Ira Moore, Woodstock.

EWES PURCHASED.

In 1865, ewe 3 was purchased of F. H. Dean; she was sired by Little Wrinkly (48); No. 4 by Gold Drop (64); No. 5 by Sweepstakes (32), and 7 and 8 by Little Wrinkly (48), were purchased of C. B. Currier; ewe 8 by Young Eureka (81) was purchased of G. Miner; in 1872 ewes 14 to 18, bred by O. H. & W. O. Bascom, were purchased of C. D. Lane; at the same time ewes 20 by Dew Drop (59) and 21 and 22 by Golden Fleece (91) were purchased of E. S. Stowell; in 1874, ewes 30 and 31 were purchased of Henry Thorp; at the same time ewe 32 was purchased of G. H. Hall; in 1875, ewes 39 to 73 were purchased of L. S. Drew and ewes 74 to 78 of Henry Thorp; in Feb., 1876, ewes 79 to 98 were purchased of Henry Thorp; of these 93 to 98 were bred by F. D. Barton. All the ewes in the several purchases were Atwood blood.

RAMS USED.

Old Genesee (161) sired ewe 10; Tariff (160) sired 11 and 12; Addison (85) sired 23, 26 to 29; Smuggler (423) sired 104, 119 to 130, 146 to 151, 153, 154; Genesee (162) sired 101, 102, 103, 105 to 114, 116 to 118, 131 to 142, 144, 145. Genesee (162), Smuggler (423) and L. P. Clark's 114 (484) were used in 1877.

----o----

FLOCK 176.

Owned by A. J. TOWNER, SHOREHAM, ADDISON CO., VT.

9 RAMS.

Marked with metallic labels in their ears, having thereon A. J. Towner and numbers 21 to 23, 30 to 35.

20 EWES.

Labeled same as rams with numbers 1 to 20.

PEDIGREE.

Descended from importations from Spain through the flocks of A. Cock, Long Island; D. Humphreys, S. Atwood. Conn; Wm. Jarvis, Weathersfield; Messrs. Rich, L. Beedle, E. R. Robinson, A. Clark, C. W. Jones, Kent Wright, F. & L. E. Moore, H. W. Jones, E. A. Birchard, Shoreham; Baker & Smith, P. Elitharp, Bridport; C. B. Cook, Charlotte.

EWES PURCHASED.

In 1874, twenty ewes were purchased of Kent Wright; they were Robinson ewes bred from stock from the flock of the late Alvin Clark, and sold to Mr. Wright by Mrs. Abbie Clark; in 1876, six Robinson ewes were purchased of C. W. Jones that were sired by Duke (275).

RAMS USED.

Fortune (475) has been used as the stock ram of the flock and the young stock were sired by him.

—o—

FLOCK 177.

Owned by LOREN TOWNER, Shoreham, Addison Co., Vt.

19 RAMS.

Marked with metallic labels in their ears, having thereon L. Towner, and numbers 1 to 15, 47 to 50.

31 EWES.

Labeled same as rams, with numbers 16 to 46.

PEDIGREE.

Descended from importations from Spain through the flocks of A. Cock, Long Island ; D. Humphreys, S. Atwood, Conn.; D. Buffum, Wm. Bailey, J. I. Bailey, Rhode Island ; Wm. Jarvis, Weathersfield ; Messrs. Rich, L. Beedle, E. R. Robinson, E. A. Birchard, F. & L. E. Moore, J. Forbes, Jr., D. & G. Cutting, J. M. Ormsbee, D. E. Robinson, Shoreham ; S. T. Baker, Whiting ; Messrs. Hammond, W. R. Remele, Middlebury ; W. R. Sanford, Orwell ; Baker & Smith, P. Elitharp, Bridport ; R. P. Hall, S. S. Gibbs, Cornwall ; C. B. Cook, Charlotte.

ORIGIN.

In 186o, twenty-nine ewes were leased upon shares of Darwin E. Robinson, and at the same time five or six ewes were purchased from the same flock ; soon after seventeen ewes were purchased of German Cutting.

RAMS USED.

Tottingham (40), Birchard & Tottingham (202), Old Ethan (67), also the stock rams of Messrs. Hammond, J. T. & V. Rich, W. R. Sanford, S. S. Gibbs, F. & L. E. Moore and J. Forbes, Jr., besides those bred within the flock.

—o—

FLOCK 178.

Owned by B. B. TOTTINGHAM, Shoreham, Addison Co., Vt.

34 RAMS.

Marked with metallic labels in their ears, twenty-one having thereon B. B. T. and numbers 111 to 130, 132, 133, 135, 136, 138 to 140, 151, 153, 156, 160, 163, 165, and one J. T. & V. R. No. 333.

134 EWES.

Labeled same as rams with numbers 1 to 110, 131, 134, 137, 141 to 150, 152, 154, 155, 157 to 169, 161, 162, 164, 166, 167.

PEDIGREE.

Descended from importations from Spain through the flocks of A. Cock, Long Island; D. Humphreys, S. Atwood, Conn.; Wm. Jarvis, Weathersfield; Leonard Beedle, Messrs. Rich, E. R. Robinson, E. A. Birchard, James Forbes, Jr., T. Stickney & Son, L. C. Remele, F. D. Douglas (now of Whiting), D. & G. Cutting, Darwin E. Robinson, J. N. North, A. C. Harris, E. D. Bush, J. M. Ormsbee, L. Catlin, Shoreham; S. T. Baker, Whiting; Ward M. Lincoln, Brandon; C. S. Rumsey, Hubbardton; W. R. Sanford, O. S. Branch, J. T. Branch, D. B. & J. N. Buell, O. H. Bascom, H. T. Cutts, H. G. Hibbard, Orwell; R. P. Hall, John Towle. Cornwall; Messrs Hammond, W. R. Remele, Oliver Severance, Middlebury; Baker & Smith, Prosper Elitharp, C. N. Hayward, W. W. Winchester, Bridport; C. B. Cook, Charlotte; N. A. Saxton, Waltham; Nelson Richards, Panton; Ebenezer Porter, Rutland; V. Wright, Weybridge; J. Sheldon, Fairhaven.

EWES PURCHASED.

In 1860, fifteen ewes were purchased of Rollin Birchard, of Shoreham; they were ewes that he had purchased the same year of J. T. & V. Rich and certified to have been bred by them; the same year ten ewes were purchased dirictly of J. T. & V. Rich; Jan. 10th, 1876, fifteen ewes were purchased of James Forbes, Jr.

RAMS USED.

Tottingham (40), Birchard & Tottingham (202), Charlie (253) and J. T. & V. R. 333 (448) successively were the stock rams of the flock; besides these a few bred in the flock have been used.

———o———

FLOCK 179.

Owned by E. N. TOWNSEND, RICHVILLE (in Shoreham), ADDISON Co., VT.

2 RAMS.

Marked with metallic labels in their ears, having thereon E. N. T., and numbers1, 2.

19 EWES.

Labeled same as rams, with numbers 3 to 21.

PEDIGREE.

Descended from importations from Spain through the flocks of A. Cock, Long Island; D. Humphreys, S. Atwood, Conn.; Wm. Jarvis, Weathersfield; Messrs. Rich, L. Beedle, L. C. Remele, T. Stickney & Son, E. Robinson, Lucius Robinson, Shoreham; Messrs. Hammond, W. R. Remele, Middlebury; W. R. Sanford, H. G. Hibbard, Orwell; J. Hinds, W. M. Lincoln, Brandon; E. Porter, Rutland; Wm. P. Wright, E. Rich, A. F. Ellsworth, C. H. Ketchum, Whiting; R. P. Hall, F. H. Dean, J. Towle, A. J. Wooster, B. S. Fields, Cornwall; W. M. Gage, Ferrisburgh; Baker & Smith, P. Elitharp, Bridport.

EWES PURCHASED.

In 1871, one ewe teg was purchased of Dr. W. P. Wright, and in Dec., 1873, three ewes were purchased of the same; in 1871, one ewe teg was purchased of J. T. & V. Rich and another was purchased in the spring of 1874; in the fall of 1873, four breeding ewes were purchased of Lucius Robinson.

RAMS USED.

In 1872, the stock ram of J. T. & V. Rich to one ewe; the produce, a ram lamb, that has since been used as a stock ram; in 1873, the same ram belonging to Messrs. Rich; in 1875, three ewes to Hibbard (214) and the rest to C. H. Ketchum's 51 (454); in 1876 and 1877, the stock rams of J. T. & V. Rich.

———o———

FLOCK 180.

Owned by URIAH D. TWITCHELL, MIDDLEBURY, ADDISON Co., VT.

50 EWES.

Marked with metallic labels in their ears, having thereon U. D. T. and numbers 1 to 50.

PEDIGREE.

Descended from the importation from Spain by Col. David Humphreys through the flocks of Stephen Atwood, Conn.; Messrs. Hammond, W. R. Remele, Willson Wright, Middlebury; N. A. Saxton, Waltham; W. R. Sanford, Orwell; V. Wright, Weybridge; R. P. Hall, E. S. Stowell, Cornwall.

EWES PURCHASED.

In 1864, eight ewes were purchased of Victor Wright that were in lamb by his Old Greasy (74) and Long Wool (112); same year four ewes were purchased of J. J. Kelsey that he purchased of W. R. Remele; same year twelve ewes were purchased of M. C. Roundy, ten of which were ewes he purchased of G. Hammond and two of them ewes bred by E. S. Stowell; six ewes bred by E. Hammond and W. R. Sanford were purchased of B. W. Couch; ten ewes were also purchased of L. A. Drake, that he purchased of E. S. Stowell; in 1868, eight ewes and three ewe tegs were purchased of William Wright; four of these were bred by Victor Wright and the rest were bred from these ewes and a ram bred by E. S. Stowell.

RAMS USED.

At first the stock rams of V. Wright; from 1865 to 1868, Little Ram (279) and Big Ram (340); ram bred by Willson Wright sired by a ram bred by E. S. Stowell, dam a V. Wright ewe; Sherman (348) was used a number of years; besides these, pure Atwood rams bred within the flock have been used.

——o——

FLOCK 181.

Owned by HENRY W. WALKER, Richville (in Shoreham), Addison Co., Vt.

27 RAMS.

Marked with metallic label in their ears, having thereon H. W. Walker and numbers 55 to 81.

54 EWES.

Labeled same as rams with numbers 1 to 54.

PEDIGREE.

Descended from importations from Spain through the flocks of A. Cock, E. Lawrence, Long Island; D. Humphreys, S. Atwood, J. N. Blakeslee, J. R. Nettleton, Conn.; D. Buffum, W. Bailey, J. I. Bailey, Rhode Island; Messrs. Rich, Z. Frost, A. Frost, T. Stickney & Son, L. Beedle, E. R. Robinson, L. C. Remele, J. M. Ormsbee, D. & G. Cutting, J. Barnum, S. L. Bissell, E. N. Bissell, H. A. Bascom, Shoreham; Messrs. Hammond, W. R. Remele, Middlebury; R. P. Hall, E. S. Stowell, J. Towle, B. S. Field, Cornwall; W. R. Sanford, Orwell; J. Hinds, Brandon; N. A. Saxton, Waltham; S. T. Baker, Whiting; C. B. Cook, Caarlotte.

ORIGIN.

This flock was inherited by H. W. Walker from his father, Wm. Walker, who commenced his flock in 1832 by a purchase of ten ewes of Abraham Frost that were descended from stock purchased of A. Cock and Effingham Lawrence in 1816, by Zebulon Frost; soon after this purchase by Mr. Walker, the Frost flock was crossed with Saxon rams, but this purchase was previous to that cross being made. Soon after this Mr. Walker purchased of J. Thurman Rich twelve or fifteen ewes of the Cock blood, and hired a ram of Mr. Rich to use in the flock. Afterwards rams bred within the flock were used until the flock was managed by the present owner, but it is believed the flock was kept entirely pure from admixture with other blood.

RAMS USED BY H. W. WALKER.

One of the first Atwood rams owned or bred by D. & G. Cutting; Wooster (16), Old Robinson (38), Monitor (142), were used to some extent; a ram was hired of E. Hammond one year; a few ewes were sent to Golden Fleece (91) and Dew Drop (59); nine in 1863 to Eureka (58) and five to the same ram in 1864; a few to Little Wrinkly (48) and a few to J. Towle's stock rams; H. W. Walker's Stowell Ram (455) was used three years; Bacon's Stickney Ram (456) was used one year; H. W. Walker's Stock ram (457) was used three years; H. W. Walker's 77 (532) was used in 1877; besides these Old Peter (264) was used in 1872 and '73, and rams were rented of T. Stickney & Son in 1870 and '71.

——o——

FLOCK 182.

Owned by JONAS A. WATTS, WHITING, ADDISON CO., VT.

9 RAMS.

Marked with metallic labels in their ears, having thereon J. A. Watts and numbers 9, 15, 17, 19, 21, 23, 25, 27, 29.

19 EWES.

Labeled same as rams, with numbers 1 to 8, 10 to 13, 14, 16, 18, 20, 22, 24, 26.

PEDIGREE.

Descended from importations from Spain through the flocks of A. Cock, Long Island; D. Humphreys, S. Atwood, Conn.; Messrs. Rich, L. Beedle, E. R. Robinson, L. C. Remele, Lucius Robinson, T. Stickney, Shoreham; Messrs. Hammond, W. R. Remele, Middlebury; W. R. Sanford, Orwell; Baker & Smith, P. Elitharp, Bridport; C. B. Cook, Charlotte; Rollin J. Smith, L. Webster, Sudbury; R. P. Hall, F. H. Dean, Cornwall; W. Gage, Ferrisburgh; V. Wright, Weybridge; S. Root, Hubbardton; J. Hinds, Brandon; J. Q. Stickney, A. F. Ellsworth, W. Baldwin, Whiting.

41

EWES PURCHASED.

In 1860, six Robinson ewes were purchased of Lucius Robinson; in 1874, one ewe lamb was added to the flock from that of J. Q. Stickney.

RAMS USED.

In 1860, Sanford Gibbs (56); Young Black Top (283) was used to the entire flock in 1865 and '66. In 1872 and '73, a Robinson ram bred by Wm. Baldwin was used; Young Fremont (215) has been used to a part of the flock, two seasons. Watts' Ellsworth ram (460) was used, and rams bred within the flock. Emerson Watts, father of J. A. Watts, owned a part of the above described flock, and his widow now owns an interest in it.

———o———

FLOCK 183.

Owned by WAGNER & SHIPLEY, McConnellsville, Ohio.

7 RAMS.

Marked with metallic labels in their ears, having thereon W. & S., and numbers 10 to 16.

19 EWES.

Five labeled same as rams, with numbers 1 to 5, and fourteen having on their labels 395, 410 to 412, 414 to 423.

PEDIGREE.

Descended from importations from Spain through the flocks of A. Cock, Long Island; D. Humphreys, S. Atwood, Conn.; Wm. Jarvis, Weathersfield; Messrs. Hammond, W. R Remele, U. D. Twitchell, Cherbino & Williamson, Willson Wright, Middlebury; V. Wright, Weybridge; R. P. Hall, C. Benedict, E. S. Stowell, Cornwall; W. R. Sanford, Orwell; Messrs. Rich, L. C. Remele, L. Beedle, E. R. Robinson, E. A. Birchard, Shoreham; Baker & Smith, P. Elitharp, C. N. Hayward, D. F. Doty, J. J. Crane, H. C. Burwell, D. E. Hill, B. Myrick, Bridport; C. B. Cook, Charlotte; N. A. Saxton, Waltham; N. Richards, Panton; A. A. Farnsworth, Brooksville.

EWES PURCHASED.

In February, 1878, fourteen Atwood ewes, 410 to 412, 414 to 423, bred by U. D. Twitchell, and one Atwood and Robinson ewe 395, bred by C. Benedict, were purchased of Cherbino & Williamson. These ewes were in lamb by Bonaparte (176), Acme (524), and a ram teg sired by Bonaparte (176); the tegs 5, 12, 13, were sired by Acme (524); 2, 4, 14, 16 were sired by Bonaparte (176), and the remainder by the ram teg.

FLOCK 184.

Owned by S. H. WESTON, WINOOSKI FALLS, CHITTENDEN Co., VT.

5 RAMS.

Marked with metallic labels in their ears, having thereon S. H. Weston and numbers 1 to 5.

28 EWES.

Labeled same as rams with numbers 26 to 53.

PEDIGREE.

Descended from the importation from Spain by Col. David Humphreys through the flocks of S. Atwood, Conn.; Messrs. Hammond, Middlebury ; V. Wright, Weybridge ; H. N. Newell, Shelburn ; C. B. Cook, H. Thorp, D. L. Spear, J. Holmes, Charlotte.

EWES PURCHASED.

Twenty-four Atwood sheep bred from the flocks of Messrs. Hammond and V. Wright, through that of H. N. Newell, were purchased of D L. Spear about the year 1867 ; Nov.,1870, fifty Atwood ewes of H. N. Newell.

RAMS USED.

The first ram used was one purchased of V. Wright ; this ram was used as long as he lived. The Holmes ram (458) was purchased and used ; two ram tegs were purchased of H. Thorp ; with these exceptions, rams bred within the flock have been used.

―――o―――

FLOCK 185.

Owned by LUTHER WEBSTER, EAST SHOREHAM, ADDISON Co., VT.

20 RAMS.

Marked with metallic labels in their ears, having thereon L. Webster, and numbers 46 to 57, 65 to 70, 72, 73.

53 EWES.

Labeled same as rams with numbers 1 to 45, 58 to 64, 71.

PEDIGREE.

Descended from importations from Spain through the flocks of A. Cock. Long Island ; D. Humphreys, S Atwood, Conn.; D. Buffum, Wm. Bailey. J. I. Bailey, Rhode Island ; Wm. Jarvis, Weathersfield ; Messrs. Rich, L. Beedle, T. Stickney, E. R. Robinson, L. Robinson, L. C. Remele, D. & G. Cutting, J. M. Ormsbee, Earl R.

Delano, Shoreham ; S. T. Baker, J. Q. Stickney, A. M. Baldwin, Wm. Baldwin, Whiting ; Lyman Webster, Sudbury ; S. Root, Hubbardton ; C. B. Cook, Charlotte ; Baker & Smith, P. Elitharp, Bridport.

EWES PURCHASED.

In the fall of 1860, ten ewes were purchased of E. R. Delano, that he had bred from ewes purchased of T. Stickney, and the Atwood rams of Messrs. Cutting, and rams bred by E. R. Robinson, including the Old Robinson (38), Lute Robinson (39), and Jenning's ram (89) ; in 1863, ten Cutting ewes were purchased of D. Cutting.

RAMS USED.

Those bred by T. Stickney & Sons, and D. & G. Cutting, and Webster's Baldwin ram (459).

———o———

FLOCK 186.

Owned by GIDEON W. WITFORD, Addison, Addison Co., Vt.

6 RAMS.

Marked with metallic labels in their ears, having thereon G. W. W., and numbers 50 to 55.

53 EWES.

Labeled same as rams, with numbers 1 to 49, 56 to 59.

PEDIGREE.

Descended from importations from Spain through the flocks of A. Cock, Long Island ; D. Humphreys, S. Atwood, Conn.; Wm. Jarvis, Weathersfield ; Messrs Rich, L. Beedle, E. R. Robinson, A. Clark, Bela Howe, Shoreham ; Baker & Smith, P. Elitharp, C. N. Hayward, H. C. Burwell, J. J. Crane, D. F. Doty, J. Hill, E. Fitch, L. S. Burwell, Bridport ; C. G. Seager, Addison ; F. D. Barton, J. H. Sprague, Waltham ; N. Richards, Panton ; Messrs. Hammond, W. R. Remele, Middlebury ; W. R. Sanford, Orwell ; R. P. Hall, E. S. Stowell, Cornwall ; C. B. Cook, Charlotte.

EWES PURCHASED.

In 1859, six ewes were purchased of J. H. Sprague ; in 1860, sixteen ewes were purchased of Alvin Clark ; a short time after this ten ewes were purchased of Bela Howe, and six of C. G. Seager ; a small part of the ewes were pure Atwoods, most of them descended direct from the Robinson flock, and a part Atwood and Robinson.

RAMS USED.

Sea Lion (83), three seasons; Eureka (58), one season to four ewes; Golden Fleece (91); Dew Drop (59), one season; Iron Sides (80), one season; except to Eureka, the numbers to each ram cannot now be given; later Silver Ring (219), Elitharp & Burwell (79) and other stock rams of H. C. & L. S. Burwell; besides those named rams bred within the flock have been used.

FLOCK 187.

Owned by E. M. WHEELER, BRIDPORT, ADDISON CO., VT.

13 RAMS.

Marked with metallic labels in their ears, having thereon E. M. Wheeler, and numbers 18, 34 to 36, 48 to 56.

32 EWES.

Labeled same as rams and numbers 1 to 12, 14 to 16, 20, 21, 23, 26, 28, 31, 37 to 47.

PEDIGREE.

Descended from importations from Spain through the flocks of A. Cock, Long Island; D. Humphreys, S. Atwood, Conn.; Wm. Jarvis, Weathersfield; Messrs. Rich. L. Beedle, T. Stickney, L. C. Remele, E. A. Birchard, Rollin Birchard. Shoreham; Messrs. Hammond, W. R. Remele, Middlebury; W. R. Sanford, Orwell; R. P. Hall, S. S. Gibbs, A. H. Sperry, Cornwall; Baker & Smith, P. Elitharp, Bridport.

EWES PURCHASED.

Sept. 27, 1860, a half-interest in four ewes was purchased of C. A. Landers, they were bred by S. S. Gibbs and were Stickney and Robinson blood; Oct. 13, 1863, four ewes were purchased of S. S. Gibbs that were bred by Tyler Stickney; in December, 1864, four Atwood ewes were purchased of P. Elitharp.

RAMS USED.

For the first three years rams of S. S. Gibbs; next a Robinson ram purchased of A. H. Sperry; since then rams bred within the flock have been used.

FLOCK 188.

Owned by ROYAL WITHERELL & SON, Shoreham, Addison Co., Vt.

10 RAMS.

Marked with metallic labels in their ears, having thereon R. W. & Son and numbers 37 to 46.

36 EWES.

Labeled same as rams, with numbers 1 to 36.

PEDIGREE.

Descended from importations from Spain through the flocks of A. Cock, Long Island ; D. Humphreys, S. Atwood, Conn.; Wm. Jarvis, Weathersfield ; Messrs. Rich, L. Beedle, L. C. Remele, E. R. Robinson, E. A. Birchard, James Forbes, Jr., T. Stickney, G. H. Hall, Shoreham ; R. P. Hall, R. J. Jones, Cornwall ; Messrs. Hammond, W. R. Remele, O. Severance, Middlebury ; W. R. Sanford, Orwell ; Baker & Smith, P. Elitharp, C. N. Hayward, W. W. Winchester, Bridport ; V. Wright, Weybridge ; C. B. Cook, Charlotte ; N. Richards, Panton.

EWES PURCHASED.

Feb. 16th, 1863, twenty ewe tegs were purchased of E. A. Birchard ; sixteen were sired by Tottingham (40) and four were sired by Old Ethan (67) ; all were from ewes bred by E. R. Robinson and leased of his estate by E. A. Birchard; (R. Witherell made this purchase in company with D. H. Sunderland).

RAMS USED.

Those from the flocks of E. A. Birchard, G. H. Hall and James Forbes, Jr., besides those bred within the flock.

FLOCK 189.

Owned by C. K. WILLIAMS, Whiting, Addison Co., Vt.

4 RAMS.

Marked with metallic labels in their ears, three having thereon C. K. Wiliams and numbers 25, 26, 27, and one M. R. Atwood and number 8.

17 EWES.

Labeled same as first three rams with numbers 1 to 17.

PEDIGREE.

Descended from importations from Spain through the flocks of A. Cock, Long Island; D. Humphreys, S. Atwood, Conn.; D. Buffum, Wm. Bailey, J. I. Bailey, Rhode Island; Wm. Jarvis, Weathersfield; Messrs. Rich, L. Beedle, L C. Remele, E. R. Robinson, M. R. Atwood, A. Frost, Z. Frost, D. & G. Cutting, T. Stickney, L. Robinson, J. M. Ormsbee, Shoreham; S. T. Baker, J. Barlow, A. H. Hubbard, A. F. Ellsworth, B. Casey, Whiting; Messrs. Hammond, W. R. Remele, D. Hooker, S. W. Remele, Middlebury; V Wright, Weybridge; A. A. Farnsworth, Brooksville; R. P. Hall, E. S. Stowell, F. Hooker, F. H. Dean, J. Towle, Cornwall; J. Hinds, W. M. Lincoln, Brandon; E. Cook, Leicester.

EWES PURCHASED.

Six ewes were purchased of L. C Remele; one of E D. Searl, bred by F. Hooker; six ewe tegs of E. Cook; four ewes from the flocks of H. W. Walker, W. M. Lincoln and D. Cutting.

RAMS USED.

The stock rams of A. F. Ellsworth, B. Casey, E. Cook, E. N. Bissell and A. H. Hubbard; one ram, No. 8, was purchased of M. R. Atwood.

——o——

FLOCK 190.

Owned by WILLIAM G. WILLSON, Richville (in Shoreham), Addison Co., Vt.

6 RAMS.

Marked with metallic labels in their ears, five having thereon W. G. Willson and numbers 56 to 60, and one with J. T. & V. R. and number 328.

80 EWES.

Labeled same as rams with numbers 1 to 55, 61 to 85.

PEDIGREE.

Descended from importations from Spain through the flocks of A. Cock, Long Island; D. Humphreys, S. Atwood, Conn.; Wm. Jarvis, Weathersfield; Messrs. Rich, L. Beedle, T. Stickney & Son, E. R. Robinson, E. N. Bissell, D. & G. Cutting, Shoreham; Messrs. Hammond, Middlebury; R. P. Hall, E. S. Stowell, J. Towle, Cornwall; N. A. Saxton, Waltham; Baker & Smith, P. Elitharp, Bridport.

EWES PURCHASED.

In 1853, five ewes were purchased of George Delano, of Whiting, who had purchased them the same season of E. A. Jennings; they were bred from ewes bred by T. Stickney and a ram bred by Stephen Atwood; in 1865, six ewes were purchased of J. T. & V. Rich, and in 1867, eight more from the same flock, and five Atwood ewes of E. N. Bissell.

RAMS USED.

Rams from the the flock of T. Stickney and T. Stickney & Son, including Gen. Fremont (126) and rams sired by him; Golden Fleece (91) was used one year; a ram from J. Towle's flock two years; Cutting's Atwood Ram, Young Sweepstakes, (50) was used to some extent; used Messrs. Hammond's stock rams to twelve ewes; in 1877, used J. T. & V. R.'s 328, purchased of V. Rich; he is now the stock ram of the flock.

—— o ——

FLOCK 191.

Owned by LEVI WOLCOTT, Shoreham, Addison Co, Vt.

7 RAMS.

Marked with metallic labels in their ears, having thereon L. Wolcott and numbers 1 to 7.

36 EWES.

Labeled same as rams, with numbers 8 to 43.

PEDIGREE.

Descended from importations from Spain through the flocks of A. Cock, Long Island; D. Humphreys, S. Atwood, Conn.; Wm. Jarvis, Weathersfield; Messrs. Rich, L. Beedle, E. R. Robinson, L. C. Remele, W. H. & R. Cook, R. N. Atwood, E. A. Birchard, Shoreham; Messrs. Hammond, Middlebury; W. M Lincoln, Brandon; Joseph Sheldon, Fairhaven; Isaac Dickinson, Benson.

EWES PURCHASED.

March 3d, 1876, twenty ewes were purchased of Isaac Dickinson; these ewes were bred by Mr. Dickinson from ewes that he purchased of the late Joseph Sheldon and by using rams from his flock (including the Lawrence Ram (24)) a ram bred by W. H. & R. Cook, sired by Tottingham (40), dam bred by J. T. & V. Rich; one bred by R. N. Atwood, sired by ram bred by L. C. Remele, dam an Atwood ewe bred by W. M. Lincoln, sired by a Hammond ram; and rams of Rich and Atwood blood bred by others and within the flock, certified to the committee to be pure from those strains of blood.

RAM USED BY L. WOLCOTT.

Since the flock was purchased by Mr. Wolcott he has used a ram leased of V. Rich.

———o———

FLOCK 192.

Owned by IRVING G. WOOSTER, WEST CORNWALL, ADDISON Co., VT.

1 RAM.

Marked with metallic label in his ear, having thereon I. G. Wooster and number 86.

85 EWES.

Labeled same as ram with numbers 1 to 85.

PEDIGREE.

Descended from importations from Spain through the flocks of D. Humphreys, S. Atwood, J. N. Blakeslee, J. R. Nettleton, Conn.; A. Cock, Long Island; D. Buffum, Wm. Bailey, J. I. Bailey, Rhode Island; Wm. Jarvis, Weathersfield; Messrs. Hammond, W. R. Remele, O. Severance, S. W. Remele, Middlebury; Messrs. Rich, L. Beedle, T. Stickney, E. R. Robinson, E. A. Birchard, Geo. & E. G. Farnham, E. B. Douglas, H. A. Bascom, S. L. Bissell, J. M. Delano (now of Ticonderoga, N. Y.), H. Birchard, J. Forbes, Jr., J. M. Ormsbee, D. & G. Cutting, Lucius Robinson, Shoreham; Baker & Smith, P. Elitharp, C. N. Hayward, W. W. Winchester, Bridport; S. T. Baker, A. F. Ellsworth, E. & J. A. Watts, F. D. Douglas, J. Q. Stickney, A. M. Baldwin, Whiting; L. Webster, R. J Smith, Sudbury; S. Root, Hubbardton; N. A. Saxton, Waltham; N. Richards, Panton; R. P. Hall, E. S. Stowell, B. S. Fields, A. J. Wooster, F. Hooker, F. H. Dean, Cornwall; C. I. Benedict, Arlington; W. R. Sanford, Orwell; V. Wright, Weybridge.

ORIGIN.

The flock was commenced about 1849 by A. J. Wooster, father of the present proprietor; he purchased a few ewes of W. S. & E. Hammond; also in 1852, six ewes of J. M. Parker, four of which were bred by T. Stickney and two by J. T. & V. Rich, also three ewes of J. T. & V. Rich direct; I. G. Wooster has purchased of E. Watts five ewes, and in company with E. A. Jennings fifty-one Robinson and Cutting sheep of E. B. Douglas; for history and breeding of these sheep see flock 108; twenty-six of these sheep were retained by Mr. Wooster at time of registration of the flock.

42

RAMS USED.

Wooster (16) was the first, afterwards other rams from the flock of Messrs. Hammond were purchased and his stock rams used to some extent; a ram sired by Sanford & Gibbs (56) and from one of the Hammond ewes was used a number of years; Comet (57) and Eureka (58) were used to some extent; Golden Fleece (91), Red Leg (109) and Little Wrinkly (48) were also used; a ram bred within the flock sired by Comet (57) and another by Little Wrinkly (48) were used each a number of years; Fremont, Jr. (215), has been used a number of seasons; Wrinkly (292) was used to some extent; and a ram bred by H. A. Bascom also. The present stock ram of the flock was sired by Fremont, Jr. (215).

——o——

FLOCK 193.

Owned by Wm. P. WRIGHT, Whiting, Addison Co., Vt.

16 RAMS.

Marked with metallic labels in their ears, having thereon W. P. Wright, and numbers 33 to 48.

41 EWES.

Labeled same as rams, with numbers 1 to 32, 49 to 57.

PEDIGREE.

Descended from importations from Spain through the flocks of A. Cock, Long Island; D. Humphreys, S. Atwood, Conn.; Wm. Jarvis, Weathersfield; Messrs. Hammond, W. R. Remele, Middlebury; Messrs. Rich, L. Beedle, E. R. Robinson, L. Robinson, T. Stickney & Son, L C. Remele, Shoreham; W. R. Sanford, H. G. Hibbard, Orwell; R. P. Hall, J. Towle, B. S. Field, A. J. Wooster, F. H. Dean, Cornwall; E. Rich, A. F. Ellsworth, A. M. Baldwin, J. Q. Stickney, C. H. Ketchum, Whiting; W. M. Lincoln, J. Hinds, Brandon; S. Root, Hubbardton; L. Webster, Sudbury; Baker & Smith, P. Elitharp, Bridport; C. B. Cook, Charlotte; N. A. Saxton, Waltham; V. Wright, Weybridge; E. Porter, Rutland.

EWES PURCHASED.

This flock has all descended on the side of the dams from two ewes purchased of L. C. Remele, one on Dec. 10, 1846, sired by Hero (4), dam a Rich ewe bred by E. R. Robinson; the other Feb. 20, 1847, was an Atwood ewe.

RAMS USED.

A minute memorandum of all the rams used outside the flock has been kept. In 1847, '48, '49 and '50 Atwood Ram (25); in

1850, Sanford (62) was used; in 1852 and three years following the stock ram of A. F. Ellsworth; and also the stock ram of T. Stickney in 1853-4; in 1856 and '57 Lute Robinson (39); in 1857 and eight following years E. Rich's Sweepstakes (53) was used; in 1858 and '59 the stock rams of T. Stickney; in 1860 Towle Ram (164); in 1861, '62, '65, '66, '68 the stock ram of A. F. Ellsworth; 1863 and '64 the stock ram of A J. Wooster, sired by Sanford & Gibbs (56); in 1864, '65, '66, Field's Eureka (450); in 1865, the stock ram of J. Towle; in 1866, the stock ram of L. Robinson, sired by Webster & Hall (163); in 1866, '68 and '71 the stock ram of T. Stickney & Son; in 1871 and '72, a ram bred by L. C. Remele, owned by C. K. Williams was used; in 1872, '73, '75, Young Fremont (215) was used; in 1873, Company ram (453); in 1873, Baldwin (449); in 1875 and '76, C. H. Ketchum's 51 (454); in 1875, Hibbard (214); besides these, rams bred within the flock have been used; of those named some years only one ewe has been served by them, but usually more.

---o---

FLOCK 194.

Owned by DON JUAN WRIGHT, Richville (in Shoreham), Addison Co., Vt.

21 RAMS.

Marked with metallic labels in their ears, having thereon D. J. Wright and numbers 2, 4, 6, 18, 26, 28 to 33, 52 to 60, 64.

31 EWES.

Labeled same as rams, and numbers 1, 3, 5, 7 to 10, 12, 13 to 17, 19 to 25, 34 to 40, 51, 61, 62, 63.

PEDIGREE.

Descended from importations from Spain through the flocks of A. Cock, Long Island; D. Humphreys, S. Atwood, Conn.; D. Buffum, Wm. Bailey, J. I. Bailey, Rhode Island; Wm. Jarvis, Weathersfield; Grant & Jennison, Walpole, N. H.; Messrs. Rich, L. C. Remele, T. Stickney, J. T. Stickney, E R. Robinson, E. A. Birchard, D. & G. Cutting, G. A. Cutting, J. M. Ormsbee, Shoreham; S. T. Baker, Whiting; Messrs. Hammond, W. R. Remele, Middlebury; V. Wright, Weybridge; N. A. Saxton, Waltham; Baker & Smith, P. Elitharp, Bridport; C. B. Cook, Charlotte; W. M. Lincoln, Brandon; E. Porter, Rutland; W. R. Sanford, H. G. Hibbard, Orwell.

HISTORY.

A part of this flock was received from the Hon M. W. C. Wright, who commenced with ten ewes purchased of W. R. Sanford, warranted by Mr. Sanford to be pure Jarvis, the blood of which he purchased of Messrs. Grant & Jennison, who certified to Mr. Sanford to be pure from the importations of Consul Jarvis; earlier than this Mr. Wright purchased one ewe of Mr. Jarvis; and four or five years after thirteen ewe tegs were purchased of J. T. & V. Rich. Besides the ewes received from the flock of the above, there has been by D. J Wright, Nov., 1875, twelve ewes descended from the flock of David Cutting, purchased of G. A. Cutting.

RAMS USED.

Those used by M. W. C. Wright were from the flocks of J. T. & V. Rich, T. Stickney, L. C. Remele, W. R. Sanford and P. Elitharp. Those used by D. J. Wright have been from the flocks of B. S. Field, L. C. Remele, R. P. Hall, J. T. Stickney and H. G. Hibbard, besides those in the former and later history of the flock, rams bred within the flock have been used. Among the number given at the commencement of the record of this flock are eight rams (52 to 59) that were purchased when tegs of L. C. Remele, Aug. 25, 1877, and Jan. 14, 1878.

——o——

FLOCK 195.

Owned by L. SILAS WRIGHT, WEYBRIDGE, ADDISON Co., VT.

11 RAMS.

Marked with metallic labels in their ears, having thereon L. S. Wright and numbers 1 to 11.

53 EWES.

Labeled same as rams with numbers 18 to 70.

PEDIGREE.

Descended from importations from Spain through the flocks of D. Humphreys, S. Atwood, Conn.; A. Cock, Long Island; Wm. Jarvis, Weathersfield; Messrs. Hammond, W. R. Remele, Middlebury; V. Wright, Weybridge; N. A. Saxton, F. D. Barton, Waltham; J. & E. D. Hinds, Brandon; W. R. Sanford, Orwell; D Giddings, P. C. Abbey, Essex; Henry Giddings, Fairfax; Rector Gage, L. P. Clark, Addison; Messrs. Rich, L. Beedle, T. Stickney, E. R. Robinson, L. C. Remele, Shoreham; R. P. Hall, J. Towle, Cornwall; Baker & Smith, P. Elitharp, Bridport; C. B. Cook, Charlotte; N. A. Saxton, Waltham.

EWES PURCHASED.

In the fall of 1866, four yearling ewes of Atwood, Rich and Jarvis blood were purchased of L. C. Remele ; in 1867, one Atwood ewe was purchased of M. T. Shackett, her sire and dam were bred by E. Hammond ; April 19, 1875, twenty Atwood ewes were purchased of Dan Giddings ; they were bred by P. C. Abbey from stock purchased of Dan & Henry Giddings, who purchased direct from Messrs. Hammond and V. Wright.

RAMS USED.

The ewes purchased in 1866 and '67 were bred to a ram called the Mason & Parmele ram, whose sire and dam were bred by J. Hinds ; the sire was by Comet (57) ; after this ram was used an Atwood ram was leased of F. D. Barton ; the next ram used was an Atwood ram bred by E. D. Hinds ; Black Top (463) and Fine Wool (464) were used next ; Rector (461) was last used and is now the stock ram of the flock.

———o———

FLOCK 196.

Owned by FREEMAN G. WRIGHT, Whiting, Addison Co., Vt.
60 EWES.

Marked with metallic labels in their ears, having thereon F. G. Wright and numbers 1 to 60.

PEDIGREE.

Descended from importations from Spain through the flocks of A. Cock, Long Island ; D. Humphreys, S. Atwood, Conn.; D. Buffum, Wm. Bailey, J. I. Bailey, Rhode Island ; Wm. Jarvis, Weathersfield ; Messrs. Rich, L. Beedle, T. Stickney, L. C. Remele, Davis Rich, E. R. Robinson, L. Robinson, L. Treadway, J. M. Ormsbee, D. & G. Cutting, Shoreham ; Messrs. Hammond, W. R. Remele, Middlebury ; V. Wright, Weybridge ; W. R. Sanford, Orwell ; R. P. Hall, J. Towle, Cornwall ; S. T. Baker, W. H. Baldwin, Whiting.

EWES PURCHASED.

In the fall of 1859, two Robinson ewe tegs were purchased of W. H. Baldwin ; Nov. 12, 1860, fifteen ewe tegs were purchased of the late Judge Davis Rich that he bred from stock he purchased of his brother, J. Thurman Rich ; in 1864, one Atwood and Robinson ewe was purchased of Lewis Treadway that he bred from a ram bred by E. Hammond and ewe bred by E. R. Robinson.

RAMS USED.

Those bred by the late German Cutting ; the Company Ram (453) ; and those bred within the flock.

FLOCK 197.

Owned by WRIGHT & JACKMAN, Vergennes (Residence Waltham), Addison Co., Vt.

15 RAMS.

Marked with metallic labels in their ears, having thereon W. & J. and numbers 51 to 57, 107 to 114.

66 EWES.

Labeled same as rams with numbers 1 to 11, 13 to 25, 27 to 37, 39 to 43, 115 to 140.

PEDIGREE.

Descended from importations from Spain through the flocks of D. Humphreys, S. Atwood, Conn.; A. Cock, Long Island; Wm. Jarvis, Weathersfield; Messrs. Hammond, J. Piper, S. Piper, Middlebury; Messrs. Rich, L. Beedle, E. R. Robinson, L. C. Remele, E. G. Farnham, A. Clark, Shoreham; C. B. Cook, D. Spear, Charlotte; N. A. Saxton, A. C. Bacon, J. H. Sprague, J. D. & F. D. Barton, Waltham; V. Wright, S. & H. B. Dodge, A. J. Stowe, Weybridge; Baker & Smith, P. Elitharp, Bridport; G. S. Harris, Panton; R. P. Hall, Cornwall.

EWES PURCHASED.

In 1864, twelve Atwood ewes were purchased of J. H. Sprague; they were a selection from thirty-two that Mr. Sprague purchased of F. D. Barton and certified to be descendants from ewes purchased of E. Hammond by John D. Barton, father of F. D.; nineteen Atwood ewes were also purchased the same year of F. D. Barton that he purchased of G. S. Harris, bred from five ewes he purchased of N. A. Saxton in 1852, and rams of the same blood; in 1874, a small number of tegs were added to the flock from that of S. Dodge and H. B. Dodge, for blood and breeding of which see flock 68 belonging to L. B. Dodge; these tegs were sired by Keeler Ram (287).

RAMS USED.

Those from the flocks of F. D. Barton, J. H. Sprague, O. C. Bacon, one purchased with the tegs purchased of S. Dodge & Son, and those bred within the flock.

FLOCK 198.

Owned by H. B. WRIGHT, MIDDLEBURY, ADDISON CO., VT.

14 RAMS.

Marked with metallic labels in their ears, having thereon H. B. Wright, and numbers 46, 51 to 63.

56 EWES.

Labeled same as rams, with numbers 1 to 45, 64 to 74.

PEDIGREE.

Descended from importations from Spain through the flocks of A. Cock, Long Island; D. Humphreys, S. Atwood, Conn.; Wm. Jarvis, Weathersfield; Messrs. Rich, L. Beedle, L. C. Remele, E. R. Robinson, T. Stickney, J. Forbes, Jr., E. D. Bush, A. C. Harris, Shoreham; Messrs. Hammond, W. R. Remele, M. C. Foote, O. Severance, Middlebury; Baker & Smith, P. Elitharp, C. N. Hayward, W. W. Winchester, Bridport; C. B. Cook, Charlotte; N. Richards, Panton; W. R. Sanford, Orwell.

ORIGIN.

In 1870 and '71, the breeding ewes of A. C. Harris (afterwards sold F. & L. E. Moore) were leased to breed from, and all the increase became the property of H. B. Wright; in Oct., 1876, thirty-five Atwood ewes were added to the flock by purchase of Manfred C. Foote; they were bred strictly from the flocks of Messrs. Hammond and E. S. Stowell.

RAMS USED.

A ram bred with the ewes leased of A. C. Harris, sired by Green Mountain (70), was used the two years the ewes were leased; all the ewes of this descent registered in this flock were sired by him; the later produce from these ewes have been sold; a ram sired by Snowflake (277), dam a Robinson ewe, was purchased of J. Forbes, Jr., and used in 1877.

——o——

FLOCK 199.

Owned by J. A. WRIGHT, MIDDLEBURY, ADDISON CO., VT.

8 RAMS.

Marked with metallic labels in their ears, having thereon J. A. Wright and numbers 83 to 90.

108 EWES.

Eighty-four labeled same as rams with numbers 1 to 47, 52 to 82, 91 to 96, and twenty-four having on their labels W. & J. and numbers 12, 26, 38, 86 to 106.

PEDIGREE.

Descended from importations from Spain through the flocks of A. Cock, Long Island; D. Humphreys, S. Atwood, Conn.; Wm. Jarvis, Weathersfield; Messrs. Rich, T. Stickney, L. Beedle, E. R. Robinson, L. C. Remele, Lucius Robinson, E. A. Birchard, A. Clark, Shoreham; Messrs. Hammond, W. R. Remele, Middlebury; N. A. Saxton, O. C. Bacon, J. H. Sprague, Wright & Jackman, J. D. & F. D. Barton, Waltham; H. Thorp, C. B. Cook, Charlotte; Baker & Smith, P. Elitharp, B. Myrick, R. Hemenway, H. S. Cross, Bridport; R. P. Hall, E. S. Stowell, S. S. Rockwell, W. McCauley, S. Benton, Cornwall; J. Q. Stickney, A. Baldwin, Whiting; A. Lawrence, J. Marsh, Hinesburgh; L. D. Gregory, V. Wright, Weybridge; A. A. Farnsworth, Brooksville; W. R. Sanford, Orwell; E. S. Rowley, E. Meach, Shelburn; S. Root, L. P. Clark, Addison.

EWES PURCHASED.

Aug. 12, 1876, thirty Robinson ewes were purchased of Alonzo Baldwin; Aug. 20, 1877, twenty-four Atwood and Robinson ewes were purchased of Wright & Jackman descended from the stock they purchased of J. H. Sprague and F. D. Barton, and not that purchased by them of S. Dodge & Son; in the fall of 1877, twenty-two Atwood ewes were purchased of the heirs of Joseph Marsh, bred by him from stock he purchased of S. Atwood, Messrs. Hammond and Almon Lawrence; Sept. 25, 1878, seventeen Atwood and Robinson ewes were purchased of J. Rice, of Bridport, that were bred by the late R. Hemenway from Robinson ewes he purchased of B. Myrick, using the stock rams of L. P. Clark and the Birchard & Tottingham Ram (202); Oct. 3d, 1878, eighteen ewes were purchased of H. S. Cross; they were bred by him from ewes he purchased of R. P. Hall and W. McCauley, certified by McCauley to be from the ewes he purchased of S. S. Rockwell.

RAMS USED.

In 1876, Wrinkly (292) was used to the Baldwin ewes; the rams 84 to 90 and ewes 91 to 96 were sired by him; in 1877, J. T. & V. R.'s 301 (470) was used to a part of the Marsh ewes, and the remainder to a ram bred by T. Stickney & Son; ram 83 was purchased of S. Benton.

FLOCK 200.

Owned by MRS. VICTOR WRIGHT, P. O. MIDDLEBURY, ADDISON Co., VT.

4 RAMS.

Marked with metallic labels in their ears, having thereon V. Wright and numbers 23, 24, 25, 27.

23 EWES.

Labeled same as rams with numbers 1 to 22, 26.

PEDIGREE.

Descended from the importation from Spain by Col. David Humphreys through the flocks of Stephen Atwood, Conn.; Messrs. Hammond, W. R. Remele, O. Severance, Middlebury; W. R. Sanford, Orwell; R. P. Hall, E. S. Stowell, Cornwall; N. A. Saxton, Waltham; L. P. Clark, Addison; C. R. Jones, Hubbardton.

HISTORY.

In 1847, Victor Wright purchased of L. C. Remele one ewe that Mr. Remele purchased of R. P. Hall; this ewe was bred by S. Atwood and purchased of him by Messrs. Hammond and Hall; in subsequent years Mr. Wright made several purchases of ewes of E. Hammond and H. W. Hammond, until they amounted in all to nearly eighty ewes; these he bred with great care until the time of his death in 1867.

RAMS USED.

Nearly all the stock rams of Messrs. Hammond were used more or less; among those bred in the flock Wright's California (43), Old Greasy (74), Black Top (90), Long Wool (112), Don Pedro (119), V. Wright's Wrinkly (173), Vanderbilt (261), Black Diamond (347) and Old Black (445) have all been celebrated stock rams and all, it is believed were used in the flock. Since the the death of Mr. Wright a ram belonging to Samuel James, bred by C. R. Jones, Hubbardton, sired by Bull Dog (104), dam an ewe purchased of G. Hammond, has been used to some extent; in 1874, Phil Sheridan (172) was used, and in 1875 Pat Henry (183) was used; in 1876 and '77, a ram bred in the flock, sired by the Samuel James ram, was used.

————o————

NOTE.—The marks and numbers of flocks 201 and 202 were not all returned to the Secretary and transferred on the records in time to have them inserted in their regular alphabetical order.

43

FLOCK 201.

Owned by J. H. CLOSE, St. Clairsville, Belmont Co., Ohio.

6 RAMS.

Four marked with metallic labels in their ears, having thereon L. P. Clark and numbers 123, 124, 134, 152, and two J. H. Close and numbers 167, 191.

113 EWES.

Labeled same as rams, ten with their labels marked J. H. Close and numbers 60, 64, 67, 71, 104, 105, 107, 110, 165, 190; fifteen marked J. W. Peck and numbers 18 to 32; five marked A. J. Towner and numbers 15 to 19; nine marked E. Townsend and numbers 105, 111, 112, 115, 120, 121, 155, 156, 157; twelve marked L. P. Clark and numbers 27, 35, 44, 48, 65, 69, 72, 74, 76 to 79; two marked L. Clark, Jr., and numbers 32, 35; three marked R. N. & O. F. A. and numbers 34, 36, 38; ten marked Mrs. M. W. M. and numbers 51 to 60; twenty marked W. H. Delong and numbers 51 to 54, 56 to 71; four marked R. J. Jones and numbers 8, 9, 58, 59; twenty marked S. G. H. & Son and numbers 51 to 70.

PEDIGREE.

Descended from importations from Spain through the flocks of D. Humphreys, S. Atwood, Conn.; A. Cock, Long Island; D. Buffum, Wm. Bailey, J. I. Bailey, Rhode Island; Wm Jarvis, Weathersfield; Messrs. Hammond, W. R. Remele, J. Piper, S. Piper, Middlebury; Messrs. Rich, L. Beedle, L. C. Remele, E. R. Robinson, D. E. Robinson, E. A. Birchard, T. Stickney & Son, A. J. Towner, G. D. Bryant, R. N. & O. F. Atwood, J. M. Ormsbee, D. & G. Cutting, J. T. Stickney, E. G. Farnham, Shoreham; N. A. Saxton, Waltham; S. G. Holyoke & Son, St. Albans; R. P. Hall, F. H. Dean, E. S. Stowell, W. H. Delong, J. W. Peck, Mrs. Mary W. Mead, Cornwall; W. R. Sanford, Orwell; L. P. Clark, Lyman Clark, Jr., Addison; Ward M. Lincoln, Brandon; S T. Baker. A. F. Ellsworth, J. E. Parker, Whiting; C. I. Benedict, Arlington; E. Townsend, Pavilion Centre, N. Y.; I. Moore, Woodstock; L. S. Drew, Burlington; C. B. Cook, H. Thorp, Charlotte; V. Wright, Weybridge.

EWES PURCHASED.

In 1877, seventy-three ewes were purchased of E. W. Eells; fifteen of these were from the flock of J. W. Peck, five from that of A. J. Towner; three from that of R. N. & O. F. Atwood; ten from that of Mrs. Mary W. Mead ; twenty from that of W. H. Delong ; and twenty in 1878 from that of S. G. Holyoke & Son ; the same year eight ewes were purchased of L. P. Clark that were bred by him, and nineteen of E. C. Eells, nine of which were bred by E. Townsend, four by R. J. Jones, four by L. P. Clark, and two by L. Clark, Jr.

RAMS USED.

In 1875, ram L. P. Clark's 123 (485) was purchased and is now owned in company with Thos. Healea ; in 1875, Pat Henry (183) was used on ewes purchased of L. P. Clark in 1875 and the ewes marked on labels J. H. Close were sired by him ; in 1877, L. P. Clark's 184 (487) was purchased and he is now owned in company with Thos. McLary, West Alexander, Pa.; in 1878, L. P. Clark's 124 (482) was purchased ; in 1878, a half interest was purchased in L. P. Clark's 152 (489.)

—o—

FLOCK 202.

Owned by THOMAS HEALEA & SON, UHRICHVILLE, TUSCARAWAS Co., OHIO.

2 RAMS.

Marked with metallic labels in their ears, having thereon L. P. Clark and numbers 123, 152.

20 EWES.

Labeled same as rams, three having on their labels L. P. Clark and numbers 22, 42, 46 ; two having on their labels L. Clark, Jr., and numbers 34, 36 ; five having on their labels R. J. Jones and numbers 3, 10, 27, 49, 69 ; and ten having on their labels J. H. Close and numbers 60, 64, 67, 71, 104, 105, 107, 110, 165, 190.

PEDIGREE.

Descended from the importation from Spain by Col. David Humphreys through the flocks of S. Atwood, Conn.; Messrs. Hammond, W. R. Remele, Middlebury ; W. R. Sanford, Orwell ; R. P. Hall, E. S. Stowell, R. J. Jones, Cornwall ; N. A. Saxton, Waltham ; L. P. Clark, Lyman Clark, Jr., Addison ; J. H. Close, St. Clairsville, Ohio.

EWES PURCHASED.

In 1877, ten ewes were purchased of E. C. Eells, three of which were bred by L. P. Clark, five by R. J. Jones, and two by L. Clark,

Jr.; ten ewes were also purchased of J. H. Close bred by Mr. Close from ewes bred by L. P.Clark ; they were sired by Pat Henry (183).

RAMS USED.

In 1876, in company with J. M. Holmes, L. P. Clark's 152 (489) was purchased of E. C. Eells; J. H. Close has since become owner of J. M. Holmes's interest; in 1878, a half interest in L. P. Clark's 123 (485) was purchased of J. H. Close.

SALES AND TRANSFERS.

The following are the reports of sales received by the Secretary and the transfers made upon the records since the several flocks were registered. Also the deaths reported :—

From Flock 1, MELVIN R. ATWOOD.

    To H. S. Brookins, Richville, Vt., Rams Nos. 9, 40, 41, 43, 44, 45, 46.
    " J. T. Stickney, Shoreham, Vt., Rams 1, 3, 4, 5. Ewes 10 to 14, 25. Sold by J. T. S. and transferred to Cherbino & Williamson, Middlebury, Vt., Rams 1, 3, 4.
    " C. K. Williams and F. G. Wright, Whiting, Vt., Ram 8.
    " E. Peck & Sons, Geneva, Ill., per E. N. B., Ram 38.
    " C. W. Mason, New Haven, Vt., Ram 50.
    " H. Williams, Hortonville, Vt., Rams 52, 53, 54. Ewes 24, 27, 55, 56.
    " A. C. Whitmore, Mukwonago, Wis., Ewes 10, 13.
    Sold by D. J. Wright for H. S. Brookins and transferred :
    " Olney Bros., Leonidas, Mich., 40, 41.
    " Southworth & Strong, Vicksburg, Mich., 43.
    " Wm. Sidler, Vicksburg. Mich., 46.
    " C. Cline, Nottaway, Mich., 44.
    " John Kellie, Fawn River, Mich., 9.

From Flock 2, R. N. and O. F. ATWOOD.

    To J. T. Stickney, Shoreham, Vt., Ram 43.
    " L. W. Wooster, West Cornwall, Vt., Ram 40.
    " E. G. Farnham, West Cornwall, Vt., Rams 44 to 47, 50 to 53.
    " George T. Dimmick, West Cornwall, Vt., Ram 41.
    " Peck & Jennings, West Cornwall, Vt., Ewes 34, 36, 38.
    " G. D. Miner, Middlebury, Vt., Ewes 35, 37, 54, 55, 57, 61, 64 to 68, 70, 71. The following have been sold by G. D. M. and transferred :
    " C. M. Inskeep, West Mansfield, O., 35, 54, 65, 66, 68, 71.
    " Jacob Killer, same P. O. address, 37, 61.
    " John H. Skidmore, same P. O. address, 70.
    " L. C. Dickinson and J. C. Smith, North Greenfield, O., 55, 57, 64.

From Flock 4, C. E. ABELL.

    To W. H. Root, Orwell, Ewes 31 to 50. There should be added to this flock, 1 Ram marked on Label C. E. Abell, 51.

From Flock 6, E. N. BISSELL.

> To A. Bagley, Salem, Wis., Ram 135.
> " Daniel Hunt, Bristol, Wis., 109, 139.
> " J. J. Clapp, Kenosha, Wis., Ram 123, 217.
> " J. J. Clapp, Kenosha, Wis., Ewes 41, 49, 51, 69.
> " D. C. Stewart. Salem, Wis., Ram 122.
> " Humbert Bros., Caldwell Prairie, Ill., ½ Ram 100, Hibbard (214).
> " M. Merchant, Alden, Ill., Ram 111.
> " H. Sullivan, Alden, Ill., Ram 119.
> " Sullivan & Fardon, Alden, Ill., Ram 131.
> " M. O. Brime, Alden, Ill., Ram 134.
> " Geo. Wakely, Harvard, Ill., Ram 138.
> " James Densmore, Markeson, Wis., Rams 103, 104, 115, 116, 117, 120, 124, 126, 142, 145, 146, 161, 184, 185, 190, 192, 204, 209, 214, and Ewes 163, 176, 179.
> " B. Thorp, Beaver Dam, Wis,, Ram 121.
> " Henry Marsh, Beaver Dam, Wis., Ram 140.
> " A. F. Burgess, Beaver Dam, Wis., Ram 136.
> " Samuel Messenger, Niles, Mich., Rams 101, 110, 114, 130.
> " G. B. Van Alta, Okemas, Mich., Ram 112.
> " Patrick Kearney, Webster, Mich., Ram 127.
> " M. Baldwin, Chelsea. Mich., Ram 107.
> " S. Weaver, Sylvan, Mich., Ram 148.
> " L. N. Tyler, Mosherville, Mich , Rams 128, 129, 141, 147.
> " Uriah Wood, Brandon, Wis., Ram 216.
> " Uriah Wood, Brandon, Wis., Ewes 160, 165, 167, 192.
> " E. Peck & Sons, Geneva, Ill., Ram 102.
> " J. Holden, Hart Prairie, Wis., Ram 108.
> " Clarence Wilber, Augusta, Kansas, Rams 118, 125, 144.
> " S. W. Andrews, Juno, Wis., Ram 180.
> " J. Sanderson, Cambria. Wis., Ram 206.
> " T. Thomas, Cambria, Wis., Ram 187.
> " J. Tanner, Cambria, Wis., Ram 205.
> " L. D. Woodworth, Pleasant Prairie, Wis., Ram 213.
> "   "   "   "   " Ewes 36, 42, 65, 72, 74, 156.
> " O. B. Knapp, Brandon, Wis., Ewes 157, 159, 168, 172, 177.
> " Geo. W. Hunt, Greenwood, Ill., Ewes 34, 35, 54, 38, 60, 66.
> " Geo. W. Everhart, Kenosha, Wis., Ewes 30 to 33, 40, 44 to 48, 50, 52, 53, 55 to 59, 61 to 63, 67, 68, 70, 71, 73, 75, 76, 88, 89, 150, 154, 155.

Of those marked J. T. & V. R. have been sold and transferred :

> To O. Cook, Whitewater, Wis , Ram 314.
> " Gregory & Hayward, Beloit, Wis, Ram 336.
> " N. S. Colby, McHenry, Ill., Ram 329.
> " E. S. Durkee, Alden, Ill., Ram 313.
> " Wm. G. Morris, Paris Center, Mich., Ram 330.
> " R. B. Curness, St. Johns, Mich., Ram 334.

To Wm. Freeman, Grass Lake, Mich., Ram 332.
" T. Robinson, Grass Lake, Mich., Ram 316.
" M. Schenks, Francisco, Mich., Ram 351.
" J. W. Clark, Pulaski, Mich., Ram 353.
" L. N. Tyler, Mosherville, Mich., Rams 340, 341, 354.
" F. H. Farrington, Brandon, Vt., Ewes, 98, 100, 109, 110, 112, 113, 116.
" J. J. Clapp, Kenosha, Wis., Ewes 157, 159, 168, 172, 177.

From Flock 7, H. C. BURWELL.

To Philips, Denver, Col., Ram 83.
" Cherbino & Williamson, Middlebury, Vt., Rams 73, 74, 75, 84 to 91. Ewes 104, 105, 109, 111, 112, 117, 118, 120, 121, 123, 124, 127, 128.
" C. P. Crane, Bridport Vt., Rams 64 to 72.
" Geo. Dimmick, by C P C, Rams 65 to 72.
" Judd Day, Addison, Ewes 7, 10, 11, 13, 23, 25, 27.
" F. S. Higbee. Ohio, Ewes 4, 40, 41, 43, 44, 53, 54, 56 to 60, 62, 63, 93, 95, 96.

From Flock 9, H MERL BOTTUM.

To Johnson, Mattison, Ram 80.
" J. M. Aldrich, Ram 92.
" W. N. Upham, Ram 93.
" Jerome Hall, Ram 94.
" Barney Dolan, Shaftsbury, Vt., Ewes 2, 7, 13, 14, 25, 28, 56, 149, 150.

From Flock 10, ALONZO BALDWIN & SON.

To J. D. Patterson, Western, N. Y., of those marked E. A. Baldwin, Ewes 33 to 36, 50.

From Flock 12, M. K. BARBOUR.

To C. W. Mason, New Haven, Vt., Rams 50, 66, 67.
" P. Bingham, for other parties, Rams 44, 46 to 48, 51 to 55.
" A. A. Fletcher, for Hemenway estate, Ram 49.
Died, Ram 45.

From Flock 16, L. S. BURWELL.

To Geo. Dimmick, West Cornwall, Vt., Rams 20, 21, 23 to 31.
Died, Ewe 19.
Since the Register of the flock was printed, at page 184, there have been added—bred from Registered Rams and Ewes— Rams 41 to 45 ; Ewes 32 to 40.

From Flock 17, H. A. BASCOM.

To Hall & Holden, Middlebury, Vt., Rams 4, 6, 7, 8, 13.
" E. N. Bissell, East Shoreham, Vt., Ram 1.
" " " " " " Ewes 16 to 21, 23, 51,
" Cherbino & Williamson, Middlebury. Vt., Ram 3.

To Jennings & Peck, W. Cornwall, Rams, 12, 14.
    "   "   "   "   Ewes 18, 28, 29, 31.
" John Hutchins, Macksburgh, Ohio, Ewes, through C. & W.,
    28, 29, 31.
" O. B. Knapp, Brandon, Wis. Ewes through E. N.B. 20, 21, 51.
" Uriah Wood, Brandon, Wis. Ewes through E. N. B. 16, 17.
    18, 19, 23.   Died, Rams 2, 9, 10, 11 ; Ewes 46, 47.

From Flock 18, H. S. BROOKINS.

To J. A. Wright, Middlebury, Vt., Ram 32.
" C. M. Fellows, Manchester, Mich., Ram. 43.
" L. B. Auterdale, Centerville,   "   " 48.
" O. Meredith, Kalamazoo,   "   " 44.
" S. Strong, Vicksburg,   "   " 47.
    Died, 43.

From Flock 19, J. G. BARKER.

To E. W. Eells, for J. H. Close, St. Clairsville, Ohio, Rams 35
    to 43, 50.
" Farnham & Birchard, Shoreham, Vt., Ewes 24 to 31.
J. H. Close has sold the following, and they have been trans-
    ferred on the Records :
To Math. Moore, Ozark, Ohio, Ram 36.
" W. G. Wright, Bealsville, Ohio, Ram 41.

From Flock 20, H. C. BROWN.

To E. N. Bissell, East Shoreham, Vt., Rams 1, 2, 11 to 18 ;
    Ewes 42, 44, 45, 47, 51, 54, 55.   Of these E. N. B. has
    sold and transferred :
To G. V. Weeks, Lyons,   Wis. Ram 12.
" Wood & Knapp, Brandon,   "   13.
" Charles Tuttle, Clinton Junction,   "   14.
" S. T. Webb, Spring Prairie,   "   18.
" J. Marris, Randolph,   "   17.
" David Carson, Cambria,   "   16.
" J. Densmore, Markeson,   "   15.
" Uriah Wood, Brandon, Wis. Ewes 47, 54, 55.
" O. B. Knapp,   "   "   42, 44, 45, 57.
H. C. Brown, has sold
To D. J. Wright, Richville, Ram 4.
" E. G. Farnham, W. Cornwall, Vt., Rams 5 to 8, 10, 25 to
    27.
" Tyylor & Dimmick, W. Cornwall, Vt., Ewes, 30 to 34, 37
    to 39.
" E. L. Needham, Whiting, Vt., Ewes 40, 59, 72, 76, 78.
Sold by D. J. Wright and transferred :
To I. M. Necsmith & Chas. S. Colby, Vicksburg, Mich, Ram 4.

From Flock 21, ELMER BARNUM.

To L. E. Moore, Shoreham, Rams 1, 2.

From Flock 22, JOHNSON S. BENEDICT.

    To Town of Castleton, Vt., Ram 113.
      " A. G. Richardson, Beeville, Bee Co., Texas, Rams 101, 103
        to 112, 114, 115 ; Ewes 5, 9, 11, 13, 18, 19, 21 to 26, 28,
        29, 37, 41 to 44. Ram 116, died.

From Flock 23, D. J. BROWN.

    To J. W. Peck, W. Cornwall, Vt., Ewe 1.
    " H. C. Brown, Whiting, Ewe 2.

From Flock 24, WILLIAM BALL.

    To J. Bamber, Highland, Mich., Ewes 10, 19, 58.
    " N. C. Branch, Williamson, Mich., Ewes 78, 79, 82, 83, 86.
    " Wm. Gage, Novi, Mich., Ewes 46, 48, 51, 57, 65, 67, 71 to
      73, 75.
    " L. Price, Bath, Mich., Ewes 62, 63, 70, 74, 76, 100, 125.
    " T. M. Southworth, Allen, Mich., Ewes 38, 39, 40.
    " E. Merithew, Howell, Mich., Ewes 8, 9, 77, 80, 81, 84, 85,
      88, 89, 90 to 99.
    " G. Lax, Howell, Mich., Ewe 47.

From Flock 27, THURMAN BROOKINS.

    To E. N. Bissell, E. Shoreham, Vt., Rams 1 to 12.
    " H. S. Brookins, E. Shoreham, Vt., Rams 13, 114 to 124.
    " J. A. Wright, Middlebury, Vt., Ewes 14 to 28, 126 to 130.
    J. A. W. has sold and transferred of these :
    To C. M. Inskeep, W. Mansfield, O., Ewe 127.
    " Jacob Killer, W. Mansfield, O., Ewe 24.
    " John H. Skidmore, W. Mansfield, O., Ewes 18, 21.
    " L. C. Dickinson and J. C. Smith, North Greenfield, O., Ewe
      15.
    T. Brookins has sold and transferred :
    To Geo. R. Waite, Shoreham, Vt., Ewes 29, 74, 103, 107, 111.
    Sold by H. S. B. and transferred :
    To Olney Bros., Leonidas, Mich., Rams 118, 124.
    " Henry Carey, Otsego, Mich., Ram 116.
    " O'Brien Bros., Mendon, Mich., Ram 121.
    " Wallace & Lockwood, Burr Oak, Mich., Ram 122.
    " J. B. Akey, Burr Oak, Mich., Ram 114.
    " Max Smalley, Vicksburgh, Mich., Ram 115.
    " A. Judson, Vicksburgh, Mich., Ram 120.
    " A. B. Chappell, Kalamazoo, Mich., Ram 119.

From Flock 33, MERRILL BINGHAM.

    To Adam Dehil, Milford, Mich., Ram 10.
    " C. W. Mason, N. Haven, Vt., Rams 2, 13.
    " C. D. Lane, Cornwall, Vt., Ram 30.
    " Chas. B. McClure, Rams 27, 28, 29, 31.
    " J. F. Randall, W. Cornwall, Vt., Ewes 163, 164, 167, 181,
      184, 185, 190, 191, 193, 196, 205, 206, 210, 211, 213, 214,
      215, 223, 226, 227.

44

Sold by J. F. R. and transferred :
To L. M. Bingham, W. Cornwall, Vt., Ewe 164.

From Flock 36, J. E. CASWELL.

To L. E. Moore, Shoreham, Vt., Ram 39 ; Ewes 36, 37, 38.
" H E. Taylor, West Cornwall, Rams 40 to 44.
" Wm. Ball, Hamburgh, Mich., Rams 22 to 24, 46, 48.
"   "   "   "   Ewes 1 to 12, 27 to 35, 51, 52.
" H. C. Caswell, Shoreham, Vt., Rams 25, 45, 47.
"   "   "   "   "   Ewe 49.
" V. Forbes, Shoreham, Vt., Ewes 13 to 21, 50.
Died, Ram 26.
Sold by Wm. Ball and transferred :
To L. Price, Bath, Mich., Ewes 31 to 34.
Sold by H. C. Caswell and transferred :
To C. Hinebaugh, Burr Oak, Mich., Ram 25.

From Flock 39, CHERBINO & WILLIAMSON.

Since the register of this flock was printed there has been added
to it twenty-one Atwood and Robinson ewes, purchased of H. W.
Jones ; before being recorded they were labeled C. & W. and numbers 549 to 569.

Sales and Transfers .

To W. R. Wickers, Washington, Co., Ohio, Ram 527.
"   "   "   "   "   "   Ewes 296, 361, 552,
    554 to 556, 569.
" Wagner and Shipley, McConnellsville, Ohio, Ewes 395,
    410 to 412, 414 to 423.
" H. Robbins, Cornwall, Vt., Ewes 137, 138, 194, 200, 209,
    227 to 229, 231, 245 to 248, 250, 253, 258, 461, 463, 468,
    480, 494, 497, 502, 504, 509, 516, 518, 519, 522.
" Jacob Clarey, Cumberland, Ohio, Ewes 327, 331, 349.

From Flock 40, ALBERT CHAPMAN.

To Cherbino & Williamson, Middlebury, Vt., Ram 3.
" E. B. Mussey, Middlebury, Vt., Ewes 1, 2, 4.
" E. Townsend, Pavilion Centre, N. Y., Ewes 8, 15, 16.
Dead, Ewe 6, 4.

From Flock 42, C. F. CHURCH.

To Taylor & Dimmick, W. Cornwall, Vt., Ewes 104.
" G. D. Bush, Orwell, Vt., Ewes 1 to 7.

From Flock 43, JOHN A. CHILDS.

To C. D. Lane, Cornwall, Vt., Ewes 34 to 50.

From Flock 44, LUMAN P. CLARK.

To E. C. Eells, Middlebury, Vt , Rams 103, 108, 111, 113, 114,
    137, 138, 140, 152, 153; Ewes 22, 42, 44, 46, 48, 55, 56,
    67, 72.

To J. H. Close, St. Clairsville, O., Rams 123, 124 ; Ewes 27, 35,
    69, 74, 76, 77, 78, 79.
'' Merrill Bingham, West Cornwall, Vt., half ram 125.
Sold by E. C. Eells and transferred :
To T. J. Close, St. Clairsville, O., Ram 103.
'' J. M. Holmes, Masterville, O., Ram 108.
'' U. Cahill, J. & B. Cahill, Wm. Hawks, and J. McMillian,
    Union Co., O., Ram 113.
'' M. G. Markham and E. Townsend, Western, N. Y., Ram 114.
'' John Allen, Bridgeport, O., Ram 111.
'' John Moore, Short Creek, O., Ram 137.
'' J. H. Close, St. Clairsville, O., Ram 140.
'' Thos. Healea & Son and J. H. Close, St. Clairsville, O.,
    Ram 152.
'' J. M. Holmes, Masterville, O., Ewes 55, 56.
'' J. Healea & Son., Urichville, O., Ewes 22, 42, 46.
'' J. H. Close, St. Clairsville, O., Ewes 44, 48, 65, 72.
Sold by J. H. Close and transferred :
To G. W. Henry, Unionport, O., Ewe 27.
L. P. Clark has sold :
To E. Townsend, Pavilion Centre, N. Y., one-half of Ram 105,
    also one-half of Ewes 3, 4, 5, 9, 12, 17, and Ewes 16. 58, 63,
    70, 71.
'' C. A. Fuller, Fairfield Centre, Me., Rams 109, 122 ; Ewe 28.

From Flock 45, CHARLES M. CLARK.

To E. C. Stewart, Rosendale, Wis., Ewes 17, 35, 61, 89, 158.
'' B. Dodd, Rosendale, Wis., Ewes 20, 199, 214, 225.
'' H. W. Goodale, Camp Baker, Meagher Co., Colorado, Ewes
    C. M. C. & C. 1, 2, 6, 7, 8, 10, 12, 13, 15, 18, 19, 22, 24 to 28,
    31, 32, 33, 37, 38, 40, 42 to 46, 47 to 51, 53, 55 to 58, 64, 67,
    68, 70, 71, 72 ; C. M. C. & B. 14, 67, 73, 75.
Died, 13, 52, 63, 87.
To the numbers of this flock should be added Ewe 193.

Flock 47, LYMAN CLARK, Jr.

To E. C. Eells, Middlebury, Vt., Ewes 32 to 36.
E. C. E. has sold and they have been transferred on the Records:
To J. H. Close, St. Clairsville, O., Ewes 22, 35.
'' J. M. Holmes, Ohio, Ewe 33.
'' Thos. Healea & Son, Urichville, O., Ewes 34, 36.

From Flock 48, ELISHA COOK.

To A. Munger, Leicester, Vt., Ram 35.
'' C. K. Williams, Whiting, Vt., Ram 41.
''   ''   ''   ''   Ewe 9.
'' C. F. Church, Whiting, Vt., Ram 46.
''   ''   ''   ''   Ewes 2, 4, 6, 10, 11, 14, 21, 28.
'' Geo. D. Bush, North Orwell, Vt., Rams 47 to 52.
''   ''   ''   ''.   ''   Ewes 1, 8, 12, 17 to 20,
    22, 24, 25, 30.

To E. G. Farnham, W. Cornwall, Vt., Rams 36 to 42.
" S. H. Rawson, Whiting, Vt., Ram 44.

From Flock 49, W. H. COOK & RALPH BROWN.

To C. W. Mason, New Haven, Vt., Rams 2, 3.

From Flock 52, O. COOK.

To N. A. Batchelder, Fort Collins, Col., Ram 97.
" Johnson Mattison, Darien, Wis., Ram 96.
" D. D. Chappell, Trempealeaw, Wis., Ram 91.
" R H. Hall, Milton Wis., Ram 93.
" George Billet, Cold Spring, Wis., Ram 87.
" Wm Lookesas, Whitewater, Wis., Ram 90.
" C. A. Sear, Garden Prairie, Ill., Ewes 65, 66, 78, 79, 82.

From Flock 53, REUBEN COOK.

To G. D. Miner, Middlebury, Vt., Ewes 72, 73, 76, 80, 85, 86,
102, 105.
The following have been sold by G. D. Miner and transferred :
To C. M. Inskeep, W. Mansfield, Ohio, Ewe 105.
" Jacob Killer, W. Mansfield, Ohio, Ewe 73.
" J. A. Killer, W. Mansfield, Ohio, Ewes 80, 85, 86.
" John W. Winer, W. Mansfield, Ohio, Ewes 72, 76.
" Dan'l Skidmore, W. Mansfield, Ohio, Ewe 102.

From Flock 55, B. W. COPE.

To R. Van Voorhis, Monongahela City, Pa., Rams 88, 89.
"   "   "   "   " Ewes 90, 94.
" J. S. & T. J. Close, St. Clairsville, Ohio, Ram 85.
" Sam'l Grubb, Mt. Pleasant, Ohio, Ram 86.
" Shep Ong, Smithfield, Ohio, Ram 87.
" C. H. Beall, Independence, Pa., Ram 101.

From Flock 58, C. P. CRANE.

To Wood & Kennedy, Mich., Rams 37 to 42. 44 to 46, 48 to 58.
" Geo. Dimmick, W. Cornwall, Vt., Ram 47 and those marked
H. C. B. 65 to 72.
" Cherbino & Williamson, Middlebury, Vt., Ewes 61 to 65, 67,
68, 70 to 87.
" J. A. Wright, Middlebury, Vt., Ewes 94, 97.
" G. D. Miner, Middlebury, Vt., Ewes 88, 105, 108, 119 to
122, 124 to 128, 130.
Dead, Ewe 66.

From Flock 59, J. J. CRANE.

To Sam'l Jewett, Independence, Mo., Rams 6 to 9 ; Ewes 33
to 51.
" H. W. Jones and J. Forbes, Jr., Shoreham, Vt., Ram 2.
" H. C. Burwell, Bridport, Vt., Ram 4 ; Ewe 7.
" Capt. Cuthill, Colorado, Rams 5, 12, 14, 17, 18, 19.
" C. D. & R. Lane, Cornwall, Vt., Rams 3, 15.

To G. D. Miner, Middlebury, Vt., Ram 16 ; Ewes 53, 56, 59, 64, 65, 66.
" C. C. Miner, Bridport, Vt., Ewe 8.
Dead, Ewes 1, 13, 19, 21.

From Flock 60, CHARLES E. CRANE.

To J. A. Wright, Middlebury, Vt., Ram 102.
" H. Jones and J. Forbes, Jr., Shoreham, Vt., Ram 103.
". J. Gillespie, Lawton, Mich, Ram 126.
" C. W. Mason, N. Haven, Vt., Rams 104, 121.
" Geo. Dimmick, W. Cornwall, Vt., Ram 112.
" D. C. Barber, Bridport, Vt., Ram 116.
" Watson Hamilton, Bridport, Vt., Ram 122.
Dead, Ewes 5, 48, 49, 52.

From Flock 61, ESTATE OF G. CUTTING.

To S. R. McFadden, Atlanta, Ill., Ewes 43, 47, 53.

From Flock 62, F. H. DEAN.

To C. W. Mason, New Haven, Vt., Rams 3 to 20.
" E. G. Farnham, W. Cornwall, Vt , Ram 2.
" " " " " Ewes 21 to 65.
" Cherbino & Williamson. Middlebury, Vt., Ewes 66 to 100.
" G. D. Miner, Middlebury, Vt., Ewes 101 to 125.
G. D. Miner has sold the following and they have been transferred on the Records :
To C. M. Inskeep, W. Mansfield, O., Ewes 101, 108, 110, 111, 113. 116, 117, 119, 122, 123.
" Jacob Killer, W. Mansfield, O., Ewes 102, 104, 109, 112, 121, 124.
" John N. Winer, W. Mansfield, O., Ewe 125.
" John H. Skidmore, W. Mansfield, O., Ewes 114, 115, 118.
" Dan'l Skidmore, W. Mansfield, O., Ewe 107.
" L. C. Dickinson and J. C. Smith, North Greenfield, O., Ewes 103, 105, 106, 120.

From Flock 63, W. H. DELONG.

To E. C. Eells, Middlebury, Vt., Ewes 51 to 54, 56 to 71.
E. C. Eells has sold the same :
To J. H. Close, St. Claireville, O.
Dead, Ewes 10, 55.

From Flock 64, H. F. DEAN.

Since the register of this flock has been printed, there has been added two ewes bred from the flock of F. H. Dean ; their numbers were 36, 37, and the same have been sold and transferred to E. C. Eells, Middlebury, Vt.

From Flock 65, DAVID F. DOTY.

> To Cherbino & Williamson, Middlebury, Vt., Rams 2 to 11.
> " J. A. Wright, Middlebury, Vt., Ewe 52.
> " G. D. Miner, Middlebury, Vt., Ewes 46 to 51, 53 to 57.
> Dead, Ewes 3, 5, 6.

From Flock 67, F. D. DOUGLAS.

> To E. D. Searl, Cornwall, Vt., Rams 4, 5, 6.

From Flock 70, A. F. ELLSWORTH.

> To C. W. Mason, N. Haven, Vt., Rams 58, 60 to 63.
> " C. W. Mason, N. Haven, Vt., Ewes 25 to 27, 69 to 83.
> . " W. H. Bingham, W. Cornwall, Vt., Ram 57.
> " H. E. Sanford, W. Cornwall, Vt., Rams 29, 42, 43, 44, 48,
> 49, 50, 52 to 55.
> Of these H. E. Sanford has sold and they have been transferred
> on the records :
> To D. H. Hill, Kalamazoo, Mich., Ram 29.
> " K. W. Barber, Alamo, Mich., Ram 55.
> " W. H. Little, Richland, Mich., Ram 46.
> " Mrs. C. Reed, Richland, Mich., Ram 53.
> " J. H. Woodman, Paw Paw, Mich., Ram 52.
> " A. Wildy, Paw Paw, Mich., Ram 50.
> A. F. Ellsworth has sold :
> To J. A. Wright, Middlebury, Vt., Ewes 67, 85.
> " R. J. Jones, West Cornwall, Vt., Ewes 35, 56, 84.

From Flock 71, MILO J. ELLSWORTH.

> To C. W. Mason, New Haven, Vt., Rams 3 to 27.
> " Capt. Cuthill, Colorado, Rams 68, 69, 71, 72, 74 to 77, 79,
> 81 to 84.

From Flock 72, FRED. H. FARRINGTON.

> To P. Bingham, Cornwall, Vt., Rams 40 to 43, 53 to 55, 59, 65
> to 68, 116.
> " C. A. Fay, Brandon, Vt., Ram 49.
> " Dea Woodcock, Pittsford, Vt., Ram 39.
> Dead, Ewes 14, 32.

From Flock 73, E. G. FARNHAM.

> To J. A. Wright, Middlebury, Vt., Rams 1 to 14.
> " Byron Farnham, Shoreham, Vt., Rams 15 to 22.
> " Farnham & Birchard, Shoreham, Vt., Rams 136 to 140.
> "   "   "   "   " Ewes 116 to 135.
> " E. H. Griswold, Brandon, Vt., Ewes 68 to 73.
> " Halliday & Farnham, W. Cornwall, Vt., Ewes 88, 89, 91, 94,
> 95, 96.
> " A. C. Fuller, Fairfield Centre, Me., Ewes 90, 92.
> " H. E. Taylor, W. Cornwall, Vt., Ewes 74 to 87.

From Flock 74, EDWIN FANNING.

To J. T. Stickney, Shoreham, Vt., Ewes 18 to 24.
By J. T. S.:
To A. C. Whitmore, Muchawongo, Wis., Ewes 20, 21, 22.

From Flock 75, A. A. FARNSWORTH.

To J. A. Wright, Middlebury, Vt., Ewes 25, 27, 28, 35, 36. 40, 41, 42, 44, 48 to 51, 54, 57, 58, 59, 61, 64, 67 to 71, 73, 74, 75.

From Flock 76, C. B. FISK.

The labels in the ears of this flock were by mistake marked C. B. Fish, and the numbers were reported to the secretary wrong; they should be seven rams 35 to 41; forty-three ewes 1 to 34, 42 to 50.

To R. I. Landon, Cornwall, Vt., Rams 35 to 41; Ewes 42 to 50.

From Flock 77, O. A. FIELD.

To Cherbino & Williamson, Middlebury, Vt., Rams 1, 76.
" L. M. Bingham, W. Cornwall, Vt. Ram 4.
" J. A. Wright, Middlebury, Vt., Ewes 19, 23 to 31.
" H. E. Taylor, W. Cornwall, Vt., Ewes 17, 18, 20, 21, 22.
" L. M. Bingham, W. Cornwall, Vt , Ewe 71.
"   "   "   "   "   Ram 78.

From Flock 78, JAMES FORBES, Jr.

To E. N. Bissell and L. E Moore, Shoreham, Vt., Rams 63 to 75; Ewes 1, 3, 4, 5, 8, 10, 13, 14.
" Farnham & Birchard, Shoreham, Vt., Rams 42 to 46.
" J. A. Wright, Middlebury, Vt., Ewes 2, 6, 7, 9, 11, 12, 14, 16, 17.
" Cherbino & Williamson, Middlebury, Vt., Ewes 56, 57, 61, 64, 83, 95, 97 to 100, 102, 104, 106 to 109, 111 to 114.
The same have been sold by C. & W. and transferred to Jacob Clary, Cumberland, O.
To V. N. Forbes, West Haven, Vt., Ewes 59, 101, 115 to 117.

From Flock 79. A. E. FULLER.

To Chas. Greenwood, Anson, Me., Rams 44, 71, 73.
" Geo. Hill, Hanover, N. H., Ram 70.
" H. B. Hough, Lebanon, N. H., Ram 72.
" E. & G. Robinson, Ostrander, O., Ram 69.
" W. E. Perkins, Pomfret, Vt., Ewes 18, 27, 29, 30, 31, 45, 66.
" W. B. Kingsley, Ewe 5.

From Flock 80, RECTOR GAGE.

To E. Townsend, Pavilion Centre, N. Y., Rams 78, 79; Ewes 10, 23, 52.
" E. C. Eells, Middlebury, Vt., Rams 24, 25, 74 to 77.

To J. E. Gilmore, Pavilion, N. Y., Ewes 15, 16, 19, 20, 34 to
    47, 50, 53.
" Guy Gage, Addison, Vt., Ewes 64, 65.

From Flock 81, HENRY GIDDINGS.

To Giddings & Ogilvie, Hartford, O., Ewes 1 to 22.

From Flock 85, DARWIN E. GROVENOR.

To Geo. Dimmick, W. Cornwall, Vt., Rams 1, 2.

From Flock 86, HENRY HARRINGTON.

To Juan Robinson, Keeler's Bay, Vt., Ram 54 ; Ewes 11, 12,
    13, 15, 19, 22, 24, 25, 50, 53.
" B. H. Brown, Alburgh, Vt., Ram 56.
" A. A. Ballard, Georgia, Vt., Ram 59.
" J. S. Landon, S. Hero, Vt., Ram 58.
" L. S. Drew, Burlington, Vt., Ewes 23, 35, 37, 39, 42, 47.
Dead 21.

From Flock 87, JOHN ORIN HAMILTON

To Cyrus H. Smith, Bridport, Vt., Ram 93.
" L. J. & J. A. Wright, Middlebury, Vt., Ram 51.
" D. E. Grovenor, Bridport, Vt., Rams 71, 73, 75, 91.
Sold by D. E. G. and transferred on records :
To Farnham & Birchard, Shoreham, Vt., Ram 71.
" Jonas Rice, Bridport, Vt., Ram 73.
Sold by J. O. Hamilton :
To E. C. Eells, Middlebury, Vt., Rams 55, 57, 59, 61, 63, 65,
    67, 69, 77, 79, 81, 85, 87, 89.
Sold by E. C. E. and transferred on Records :
To John Kirkbride, Bealsville, O., Ram 61.
" T. J. Williamson, Ozark, O., Ram 85.
" R. R. Driggs, Ozark, O., Ram 77.
J. O. H. has sold :
To Wilbur Hamilton, Bridport, Vt., Ewes 1 to 5, 10, 12, 24,
    27, 37.
" Cherbino & Williamson, Middlebury, Vt., Ewes 6, 26, 44 to
    50, 52, 54, 56, 58, 62, 64.

From Flock 88, MRS. C. N. HAYWARD.

To J. A. Wright, Middlebury, Vt., Ram 77.
" Milo B. Williamson, Middlebury, Vt., Ram 15.
" H. C. Burwell, Bridport, Vt., Ewes 12, 64 to 67, 69 to 71,
    73, 74.
Dead, Ewe 3.

From Flock 89, J. B. HAMBLIN.

To Cherbino & Williamson, Middlebury, Vt., Rams 2 to 13.
Sold by C. & W. and transferred :
To Jacob Clary, Cumberland, O., Ram 12.

Sold by J. B. Hamblin :

To E. C. Eells, Middlebury, Vt., Rams 1, 81 to 86, 88, 89, 90.
" G. D. Miner, Middlebury, Vt., Ewes 47, 48, 52, 55, 57 to 62,
    69, 70, 74, 76, 77, 79.
Dead, Ewe 64.

From Flock 90, O. & E. S. HALL.

To D. C. Stearns, Royalton, Vt., Ram 43.
" James Farnum, Tunbridge, Vt., Ram 92.
" J. H. & A. W. Peters, Bradford, Vt., Rams 42, 90 ; Ewes
    7, 75.
" E. S. Fulsom, Tunbridge, Vt., Ram 44 ; Ewe 103.
" A. Chapman, Middlebury, Vt., Ewe 28.
Dead, Rams 33, 124.

From Flock 93, E. J. & E. W. HARDY.

To A. W. Baker, Highland, Mich., Ewes 115 to 129.
" W. R. Hedder, Clyde, Mich., Ewes 178, 185. 186, 188, 192
    to 196, 198, 199.

From Flock 96, A. HITCHCOCK & SON.

To E. G. Farnham, W. Cornwall, Vt., Rams 1 to 5.
Sold by E. G. F. and transferred :
To Farnham & Birchard, Shoreham, Vt., Ram 1.

From Flock 97, H. G. HIBBARD.

To E. G. Farnham, W. Cornwall, Vt., Rams 1 to 5.
" A. C. Martin, Benson, Vt., Rams 3, 7.
" H. S. Brookins and D. J. Wright, Richville, Vt., Rams 4,
    5, 6.
Sold by H. S. B. & D. J. W. and transferred :
To Peter Butler, Sturgis, Mich., Ram 4.
" A. Harvey, Mendon, Mich., Ram 6.
" S. Strong, Vicksburgh, Mich., Ram 5.

From Flock 98, E. D. HINDS.

To C. H. Pinckney, Pittsford, Vt., Rams 51 to 53, 57 to 63.
" F. Hooker, Cornwall, Vt., Ram 54.
" C. E. Bush, Shoreham, Vt., Ewes 64 to 76.

From Flock 99, FRANKLIN HOOKER.

To Hall & Kingsbury, Walpole, N. H., Rams 42. 53.
" C. W. Mason, N. Haven, Vt., Rams 40, 41, 43 to 52, 38.
Sold by C. W. M. and transferred :
To J. A. Wright, Middlebury, Vt., Ram 38.
F. Hooker has sold :
To J. A. Wright, Middlebury, Vt., Ewes 27, 28.
" J. M. Norton, E. Bloomfield, N. Y., Ewe 72.
Sold by J. A. Wright and transferred :
To W. E. Bosman, Blue Rock, Ohio, Ewes 27, 28.

45

From Flock 100, S. G. HOLYOKE & SON.

To E. W. Eells, Middlebury, Vt., Rams 85 to 98; Ewes 51 to 70.
Sold by E. W. E. and transferred :
To J. H. Close, St. Clairsville, O., Ewes 51 to 70.

From Flock 101, JEROME HOLDEN.

To Mills & Condit, Condit, Ohio, Rams 97 to 100; Ewes 51 to 95.

From Flock 103, FRANK S. HOLLEY.

To J. A. Wright, Middlebury, Vt., Ewes 26, 34, 36, 38, 42, 43, 46, 47, 50, 51, 55.

From Flock 104, A. H. HUBBARD.

To R. J. Jones, W. Cornwall, Vt., Ewes 34, 36 to 39.
" L. C. Smith, Whitewater, Wis., Ram 51.
" J. E. Smith, Ypsilanti, Mich., Ram 52.
" J. Q. Smith, W. Rutland, Vt , Ram 53.
" C. W. Mason, N. Haven, Vt., Rams 55, 57 to 59.
" E. A. Baldwin, Whiting, Vt., Ewes 33, 35, 40 to 50.

From Flock 106, CURTIS H. JAMES.

To James Murdock, Crown Point, N. Y., Ram 76.
" H. C. Burwell, Bridport, Vt., Ram 77.

From Flock 107, SAMUEL JEWETT.

To T. C. Lippitt, Monmouth, Ill., Ewe 101.
" W. M. Gentry, Sedalia, Mo., Ewe 191.

From Flock 109, C. H. KETCHUM.

To G. D. Miner, Middlebury, Vt., Rams 52, 54 to 58.
Sold by G. D. M. and transferred :
To C. M. Inskeep, W. Mansfield, O., Ram 52.
" L. C. Dickinson and J. C. Smith, North Greenfield, O., Ram 55.

From Flock 113, C. D. LANE.

To C. P. Crane, Bridport, Vt., Ewes 1, 2, 4, 13, 14, 20, 24, 45, 57, 69, 71, 76, 164, 170, 262.
" B. W. Crane, Bridport, Vt., Ewes 12, 15, 21, 23, 26, 28, 31, 34, 37, 43, 47, 49, 50, 53, 64, 65, 68, 104, 111, 168, 174, 176, 258, 261, 264.

From Flock 114, HENRY LANE.

To Mills & Condit, Condit, O., Ram 45.

From Flock 115, ROLLIN LANE.

To Cherbino & Williamson, Middlebury, Vt., Rams 88, 89, 94.
" J. A. Wright, Middlebury, Vt., Rams 13, 85.
" Mills & Condit, Condit, O., Ram 87.
" H. Kirkpatrick, Utica, O., Ram 14.

To Jonas Rice, Bridport, Vt., Rams 130, 132, 133, 134, 142 to
146, 148, 150, 151.
" Wood & Kennedy, Mich., Ewes 11, 40, 55, 56, 62, 63, 66 to
68,.74. 75, 77, 79, 81, 82, 84, 86, 101 to 107.
" A. M. Evarts, Salisbury, Vt., Ewes 18, 22, 23, 27, 29, 31, 35,
36, 42 to 44, 46, 48, 49, 50.
" Mills & Condit, Condit, O., Ewes 51, 52, 53.
" Cherbino & Williamson, Middlebury, Vt., Ewes 20, 24, 33,
59, 60, 65, 78, 83, 95, 96, 97.

From Flock 116, H. S. LANGDON.

To H. C. Burwell, Bridport, Vt., Rams 14, 16, 19, 23, 26.
" Will Nash, New Haven, Vt., Rams 21, 22.
" Z. & S. R. McFadden, Atlanta, Ill., Ewe 15.
Dead, Ram 27 ; Ewe 17.

From Flock 117, OTIS P. LEE.

To J. A. Wright, Middlebury, Vt., Ewes 33 to 49.
" H. B. Wright, Middlebury, Vt., Ewes 4, 7, 8, 10, 15, 19, 24,
25, 28, 51, 53, 65, 67.

From Flock 119, ORSON C. MARTIN.

To L. J. Wright, Weybridge, Vt., Ram H. G. Hibbard 8.
" J. Forbes, Jr., Shoreham, Vt., Ram H. G. Hibbard 7.

From Flock 121, G. F. MARTIN.

No sales have been reported from this flock, but the report of
number of ewes as published was wrong ; since it was printed in
the Register, the following corrected list of numbers has been
received: 2, 5, 21, 25, 48, 96, 100, 105 to 108, 114, 119 to 122, 127,
128, 134, 135, 137 to 139, 149 to 152.

From Flock 123, WILLIAM McCAULEY.

To L. M. Bingham, W. Cornwall, Vt., Ram 3.
" Joel Randall, W. Cornwall, Vt., Rams 3 to 6.
" J. B. Hamblin, Cornwall, Vt., Ewe 43.
" F. Hooker, Cornwall, Vt., Ewes 38, 40.
" R. J. Jones, Cornwall, Vt., Ewes 17, 42.
Dead, Ewes 18, 35.

From Flock 124, JOHN H. MEAD.

To C. H. Hubbard, Springfield, Vt., Ram 92, now dead.
" Ora Paul, Pomfret, Vt., Ram 74.
" John Pearrows, York Station, O., Ram 86.
" D. G. Spaulding, Taftsville, Vt., Ram 79.
" A. B. Blood, Brownsville, Vt., Ram 89.
" Geo. Ashley, Otisco, Mich., Ram 84.
" J. F. Hemenway, Chelsea, Vt., Ram 88.
" B. F. Van Vliet, Shelburn, Vt., Ram 103.

To J. & F. Smith, W. Rutland, Vt., Ewes 3, 9, 10, 17, 25, 27, 35, 43, 47, 69.

" A. G. Richardson, Beeville, Texas, Rams 75, 76, 78, 80 to 83, 85, 87, 90, 93 to 100, 102, 104 to 106, 109 to 112, 114 to 119 ; Ewes 123, 125, 133 to 137, 140, 141.

From Flock 125, MRS. MARY W. MEAD.

To E. W. Eells. Middlebury, Vt., Ewes 51 to 60.
Dead, Ewes 14, 21, 40.

From Flock 126, C. C. MINER.

To J. A. Wright, Middlebury, Vt., Ewe 14.

From Flock 127, F. & L. E. MOORE.

To Charles M. Clark, Whitewater, Wis., Rams 114, 116.
" E. Gardner, Whitewater, Wis., Ram 118.
" J. C. George. Harvard, Ill., Ram 99.
" Thos. Sicklon, Wilmot. Wis., Ram 115.
" Dr. F. Paddock, Salem, Wis., Ram 113.
" Robert Fleming, Salem, Wis , Ram 105.
" L. W. Tyler, Mosherville, Wis , Ram 119.
" James Densmore, Markeson, Wis., Rams 101, 104, 106, 107, 108, 112.
" D. A Hussey, Columbus, Wis., Ram 100.
" Clapp Miner, Salem, Wis., Ram 109.
" E. G. Farnham, W. Cornwall, Vt., Ram 111.
" H. Wellman, Western, N. Y., Ewes 1 to 13.
" J. A. Wright, Middlebury, Vt., Ewes 63 to 80.
" A. J. Towner, Shoreham, Vt., Ewes 130 to 135.
" Wm. Ball, Hamburgh, Mich., Ewe 35.

From Flock 129, E. B. MUSSEY.

To Dan'l Sullivan, Middlebury, Vt., Ram 7.

From Flock 131, E. L. NEEDHAM.

To C. M. Inskeep, West Mansfield, Ohio., Ewe 5.
" L. C. Dickinson and J. C. Smith, North Greenfield, O., Ewes 1, 2, 3.
" H. Skidmore, Ewes 4, 6.

The above were sold through G. D. Miner, Middlebury, Vt.

From Flock 132, GEORGE N. PAYNE.

To J. J. Crane. Bridport, Vt., Ewes 40, 41, 42.

From Flock 133, LYMAN H. PAYNE.

To R. J. Jones, W. Cornwall, Vt., Ewes 61 to 70.

From Flock 135, J. W. PECK.

To E. W. Eells, Middlebury, Vt , Ewes 18 to 32.
" H. E Taylor, W. Cornwall, Vt., Ewes 60 to 71.

Since the Register of this flock was printed, page 285, there has been added to the flock tegs sired by Delong's Remele Ram (361) and registered ewes of the flock; Rams 74, 76, Ewes 60 to 73.

From Flock 136, H. M. PERRY & G. A. CUTTING.

To H. Kirkpatrick, Utica, O., Rams 87, 89 to 91, 93.
" A. Harding, Pomfret, Vt., Ram 95
" H. R. Holden, Middlebury, Vt., Rams 51, 92, 96 to 98, 100, 102, 103, 105 to 108.
" E. G. Farnham, W. Cornwall, Vt., Ewes 60 to 72.
" W. H. Gibbs, Hart Prairie, Wis., Ewes 40, 43, 46, 49, 50, 54 to 58.
" C. R. Gibbs, Hart Prairie, Wis., Ewes 36, 37, 44.
" R. J. Williams, Hart Prairie, Wis., Ewes 35, 38, 39, 42, 45, 47, 48, 52, 53.

From Flock 137, E. PECK & SONS.

Died, Ewes 10, 17, 31, 35.

From Flock 140, ALBRO E. PERKINS.

To John D. Patterson, Geneva, N. Y., Rams 2, 4, 5, 6.
" Randall Waugh, Starks, Me., Ram 11; Ewes 12, 13, 14.
Dead, Ram 7; Ewe 33.

From Flock 143, J. F. RANDALL.

To L. M. Bingham, W. Cornwall, Vt., Ewes 3, 4, 5, 7, 8, 9.
" H. Manchester, W. Cornwall, Vt., Ewes 11, 13, 15, 23 to 25, 32, 35.
Dead, Ewe 33.

From Flock 145, L. C. REMELE.

To H. S. Brookins, Richville, Vt., Rams 5, 6, 7.
" G. D. Bush, N. Orwell, Vt, Ram 8.
" E. L. Needham, Whiting, Vt., Ram 4.
" C. K. Williams, Whiting, Vt., Ram 2.
" E. B. Pond, Whiting, Vt., Ewes 9, 11, 12, 17, 23, 32, 37, 38, 47, 48.

From Flock 147, W. R. REMELE.

To B. Linville & Son, Cloverdale, Cal., Ram 15.
" J. D. Patterson, Geneva, N. Y., Ram 14.
" E. N. Bissell, E. Shoreham, Vt., Ewes 1, 2, 3, 4.
Sold by E. N. B. and transfered:
- To John Lockin, Brandon, Wis., Ewes 1, 2, 3, 4.

From Flock 151, VIRTULAN RICH.

To J. A. Wright, Middlebury, Vt., Ram 301.
" Wm. G. Willson, Richville, Vt, Ram 328.
" Geo. W. Hunt, Greenwood, Ill., Ram 350.
" G. M. Everhart, Kenosha, Ill., Ram 358.
" A. Lobin, Grand Blanc, Mich., Ram 367.

To Hayward & Gregory, Beloit, Wis., Ram 360.
" W. W. Dykeman, Fairfield, Wis., Ram 361.
" I. Kedditch, Fairfield, Wis., Ram 362.
" C. R. Gibbs, Whitewater, Wis., Ram 368.
" E. Lewis, Randolph Center, Wis., Ram 348.
" Wm. Thomas, Randolph Center, Wis., Ram 371.
" M. F. Arms, Randolph Center, Wis., Ram 369.
" C. D. McConnell, Ripon, Wis., Rams 357, 390.
" S. D. Woodworth, Pleasant Prairie, Wis , Ram 366.
" James Densmore, Markeson, Wis., Rams 338, 339, 344, 347, 355.

With the exception of the first two, the above were sold by E. N. Bissell, who had purchased a half interest in them of Mr. Rich.

Sold by Mr. Rich :

To George Campbell, Westminster West, Vt., Rams 382, 385, 386.
" J. T. Rich, Elba, Mich., Ram 359.
" G. D. Bush, W. Orwell, Vt., Ram 363.

Sold by J. A. Wright and transferred :

To Brant, Thompson, Kelly & Co., Ohio, Ram 301.

Dead, Ewes 6, 77, 156.

From Flock 152, JOHN T. RICH.

To John Flanagan, Elba, Mich., Ram 2.
" J. C. Smith, Farmer's Creek, Mich., Ram 26.
" Albert Howman, Lapeer, Mich., Ram 3.
" James Henderson, Lapeer, Mich., Ram 4.
" John C. Henderson, Lapeer, Mich., Rams 13, 23.
" Philo J. Bristol, Lapier, Mich., Ram 20.
" George Barber, Elba, Mich., Ram 5.
" Oscar Moore, Elba, Mich., Ram 9.
" Joseph Ognolle, Elba, Mich., Ram 18.
" David Godfrey, Elba, Mich., Ram 21.
" James Arnold, Oakwood, Mich., Rams 7, 15.
" Perry Stimson, Hadley, Mich., Ram 10.
" Wm. McCoy, Tuscola, Mich., Ram 24.
" H. D. Phelps, Ortonville, Mich., Ram 28.
" George H. Stuart, Grand Blanc, Mich., Ewes 12, 13, 21, 27, 28, 32, 36, 41, 42, 49, 67.
" Morris Moore, Elba, Mich., Ewes 19, 26.

Dead, Ram 1 ; Ewe 71.

From Flock 153, S. S. ROCKWELL & H. E. SANFORD.

To L. P. Clark, Addison, Vt., Rams 85 to 90 ; Ewes 24 to 50, 100 to 105, 263.
" Z. & S. R. McFadden, Atlanta, Ill., Ram 52 ; Ewe 209.
" C. W. Mason, N. Haven, Vt., Rams 53, 54, 55, 57 to 61, 65 to 68, 70 to 75, 77 to 83.
" Peck & Jennings, W. Cornwall, Vt., Ram 62 ; Ewes 19, 110, 167.

To J. A. Hendrix and A. H. Williams, Decatur, Mich., Ram 51.
" G. F. Harrington and D. W. Broadhead, Paw Paw, Mich.,
    Rams 233, 262.
" J. N. Ransom, Alamo, Mich., Ram 237.
" O. Snow, Kalamazoo, Mich., Ram 69.
" A. W. Cooper, Cooper, Mich., Ram 76.
" W. F. Willson, Cooper, Mich., Ram 84.
" H. Dowd, Hartford, Mich., Ram 243.
" C. Limberk, Keller, Mich., Ram 241.
" A. N. & W. Spaulding, Hartford, Mich., Rams 235, 239.
" J. A. Wright, Middlebury, Vt., Ewes 1 to 10.
" John Gillespie, Lawton, Mich., Ewes 64, 172, 175, 177 to
    181, 183, 187, 198, 199, 202, 203, 205, 206, 207, 211, 229,
    232.
" J. F. Randall, W. Cornwall, Vt., Ewes 17, 18, 22, 99, 122,
    134, 136, 137, 142, 145, 149, 157, 161, 328, 329.
Dead, Rams 56, 313; Ewes 11, 13, 20, 21, 55, 96, 97, 111, 117,
    124, 127, 141, 148, 191.

From Flock 155, MARCELLUS ROYCE.

To Farnham & Birchard, Shoreham, Vt., Ram 34.

From Flock 157, C. B. RUSSELL.

To B. F. Rugg, St. Albans, Vt., Rams 1 to 4, 6, 8 to 14.
" J. A. Dodds, North Hero, Vt., Ram 5.

From Flock 158, EDGAR SANFORD.

To Chas. W. Mason, N. Haven, Vt, Rams 53, 55.
" Capt. Cuthill, Colorado, Rams 51, 52, 77, 78, 80 to 82, 84.
" H. E. Taylor, W. Cornwall, Vt., Ram 83; Ewes 39 to 47.
" H. J. Manchester, Cornwall, Vt., Ewes 1, 7, 31, 35.

From Flock 159, W. R. & CHAS. SANFORD.

To H. E. Taylor, W. Cornwall, Vt., Rams 280 to 282, 284, 285.

From Flock 160, E D. SEARL.

To Hall & Holden, Middlebury, Vt., Rams 11 to 16.
" E. G. Farnham, W. Cornwall, Vt., Rams 19 to 32.
" Chas. Church, Whiting, Vt., Rams 17, 18.
" Capt. W. A. Cuthill, Colorado, Rams 36, 38, 39, 41.
" J. A. Wright, Middlebury, Vt., Ewes 33, 34, 35.

From Flock 161, MOSES S. SHELDON.

To Cherbino & Williamson, Middlebury, Vt., Rams 33, 34.
Sold by C. & W. and transferred:
To Jacob Clarey, Cumberland, Ohio, Ram 33.

From Flock 165, A. H. SPERRY.

To John S. Beecker, Livonia, N. Y., Ram 49.
" H. B. Williamson, Starks, Me., Ram 41.
" E. S. Stowell, Cornwall, Rams 43, 47.
" Chas. A. Sperry, Grayson, Cal., Ram 60; Ewes 61, 62.

From Flock 166, JOSEPH T. STICKNEY.

To Cherbino & Williamson, Middlebury, Vt., Rams 22, 43, 128, 129, 130, 132 to 135, 139, 140.
" Farnham & Birchard, Shoreham, Vt., Rams 234, 235.
" J. A. Wright, Middlebury Vt., Ewe 42.
" E. N. Bissell, Shoreham, Vt., Ewes 10, 13, 20, 21, 22, 46, 53, 54, 62, 89, 91.
Sold by E. N. B. and transferred:
To A. C. Whitmore, Mukwongo, Wis., Ewes 46, 53, 54, 62, 89, 91.

From Flock 167, T. STICKNEY & SON.

To Farnham & Birchard, Shoreham, Vt., Rams 210, 212, 223, 227, 232, 234, 329.
" J. A. Wright, Middlebury, Vt., Ewes 107, 241.

From Flock 168, JOHN Q. STICKNEY.

To Capt. W. A. Cuthill, Colorado Springs, Col., Rams 16 to 31.
" Wm. Ball, Hamburgh, Mich., Ewes 40 to 43, 45 to 59.

From Flock 169, S. L. STEVENS.

To E. G. Farnham, West Cornwall, Vt., Ewes 4, 46, 61, 62, 71, 101, 112, 116, 120, 126, 141, 144, 151, 154, 155, 165 to 167, 171, 172, 177 : these were marked Wm. M. Holmes; to same, Ewes 42, 43, 44, 45, 47, 48, 49, 51 to 58, 69, marked C. L. Stevens.

From Flock 170, W. C. STURDEVANT.

To H. B. Sturdevant, Sherman, Mich., Ewe 13.

From Flock 173, HENRY THORP.

To Henry Giddings, Fairfax, Vt., Ram 22.
" C. B. Russell, North Hero, Vt., Ram 37.
" E. Townsend, Pavilion Center, N. Y., Ewes 2, 3, 6, 7, 21, 24, 25, 47, 50, 73, 85, 130, 141, 177, 178; these had L. J. O. on their labels.
" Same Ewes 5 to 9, 11, 12, 13, 15, 16, 17, 19, 20. Marked H. Thorp.

From Flock 174, JOHN TOWLE.

To E. S. Casey, Whiting, Vt, Ewes 54 to 56, 58, 60 to 63, 65, 67, 69.

From Flock 176, A. J. TOWNER.

To L. E. Moore, Shoreham, Vt., Rams 30 to 35.
Sold by L. E. M. and transferred:
To Herbert Bros., Belvidere, Ill., Rams 34, 35.
" A. Patterson, Fox River, Ill., Ram 30.
" Dyke Bros., Crystal Lake, Ill., Ram 32.
" W. N. Batchelder, Fort Collins, Col., Ram 31.
" G. W. Patterson, Salem, Wis., Ram 33.
" E. W. Eells, Middlebury, Vt., Ewes 15 to 19.

From Flock 177. LOREN TOWNER.

To E. G. Farnham, W. Cornwall, Vt., Rams 1 to 15.

From Flock 178, B. B. TOTTINGHAM.

To H. W. Jones, Shoreham, Vt.,Ewes 12, 13, 27, 28, 59, 60, 96, 106, 142.

" Cherbino & Williamson, Middlebury, Vt., Ewes 38, 41, 46, 55, 93, 100, 101, 104, 105, 109, 143, 148, 164, 166, 167, and these have been sold by C. & W. and transferred to J. W. Lippet, Nobles Co., O.

From Flock 179, E. N. TOWNSEND.

To H. S. Brookins and D. J. Wright, Richville, Vt., Rams 1, 2.

" George Waite, Shoreham, Vt., Ewes 11, 12, 13.

Sold by H. S. B. & D J. W. and transferred :

To P. Butler, Sturgis, Mich., Ram 1.

" S. Strong, Vicksburg, Mich., Ram 2.

Dead, Ewe 10.

From Flock 182, J. A. WATTS.

To J. Q. Stickney, Whiting, Vt , Rams 15, 17, 19 ; Ewe 14.

From Flock 185, LUTHER WEBSTER,

To H. Kirkpatrick, Utica, O., Ram 57.

" Farnham & Birchard, Shoreham, Vt., Ram 52.

" J. A. Wright, Middlebury, Vt , Ewes 22, 24, 25, 27 to 29, 31 to 34, 36 to 40.

" C. F. Church, Whiting, Vt., Ewes 35, 41 to 45.

From Flock 186, G. W. WHITFORD.

To L. W. Frost, Bridport, Vt., Ewes 56 to 59.

From Flock 187, E. M. WHEELER.

To Cherbino & Williamson, Middlebury, Vt., Ewes 2, 5, 7, 11, 12, 15, 16, 21, 23.

From Flock 188, R. WITHERELL & SON.

To Capt. Cuthill, Colorado, Rams 37 to 46.

" J. Forbes, Jr., Shoreham, Vt., Ewes 30 to 36.

Sold by J. F., Jr., and transferred :

To V. N. Forbes, West Haven, Vt., Ewes 30, 32 to 34.

From Flock 191, LEVI WOLCOTT.

To W. E. Rich, Shoreham, Vt., Rams 1 to 7, 18 to 24.

" Wood & Kennedy, Saline, Mich., Ewes 8 to 17.

" A. Christman, Pavilion, N. Y., Ewes 36 to 59.

The report of numbers of this flock made to the Secretary as printed on page 232 is incorrect ; it should be 15 Rams, 1 to 7, 18 to 24 ; 44 Ewes, 8 to 17, 26 to 58, 60.

From Flock 192, J. G. WOOSTER.

> To J. A. Wright, Middlebury, Vt., Ewes 67 to 85.
> " Rollin Lane, Cornwall, Vt., Ewes 13, 18, 25, 33, 41, 55, 61.
> " James Cook, Richville, Vt., Ewe 50.
> Sold by J. A. Wright, and transferred:
> To W. E. Bozman, Blue Rock, Muskingum Co., O., Ewes 67 to
> 72, 74 to 85.

From Flock 193, WILLIAM P. WRIGHT.

> To A. Ketchum, Whiting, Vt., Ram 33.
> " H. S. Brookins and D. G. Wright, Richville, Vt., Rams 34
> to 37.
> " F. G. Wright, Whiting, Vt., Ram 44.
> " C. K. Williams, Whiting, Vt., Ram 45 ; Ewe 24.
> " J. H. Kirkpatrick, Utica, O., Ewes 1, 10, 12 to 16.
> " Geo. D. Bush, N. Orwell, Vt., Ewes 55 to 57.
> " A. & T. J. Ketchum, Whiting, Vt., Rams 38 to 43.
> Sold by H. S. B. & D. J. W. and transferred:
> To S. Perham, Flowersfield, Ind., Ram 34.
> " Chas. Stroud, Sturgis, Mich., Ram 36.
> " A. B. Smith, Vicksburg, Mich., Ram 37.

From Flock 194, DON JUAN WRIGHT.

> To Michael Schenk, Francisco, Mich., Ram 4.
> " A. A. Corwin, Grass Lake, Mich., Ram 60.
> " John Cook, Chelsea, Mich., Ram 6.
> " Z. Robinson, Mosherville, Mich., Rams 2, 64.
> " L. N. Tyler, Mosherville, Mich., Ram 18.
> " John McKerlie, Sturgis, Mich., Ram 54.
> " Hiram Draper, Calon, Mich., Ram 59.
> " Chas. Loud, Vicksburg, Mich., 30.
> " Max Smaley, Vicksburg, Mich., Ram 55.
> " W. Strong, Vicksburg, Mich., Ram 57.
> " C. Johnson, Centerville, Mich., Ram 28.
> " Orin Meredith, Kalamazoo, Mich., Ram 29.
> " Norton Pomeroy, Kalamazoo, Mich., Ram 26.
> " J. Woodward, Kalamazoo, Mich., Ram 56.

From Flock 196, F. G. WRIGHT.

> To W. P. Nash and L. M. Bingham, Cornwall, Vt., Ewes 9, 44,
> 45, 51 to 60.

From Flock 198, H. B. WRIGHT.

> To Kent Wright, Shoreham, Vt., Ewes 1 to 45.

From Flock 199, J. A. WRIGHT.

> To G. D. Miner, Middlebury, Vt., Ewes 73 to 82.
> Sold by G. D. M. and transferred:
> To C. M. Inskeep, West Mansfield, O., Ewe 80.
> " Jacob Killer, West Mansfield, O., Ewe 73.

" J. A. Killer, West Mansfield, O., Ewes 74, 75, 76.
" John W. Winer, West Mansfield, O., Ewes 78, 82.
" Dan'l Skidmore, West Mansfield, O., Ewes 77, 79, 81.

From Flock 201, JOSEPH CLOSE.

To N. J. Israel, Bealsville, O., Ram 167.
" Thos. Healea, Urichville, O., Ewes 60, 64, 67, 71, 104, 105, 107, 110, 165, 190.

AWARDS ON VERMONT WOOLS AT PARIS.

Too late to be inserted in our account of Improvement of Merinos we received the following Report of Awards received on the exhibits of wools at Paris. It will be seen that Vermont wools received two out of four Gold Medals, seven out of twelve Bronze Medals and two Honorable Mentions out of seven accorded to Exhibitors of Wool from the United States. It will also be noticed that one Gold Medal to Mr. E. Townsend and one honorable mention to Mr. G. F. Martin were awarded to wools from Flocks that are recorded in this Register. All the Vermont Wools were from sheep that are recorded in this Register and were prepared and forwarded at the expense of the Vermont Merino Sheep Breeders' Association.

The awards were as follows:

GOLD MEDALS—George Campbell, Westminster West, Vt; Luman P. Clark, Addison, Vt.; E. Townsend, Pavilion Centre, N. Y.; E. Hammond, Cadiz, Ohio.

SILVER MEDALS—Sam'l Archer, Kansas City, Mo.; E. J. Hiatt, Chester Hills. Ohio.

BRONZE MEDALS—H. C. Burwell, Bridport, Vt.; H. M. Bottum, Shaftsbury, Vt.; Edgar Sanford, West Cornwall, Vt.; Henry Harrington, Keeler's Bay, Vt.; J. T. Stickney, Shoreham, Vt.; C. B. Russell, North Hero, Vt.; Milo J. Ellsworth. Cornwall, Vt.; W. G. Markham, Avon, N. Y.; and E. Cleland, W. Pierpoint, J. H. Blaine, Mark Durand, no residence given.

HONORABLE MENTION—C. P. Morrison, Addison, Vt.; T. Stickney & Son. East Shoreham, Vt; G. F. Martin, Rush, N. Y.; Bronson & Mariner, West Bloomfield, N. Y.; and J. Forbes, J. C. Short, ——— Simmonds, residences not given.

INDEX TO STOCK RAMS.

INDEX TO FLOCKS.

GENERAL INDEX.

PAGE.

ERRATA.

The following errors of the printer and editor are noted since the work was printed :

At Page 20, Worrell should read Morrell.
" 22, Hale should be Hall.
" 46, Sampson should be Lampson.
" 47, Miss Cutts should be Mrs. Cutts.
" 48, Claueau should be Cleaveau.
" 54, Wm. Samson should be Wm. Lampson.
" 58, Joseph J. Bailey should be Joseph I. Bailey.
" 93, in 22d line for rams read dams.
" 101, in pedigree of ram 52, for O. S. Hamilton read J. O. Hamilton.
" 102, in pedigree of ram 58, for 1860 read 1868.
" 120, in pedigree of ram 224, for H. B. Burwell read H. C. Burwell.
" 122, in pedigree of ram 247, for J. C. Cherbino read J. B. C.
" 123, in pedigree of ram 258, for B. F. Field read B. S. Field.
" 132, in pedigree of ram 342, for J. C. Cherbino read J. B. C.
" 134, in pedigree of ram 359, for S. Stickney read T. Stickney.
" 134, in pedigree of ram 362, for A. W. Baldwin read A. M. B.
" 134, in pedigree of ram 363, for Dowd read Doud.
" 135, in pedigree of ram 369, for Lane Ram read Lame Ram.
" 136, in pedigree of ram 385, for J. T. Rich read J. W. Rich.
" 137, in pedigree of ram 386, for S. W. Rich read J. W. Rich.
" 140, in pedigree of ram 419, for W. P. Porter read W. B. Porter.
" 183, in Ewes purchased for flock 13, for Messrs. E. R. Robinson read Mrs. E. R Robinson.
" 196, in pedigree of flock 29, H. H. should be omitted between R. P & Hall.
" 232, in pedigree of flock 71, the I should be omitted in the name of Charles Benedict, Cornwall.
" 234, in pedigree of flock 73, for D. Hooker read F. Hooker.
" 235, " " 74, for B. F. Fields read B. S. Fields.
" 258, in pedigree of flock 102, for E. R. Sanford read W. R. Sanford.

The flock of W. R. Remele should be in the pedigrees of the following flocks : 1, 4, 5, 7, 8, 11, 12, 14, 15, 16, 17, 20, 22, 23, 25, 28, 29, 30, 31, 32, 35, 36, 37, 38, 41, 42, 43, 44, 45, 46, 47, 48, 50, 51, 52, 54, 55, 57, 58, 59, 60, 61, 62, 63, 64, 65, 68, 69, 100, 119, 158, 156, 175, 197.

The following flocks lack the flock of R. P. Hall in their pedigrees:
4, 5, 7, 12, 16, 35, 36, 38, 44, 46, 47, 48, 52, 54, 55, 56,
58, 60, 65, 69, 70, 84, 85, 88, 92, 100, 110, 119, 148,
150, 159, 163, 164, 173, 175, 194, 198.

The following flocks need the flock of W. R Sanford in their pedigrees to make them complete: 25, 29, 41, 82, 88, 111,
114, 116, 117, 118, 121, 126, 160, 161, 164, 197.

Flock 33 should have the flock of H. Cross in its pedigree.

" 67 should have that of Wm. Jarvis in its pedigree.

" 158 should have that of Messrs. Hammond in its Pedigree.

At page 328 the owner of the flock is Whitford.

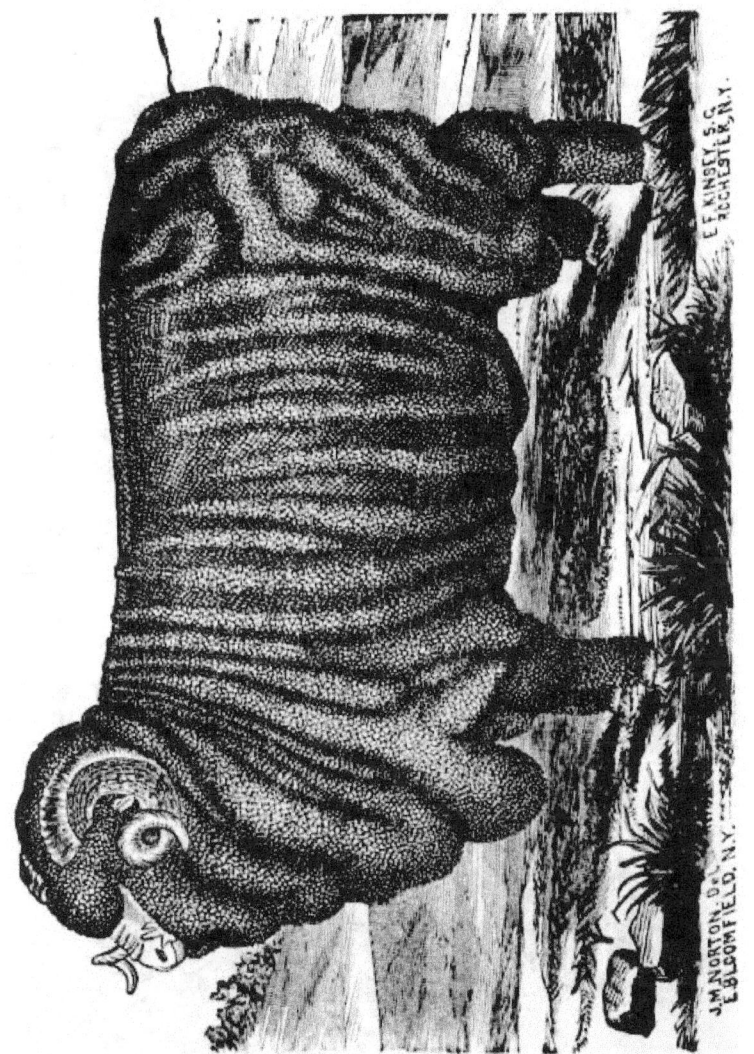

FREMONT, Jr.

BRED AND OWNED BY J. Q. STICKNEY, WHITING, VT.

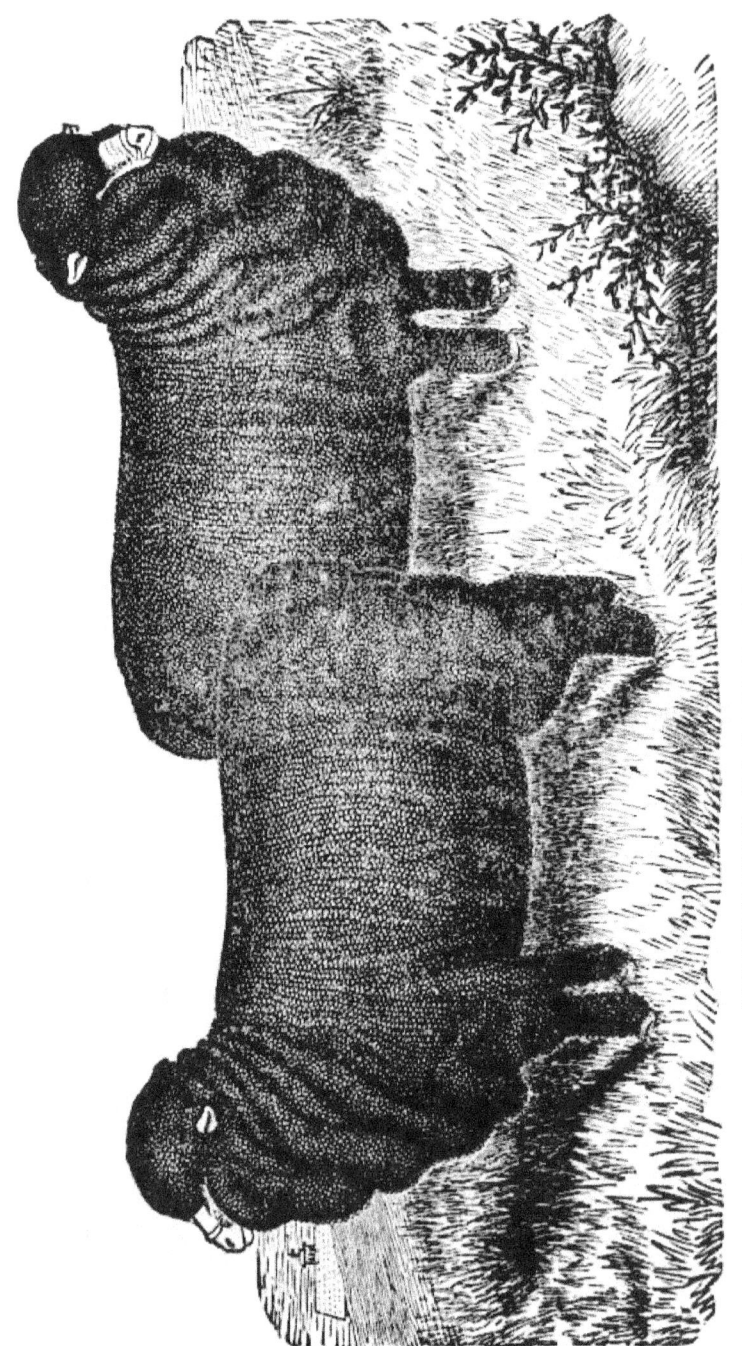

IMPROVED SPANISH MERINO EWES.

Bred By H. E. Sanford, West Cornwall, Vt.

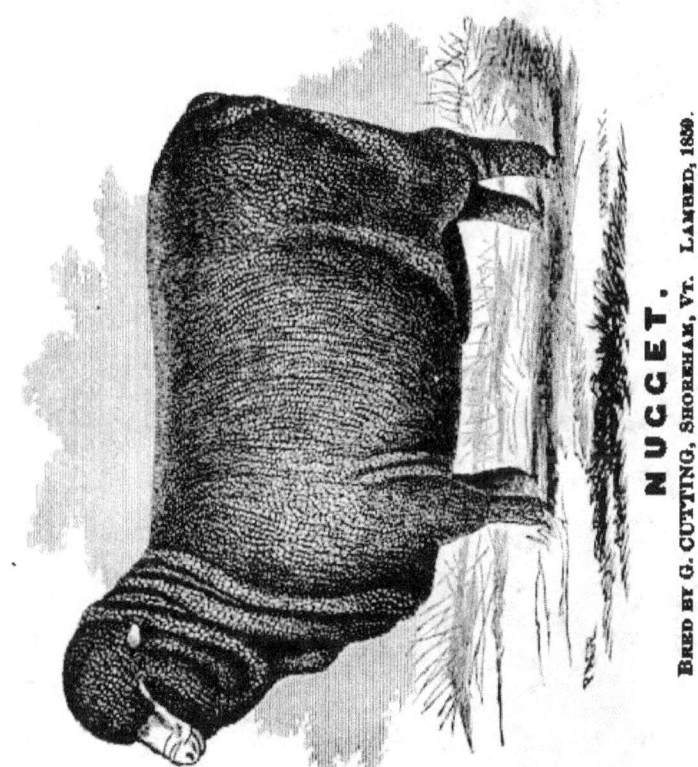

NUGGET.

BRED BY G. CUTTING, SHOREHAM, VT. LAMBED, 1859.

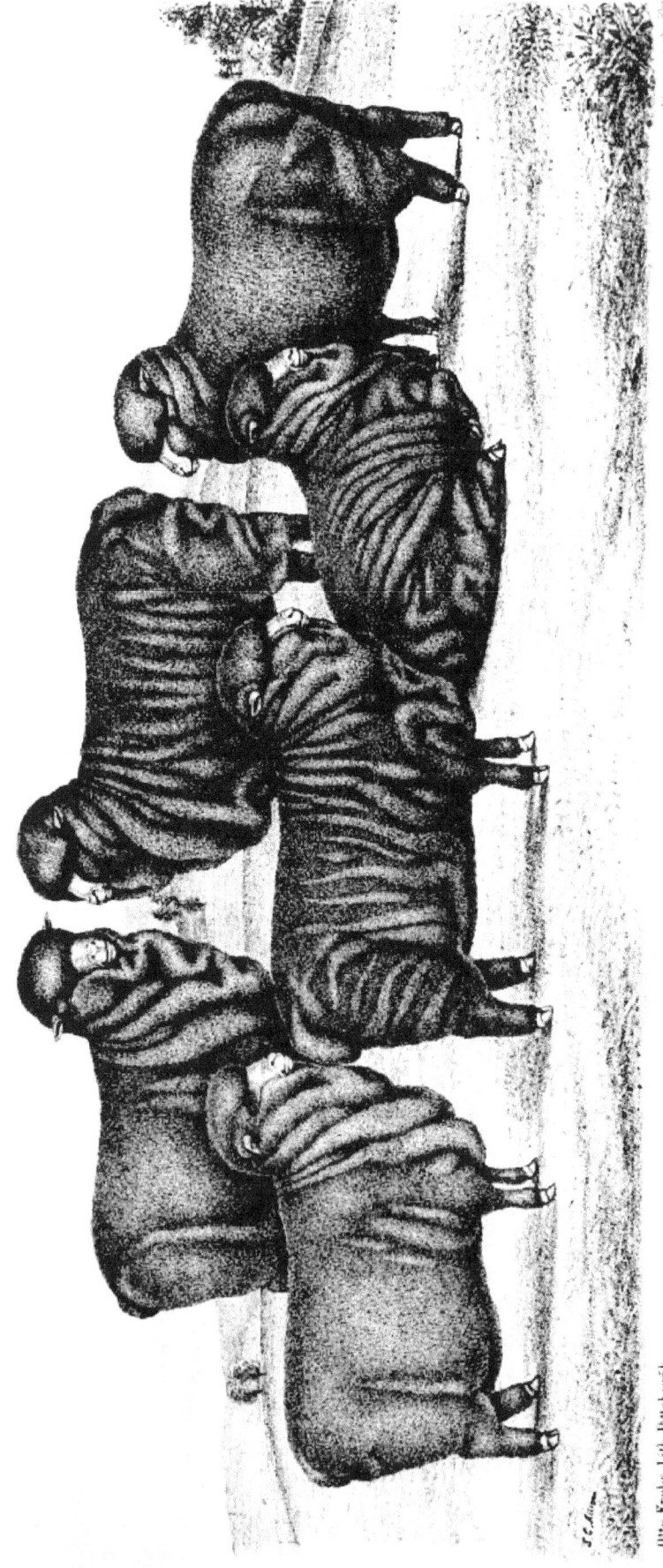

Otto Krebs Lith. Pittsburgh

"AMERICAN MERINOS"

Property of B.W.COPE, Breeder of Thorough Breds exclusively. Smithfield, Jefferson Co Ohio

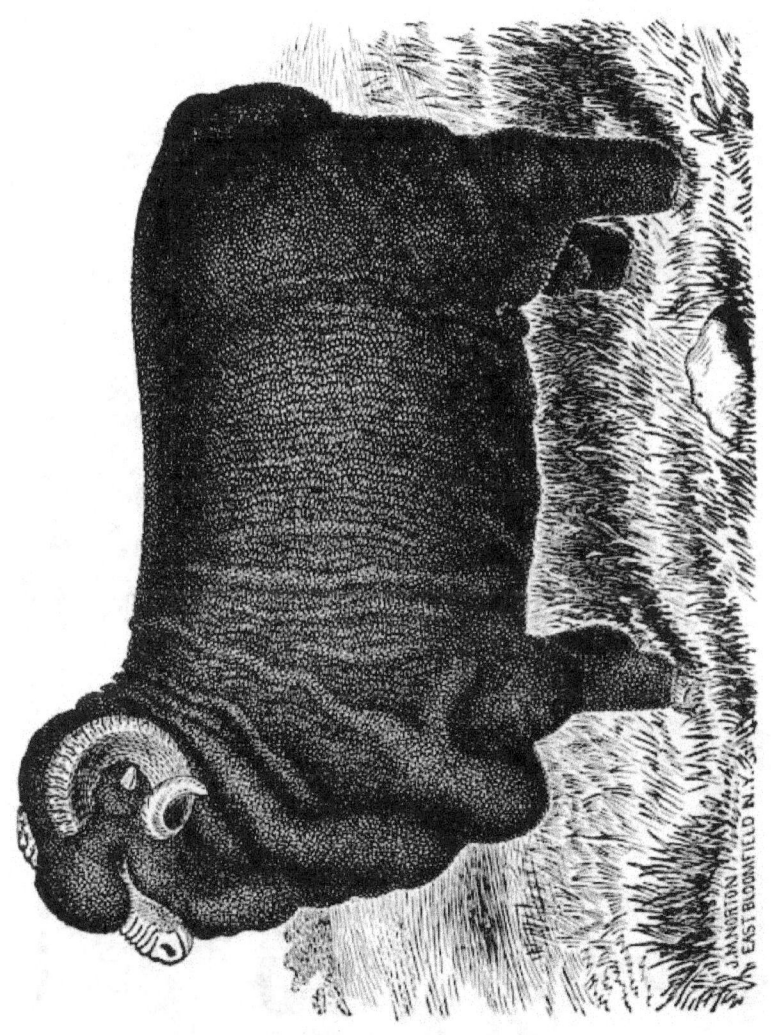

STOGA.

BRED BY M. R. ATWOOD, RICHVILLE, VT. OWNED BY E. N. BISSELL, EAST SHOREHAM, VT.

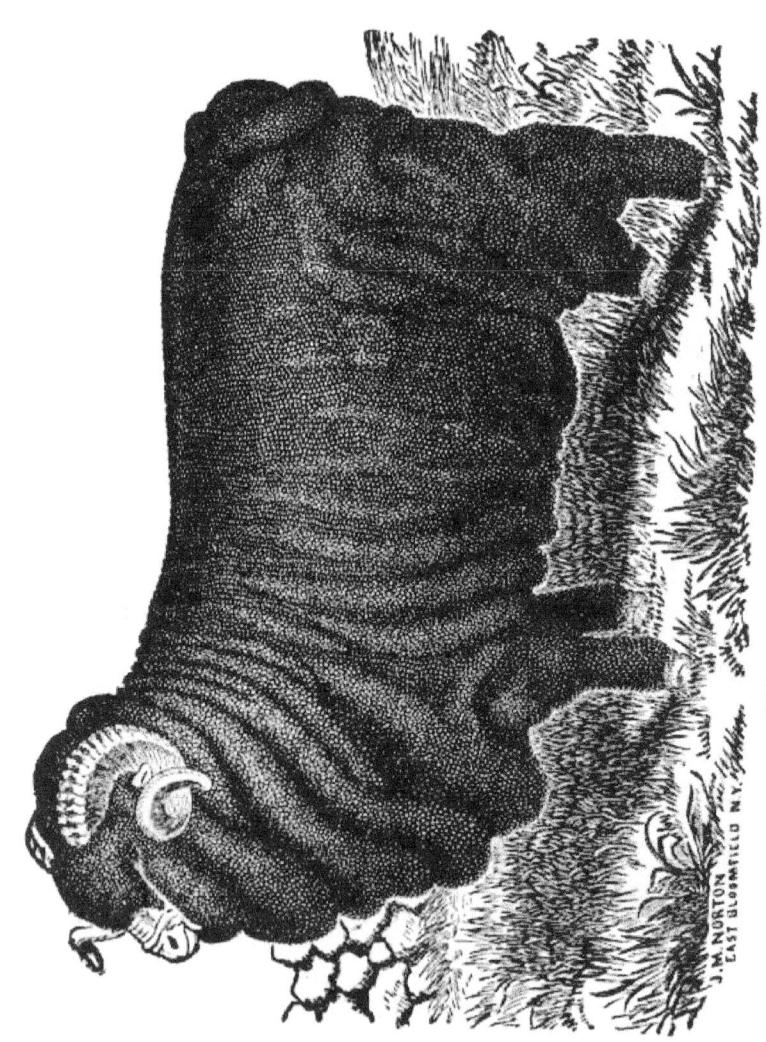

SILVER HORNS, JR.

Bred By H. C. BURWELL, Bridport, Vt. Owned By N. T. SPRAGUE, Brandon, Vt.

J. M. NORTON
EAST BLOOMFIELD, N.Y.

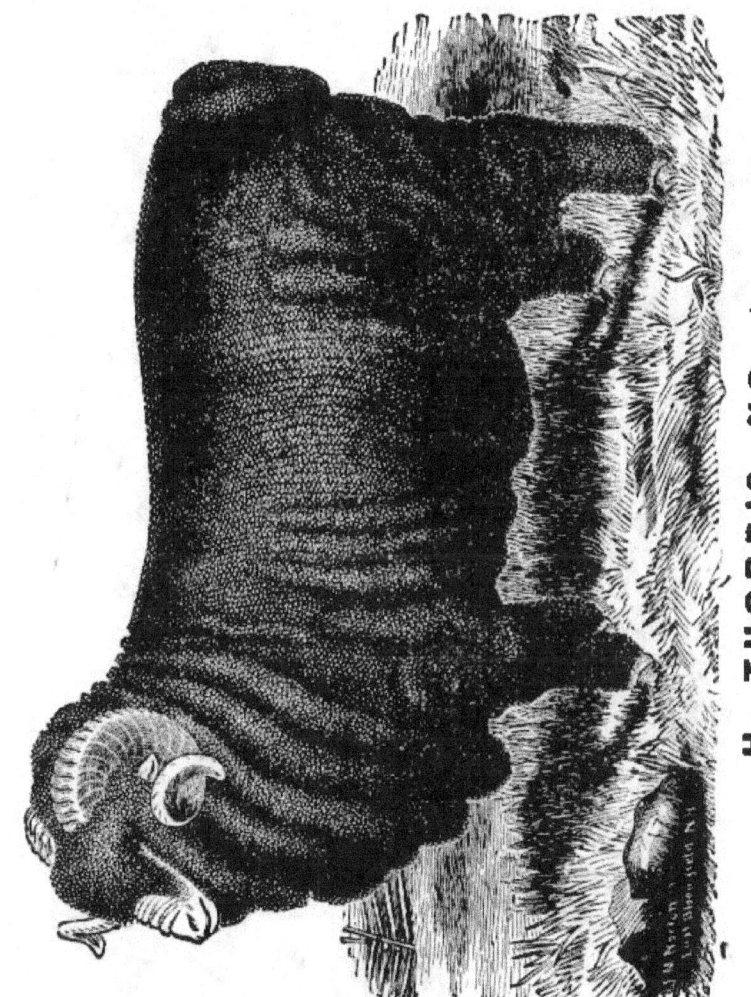

H. THORP'S, NO. 1.

BRED AND OWNED BY HENRY THORP, CHARLOTTE, VT.

WRINKLEY.

BRED AND OWNED BY FRANKLIN HOOKER, CORNWALL, VT.

SNOW FLAKE 1091 AT 6 YEARS.

CARCASS 15 FLEECE 25 Lbs.

PROPERTY OF MESSRS. O. BLACKMER & F. H. FARRINGTON, BRANDON, VT.

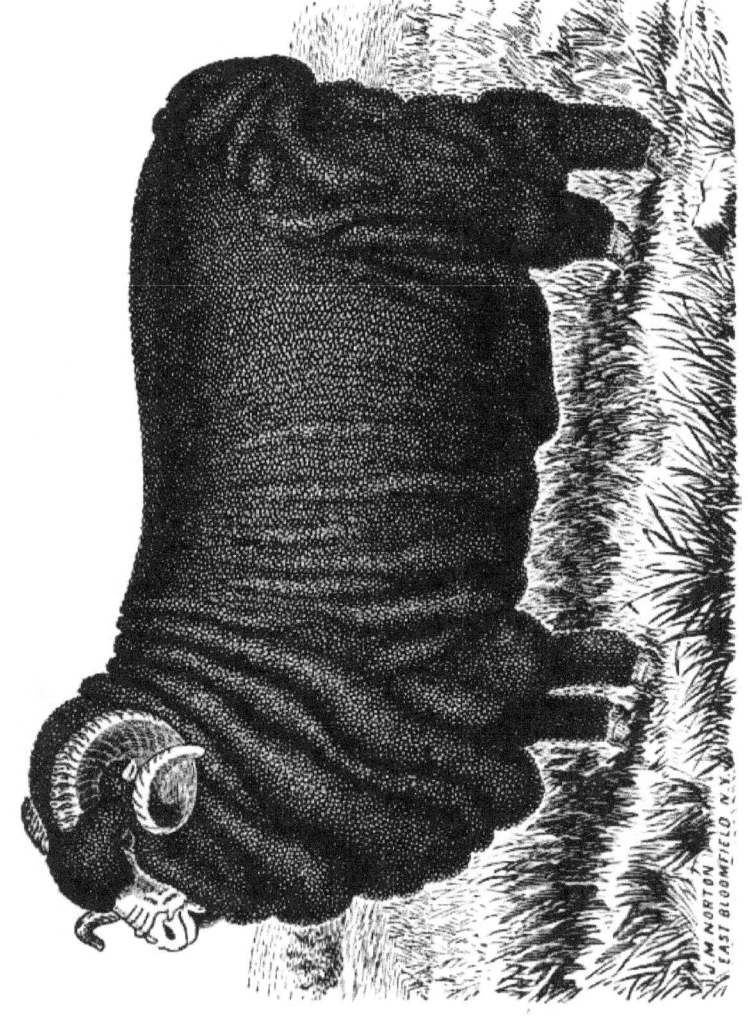

EUREKA 3d.

J. M. NORTON
EAST BLOOMFIELD, N.Y.

BRED BY J. J. CRANE, BRIDPORT, VT. OWNED BY J. J. CRANE AND H. C. BURWELL,
BRIDPORT, VT.

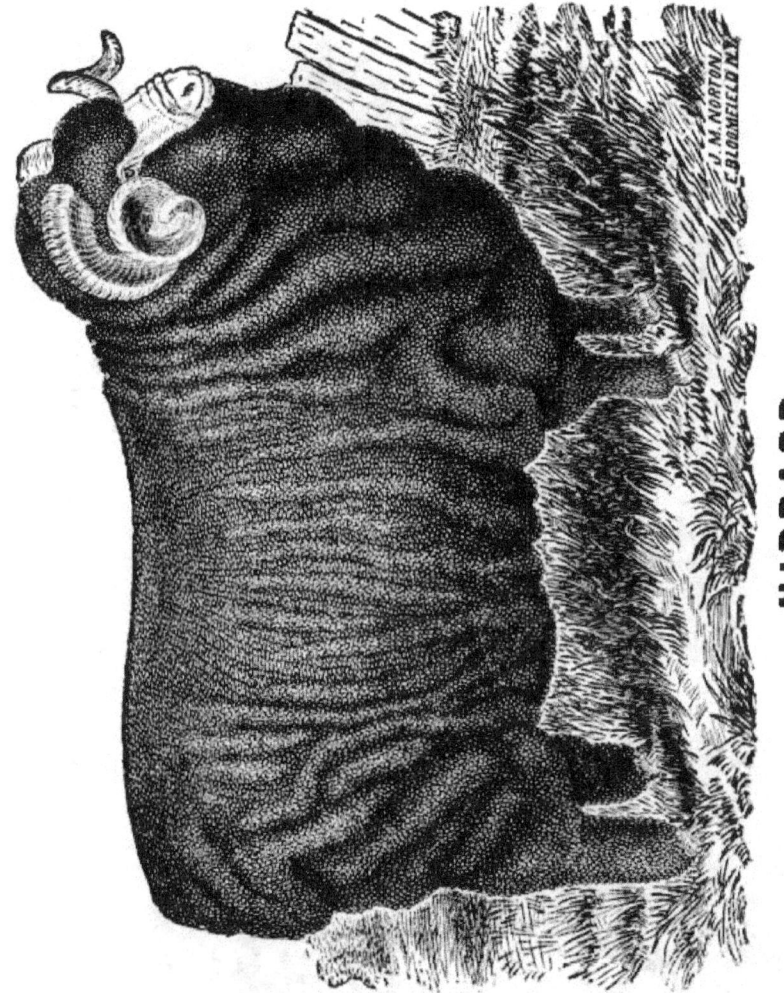

HIBBARD.

BRED BY H. G. HIBBARD, ORWELL, VT. OWNED BY E. N. BISSELL, SHOREHAM, VT., AND HUMBERT BROS., CALDWELL PRAIRIE, ILL.

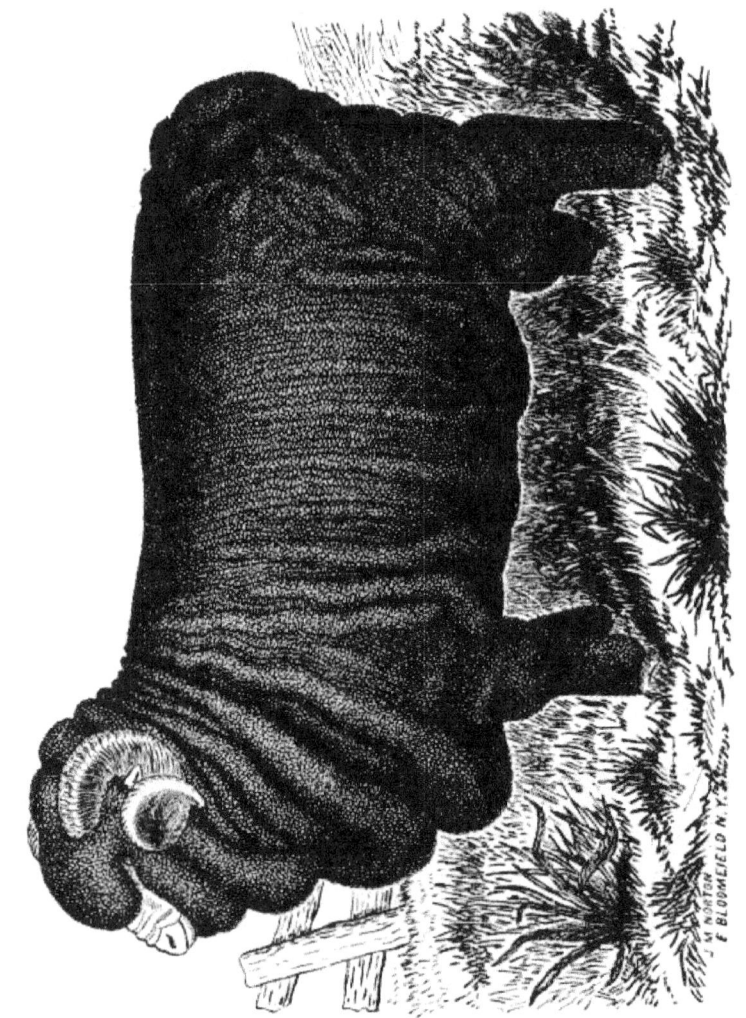

GOLIATH.

BRED AND OWNED BY E. S. STOWELL, CORNWALL, VT.

PATRICK HENRY. (183).

BRED AND OWNED BY L. F. CLARK, ADDISON, Vt.

GENESEE.

BRED AND OWNED BY E. TOWNSEND, PAVILLION CENTRE, NEW YORK.

GEN. FREMONT.

BRED AND OWNED BY T. STICKNEY & SON, SHOREHAM, VT.

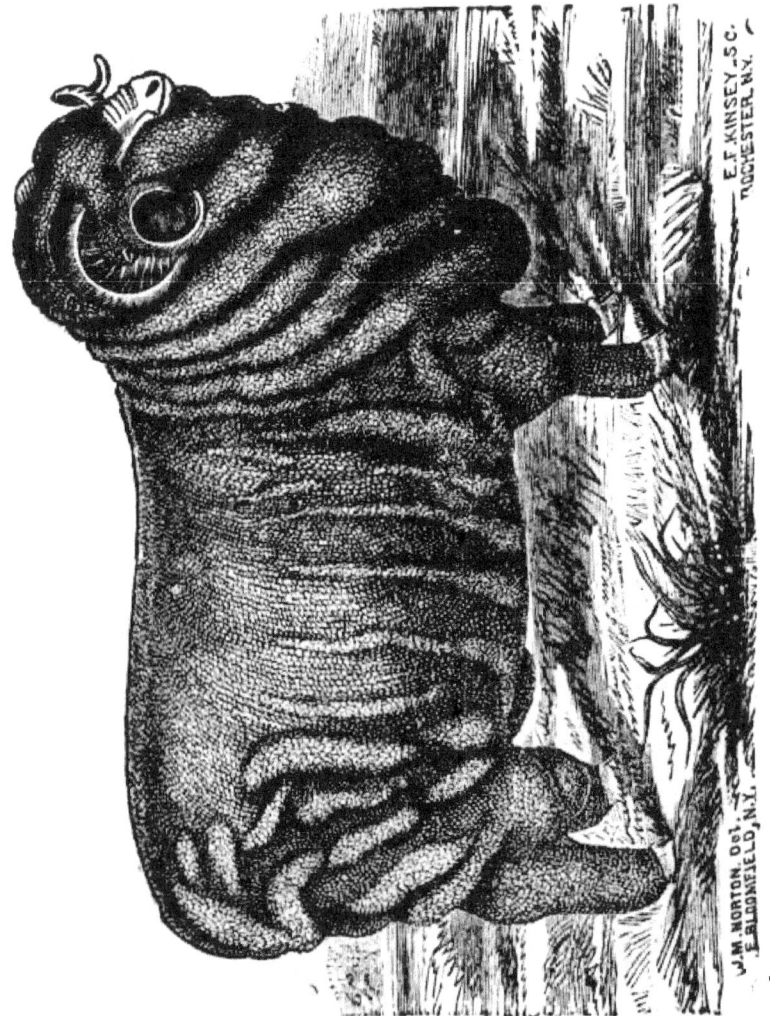

TRIUMPH.

BRED AND OWNED BY F. & G. V. MARTIN, RUSH, NEW YORK.

OLD TOTTINGHAM RAM.

Bred by E. A. Birchard. Lambed 1868.

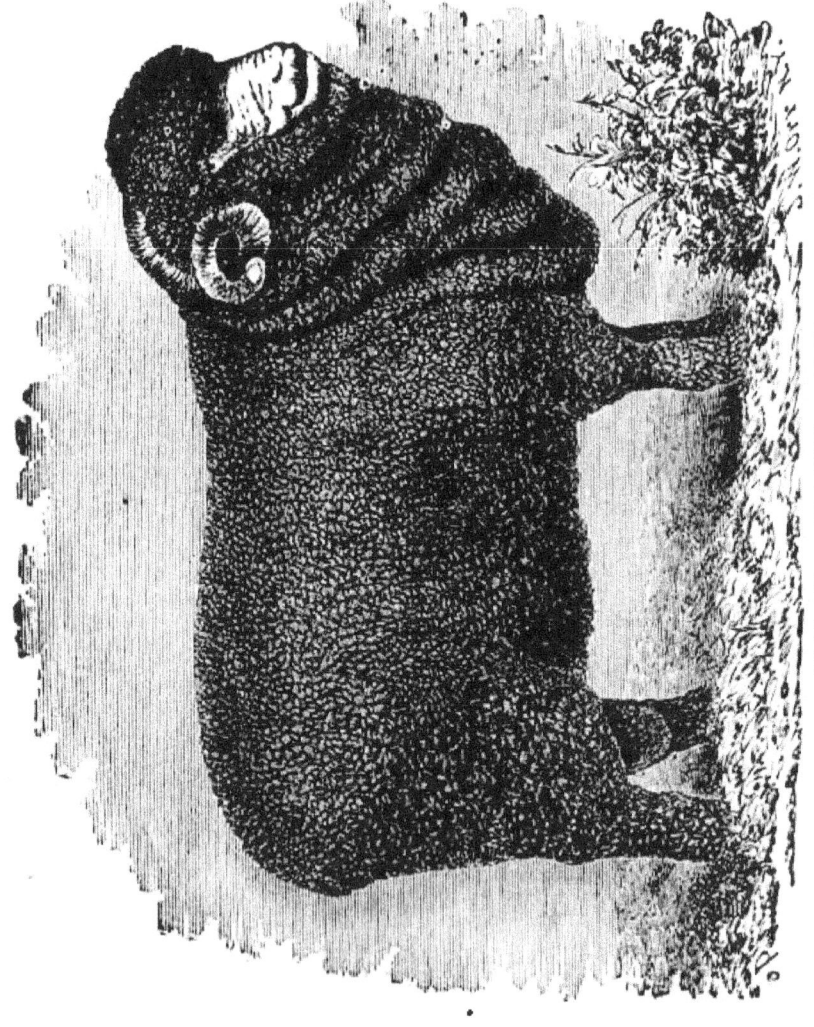

SWEEPSTAKES, (32).

BRED BY W. S. & E. HAMMOND, MIDDLEBURY, VT. ONE OF THE VERY BEST IMPROVED
SPANISH MERINO RAMS OF TWENTY YEARS AGO.

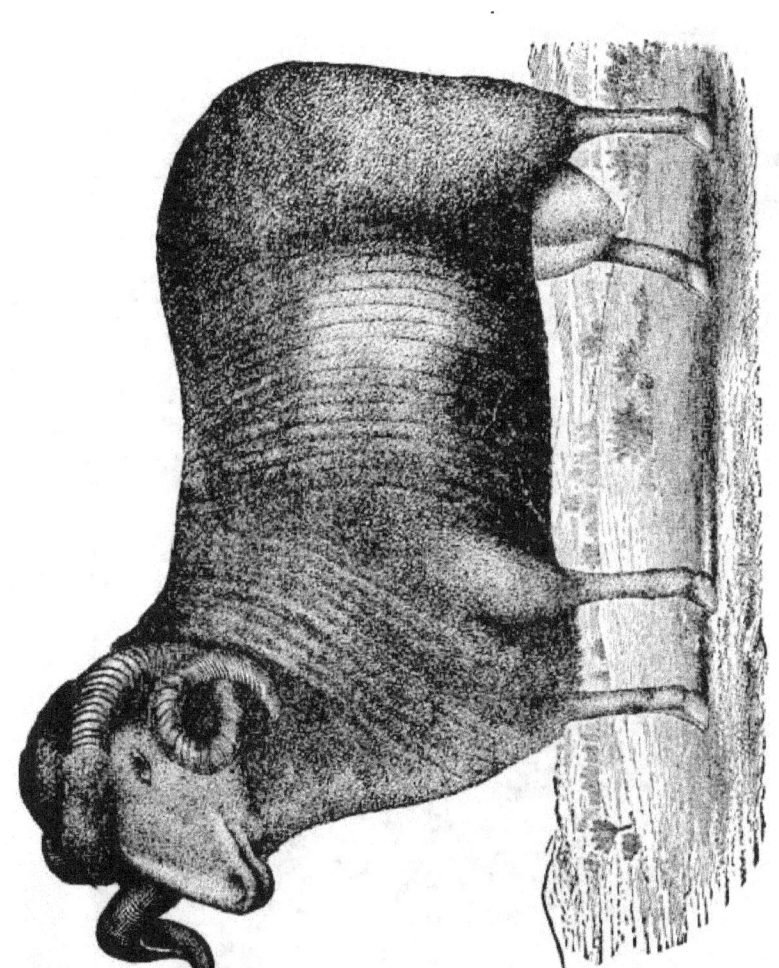

CONSUL.

BRED BY WM. JARVIS, WEATHERSFIELD, VT.

An Improved Spanish Merino of Forty Years age.

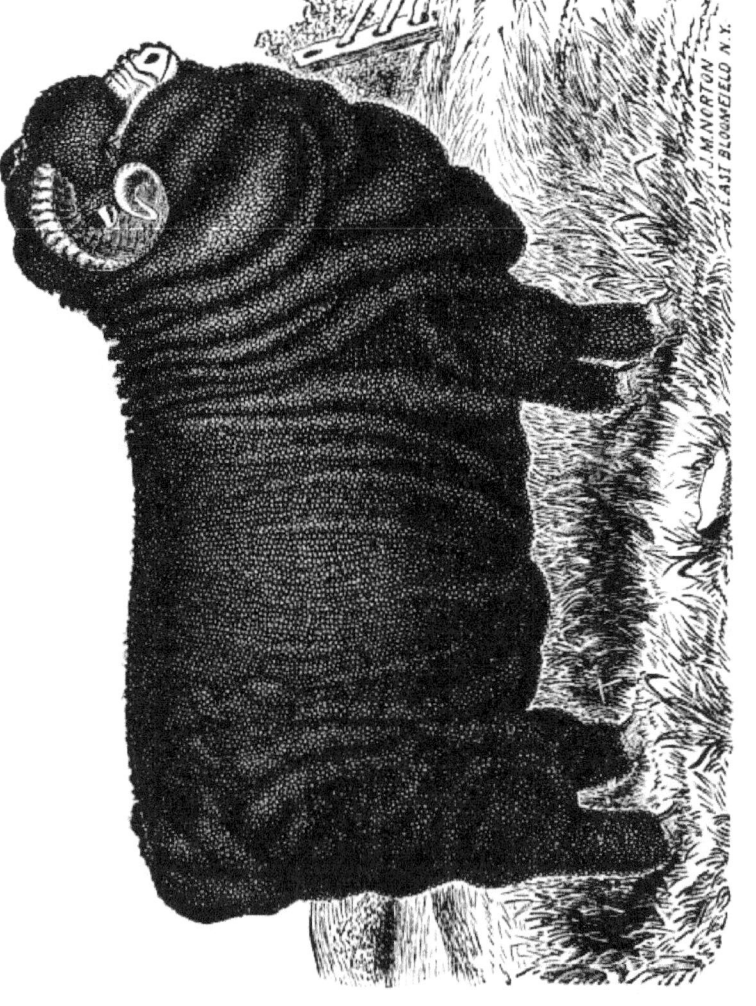

BISMARCK.

BRED AND OWNED BY H. C. BURWELL, BRIDPORT, Vt.

Winner of the SWEEPSTAKES at the Centennial for "the best American Merino Ram of any age," in a competition of twenty-seven.

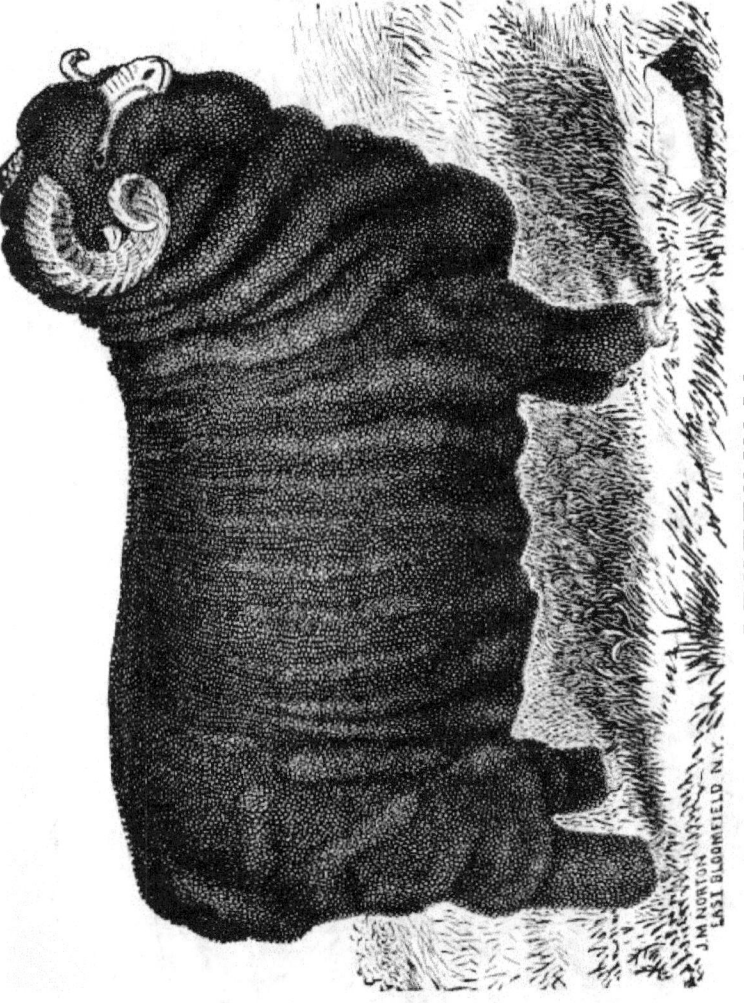

J.M. HORTON
EAST BLOOMFIELD N.Y.

CENTENNIAL.

BRED AND OWNED BY JOSEPH T. STICKNEY, SHOREHAM, VT.

Centennial with four ewes was awarded the prize of $100, offered by the Pennsylvania State Agricultural Society, for "the best flock of American Merino Sheep of any age," exhibited at the Centennial.

CENTENNIAL PRIZE EWES.

BRED AND OWNED BY JOSEPH T. STICKNEY, SHOREHAM, VT.

These ewes with the ram "Centennial" were awarded the prize of $100, offered by the Pennsylvania State Agricultural Society for "the best flock of American Merino Sheep of any age," exhibited at the Centennial.

J.M. MORTON
EAST BLOOMFIELD. N.Y.

Should any of the Sheep die
on the passage the Skin is
to be taken off from the
tip of the nose to the extrem
-ity of the Heels, the hind
feet as high as the first joint
above the hoofs to remain
attached to the Skin which
is to be dried & delivered to
the Consignees —

A gratification of fifty cents ahead
is to be paid to the Captain &
twenty five cents a head to the
two mates, if not more than
ten ♯ Cent die on the passage

Fac-Simile of Bill of Lading taken by Consul Jarvis on Merino Sheep shipped at
Lisbon, Oct. 18, 1810, with this endorsement on the back in his own handwriting. The
original certificate was kindly presented to the Secretary of this Association by Consul
Jarvis' daughter, Mrs. Hampden Cutts, of Brattleboro, Vt.

Shipped by the Grace of God, in good Order and well conditioned, by Wm Jarvis in and upon the good Ship called the *Alfred*, whereof is — Master, under God, for this present Voyage Josh S. Patch and now riding at Anchor in the River Tagus and by God's grace bound for

Boston

forty eight Spanish Merino Ewes & six Spanish
Sheep Rams for Account of Risk of the
Shippers, a Native Subject of the United
States of America

being marked and numbered as in the Margin, and are to be delivered in the like good Order and well conditioned, at the aforesaid Port of Boston (the Danger of the Seas only excepted) unto Mr Wm Lewis or in his absence to Doctor George Patch to they, or their assigns, Freight for the said Goods, ... with Primage and Average accustomed. In Witness whereof the Master or Purser of the said Ship hath affirmed to Bills of Lading, all of this Tenor and Date the one of which Bills being accomplished the other to stand void. And so God send the good Ship to her desired Port in Safety. Amen. Dated in Lisbon, Girl and this 12 Dec 1810

Joseph Patch

48 Agunes Ewes
6 Ditto Rams
— — —
54 in all

www.ingramcontent.com/pod-product-compliance
Lightning Source LLC
Chambersburg PA
CBHW081714220526
45468CB00008B/1837